THE STRONGER SEX

THE STRONGER SEX

WHAT SCIENCE TELLS US ABOUT THE POWER OF THE FEMALE BODY

STARRE VARTAN

SEAL PRESS

New York

Copyright © 2025 by Starre Vartan
Cover design by Ann Kirchner
Cover images © vm via Getty Images; © Firax / Shutterstock.com
Cover copyright © 2025 by Hachette Book Group, Inc.

Hachette Book Group supports the right to free expression and the value of copyright. The purpose of copyright is to encourage writers and artists to produce the creative works that enrich our culture.

The scanning, uploading, and distribution of this book without permission is a theft of the author's intellectual property. If you would like permission to use material from the book (other than for review purposes), please contact permissions@hbgusa.com. Thank you for your support of the author's rights.

Seal Press
Hachette Book Group
1290 Avenue of the Americas, New York, NY 10104
www.sealpress.com

Printed in Canada

First Edition: July 2025

Published by Seal Press, an imprint of Hachette Book Group, Inc. The Seal Press name and logo is a registered trademark of the Hachette Book Group.

The Hachette Speakers Bureau provides a wide range of authors for speaking events. To find out more, go to hachettespeakersbureau.com or email HachetteSpeakers@hbgusa.com.

Seal Press books may be purchased in bulk for business, educational, or promotional use. For more information, please contact your local bookseller or the Hachette Book Group Special Markets Department at special.markets@hbgusa.com.

The publisher is not responsible for websites (or their content) that are not owned by the publisher.

Library of Congress Cataloging-in-Publication Data has been applied for.

ISBNs: 9781541604421 (hardcover), 9781541604438 (ebook)

MRQ-T

10 9 8 7 6 5 4 3 2 1

*For my grandmother, Doris May Ross,
the strongest woman I have ever known*

CONTENTS

Introduction		1

PART ONE MUSCLES: POWER
Chapter One	The Culture of Women's Muscles	13
Chapter Two	Centering Women's Muscles	31

PART TWO THE CYCLE: FLEXIBILITY
Chapter Three	The Power of Periods	55
Chapter Four	Hacking the Female Body to Win at Sports	77

PART THREE PERSISTENCE: DURABILITY
Chapter Five	Female Fat Is Fundamental	99
Chapter Six	Endurance, the Female Superpower	119
Chapter Seven	Running on Empty	139
Chapter Eight	Female Pain and Disempowerment	153
Chapter Nine	Pain and Perseverance in Athletics	177

PART FOUR OVERLOOKED ABILITIES: SENSITIVITY
Chapter Ten	Held Back, Hamstrung, and Hobbled, Women Still Nail It	199
Chapter Eleven	A Well-Balanced Life and Sporting Culture	215

PART FIVE	**IMMUNITY: PROTECTION**	
Chapter Twelve	Female Bodies: Great at Defense	235
Chapter Thirteen	Estrogen the Pathogen Slayer	245
PART SIX	**THE LONG GAME: LONGEVITY**	
Chapter Fourteen	Living Longer, Living Better	265
Chapter Fifteen	A More Exciting Future for Women's Bodies	287
Epilogue		309
	Acknowledgments	*313*
	Notes	*321*
	Selected Bibliography	*363*
	Index	*365*

INTRODUCTION

For as long as women have supposedly been "the weaker sex," physical strength has been defined by what men's bodies can do. Men have built the idea of what bodily power means in their own image, and they have defined the terms.

But what if we defined strength by what the female body does best? What if strength were measured by a body's capacity to fight off bacteria and viruses, the durability of its muscles, the flexibility of its form, its endurance in swimming or running, the length of its life, its toughness in the face of pain? In these cases, and many others, the female body is physically stronger than the male. Of course, male bodies can generally do some things well too: Lifting heavy weights over one's head or sprinting a short distance quickly are clearly kinds of physical strength. They are just not the only ones.

For the four-plus decades I've been alive, I've been told not only that women aren't as strong as men, but that those things inherent to the female body are what make us delicate and fragile. How many times have I heard that female fat is gross and holds us back in athletic pursuits? That having a period, and the associated hormones, makes us less able to do physically demanding jobs and makes it too hard for scientists to study our bodies? That we are too sensitive? That being smaller makes us inferior? That all the qualities that make us female make us weak? The sometimes-said, mostly inferred implication of those ideas is that physical weakness means we get to have less power in the world, and certainly less power over our own bodies.

The science shows that none of the above is true. As I detail in *The Stronger Sex*, women's muscles retain strength over time better than men's, women's fat and metabolism offer a huge advantage for any pursuit that requires endurance, menstruation might be a pain, but it has lots of unique powers, sensitivity is a key to strength, and the very biology of women's nervous systems provides significant advantages when it comes to pacing in physical challenges, durability, and pain tolerance.

The Stronger Sex is a journey from muscles to ovaries and uterus, through metabolic systems and fat stores, into nerve endings, and deep into immune cells and the warrior hormone, estrogen. It's a voyage into the lives of a Gen Z woodchopper in Oregon, a firefighter in New York, and a teenage wrestler in Arizona. I watched young korfball players in the Netherlands, learned about judo from a brown belt, saw Megan Rapinoe retire her jersey, and discovered the secrets of ice swimming. I met with sports scientists and surfers in Australia, revisited getting kicked in the face by a mule in the mountains, and sipped martinis in a bar in Okinawa, Japan, to learn the secrets of the oldest women in the world. I delved into the bowels of an art museum in Antwerp to interrogate the beauty of muscles, meditated on the power of fat with a Brooklyn marathoner, and spent the day in a lab in Seattle with an immunity researcher—and way too many mosquitoes.

I interviewed scientists—from ob-gyns in the UK and US to experts in longevity, pain, immune systems, bones, menopause, menstruation, biological anthropology, physiology, sports science, and more. I read through hundreds of research papers by these experts and many others, some from early science history and some published the week this book was due. My reporting uncovered the myriad ways that female bodies outperform male bodies—and just as importantly, how much human bodies' abilities overlap when considering training and body size, not just sex. There are many areas where bodies do well, or not, based on height, where weight is located, life experience, and practice—and where sex has little relevance.

Female bodies are far stronger and more capable than most of us have been taught, to the extent that anthropologists have been surprised by how physically powerful our female ancestors really were—stronger than even female athletes are today. Our great-great-great-great-great-great grandmothers would likely laugh at some of the stories we have swallowed and reiterate about weaker female forms.

So how did we get to be living under these pervasive myths about what the female body is capable of? The people who created and tell the stories about all of our bodies have been, until recently, men. They have long looked at the world from the point of view of male bodies and minds, talked to and worked with other men (often to the purposeful and systemic exclusion of women), and been interested in and experimented on their own anatomy. Female bodies, to those who don't have them, seem complicated, a little "too much" to deal with. If there aren't any female researchers asking questions, making assertions, and testing ideas, female bodies are easy to ignore in the name of ease, efficiency. Stated more bluntly, female bodies haven't been of interest to many male researchers due to disrespect, dislike, disgust, and disapproval.

The fact that women's bodies have been understudied in science is well known and reported on, and continues into the current day, despite a 1993 National Institutes of Health–mandated increased enrollment of women in clinical trials in the United States. Researchers at the University of California, Berkeley, found that in 2010, women were still significantly underrepresented in scientific studies, and a follow-up study by a team from Northwestern University and Smith College in 2020 found that not much has changed. A 2023 study from the University of New South Wales in Australia found similar results, and also pointed out that male researchers did a poor job of including female bodies in their studies, whereas female researchers usually included both male and female test subjects.[1]

It's not just human females that have been and still are excluded in science. This bias has resulted in dramatically less research conducted on female animals, so the disparity starts early in the knowledge process. Whether it's to study new medications, surgical procedures, or sports medicine, basic science too often leaves female bodies out. Some fields do better than others—pharmacology has very low rates of female inclusion, while psychedelics research, according to scientists I talked to for an article I wrote on the subject, said it was equitable.

Similarly, research into women's athletics is lacking, and female athletes are trained as "smaller men," to their detriment. We just don't really know what female bodies are capable of, because they have been limited, constrained, oppressed, or just ignored for generations.

Still, there's more data on female bodies than there used to be. That's thanks in large part to female researchers, many of whom have made the breakthrough discoveries and are doing the arduous basic science on diseases of the reproductive system, but also in areas like immunology.

That uptick in research means all kinds of new information is being revealed every year, which is exciting, and which drove my interest in reporting it to you in *The Stronger Sex*.

Female bodies are more than half of all bodies on earth. The disparate treatment of those bodies in medicine, physiology, and athletics is deeply frustrating for the researchers, coaches, competitors, and fans who know better. As many of the examples and stories that unfold in this book prove, when we know more about female bodies, we find solutions to problems and illnesses that can apply to all humans.

That's why this work matters. It's not just about being fair and inclusive of more than half the population (though that would be reason enough to do the work). It's that we are missing out on key information about humanity when we don't consider everyone.

Chances are, whatever your gender, sexual orientation, age, or sex, you have complicated feelings around your body. Maybe you were taught to hate it but have grown to appreciate it, or you like it when it's running or lifting things, but feel a sense of disappointment otherwise. Maybe you feel that your body has betrayed you in some way aesthetically or functionally, or maybe you grudgingly accept it but dream of how you would alter it if you could. Maybe you just feel pretty detached from your corporeal form.

Rarely have I met anyone who has had a truly uncomplicated relationship to their designated meat sack. Feeling weird about your body is Human 101.

Like many people, and especially those who grew from girls to women anytime in the last hundred years, I have received countless messages about how my body was not enough or was too much, was shaped wrong, had fat in the wrong places, or was weak or lame or gross.

I no longer feel any of those things. Researching and writing *The Stronger Sex* changed how I feel about and see my body. It's my hope that this book changes how you understand the female body too—whether you have one, want one, are transitioning to one, or love someone who lives in one. I hope it challenges your ideas about what the female body is capable of and serves as a reset to what you think you know. We have all heard, seen, considered, dismissed, and absorbed ten million messages about our bodies, most of them negative, too many of them from our closest friends and family. *The Stronger Sex* will undo some of that negativity, some of those messages, and hopefully some of the self-hate that can come along with them.

The Stronger Sex addresses many aspects of the female body—from athletics to basic biology, from lived experience to experimental ideas—and you will learn quite a lot of fascinating details. But I have a larger mission, which is to upend what you've been taught about the familiar idea of women's bodies and how you see them. Writing this book certainly changed how I perceive mine.

We might live in a world increasingly dominated by virtual realities of various kinds, but we are still embodied humans. From our bodies we all derive pain, pleasure, joy, anger, life, and eventually death, which Jane Goodall recently called her "next great adventure." Let's get the truth about them.

A NOTE ON SEX AND GENDER

This book talks a lot about sex and gender, and there are some people who argue that sex is the same thing as gender. But I've been living here on earth for a bunch of decades, and that's not my experience or the experience of many others. Gender has physical aspects to it, but it's mostly what we are taught and how we express ourselves considering that education. We know this because what makes someone a woman or man has changed over time, and it also differs across cultures.

Like most systems in most human societies, gender rules are nuanced and far from universal. "Gender extends far past the physical realm. There are behavioral, social, cognitive, and other markers of what we would consider masculine and feminine," said Amethysta Herrick, a transgender woman, chemist, and gender theorist that I talked this over with.[2] Gender is many things, but it's not simple, and it's not binary. There are too many people for whom the binary doesn't fit—and too much variety in cultural definitions, including plenty of societies that have third, fourth, or more genders—for it to be the simple thing too many people insist it is.

In this book, I'm not going to simply state that I think transgender women are women, and they therefore merit inclusion. I am going to tell you precisely why I believe that. It's rooted in both my personal life experience and what I've learned while researching this book. First, I have had several trans women friends, the first of whom I met when I was a freshman in college, in 1996, before anyone had ever told me how to think about trans people or the current divisive conversations. My grandma, who raised me, had taught me that everyone deserves respect and kindness until they merit different treatment. For my grandma, this idea was rooted in her Episcopalian faith and her experience having lived alongside a huge variety of people throughout most of her life in New York City.

So when I met Maria as a sophomore at Syracuse University and asked her to go thrift shopping with me, it's because she was smart and interesting and funny—my qualifications for friendship then as now. I didn't like or dislike her because she was trans, but because of who she was as a person. When she was harassed one day by campus police while we were walking past the Hall of Languages, I was confused and then angry (I was nineteen, it was 1996, and I hadn't known that trans people were and are often targeted by law enforcement). I loved my undergraduate experience, which was overwhelmingly positive, but that experience with Maria was as much a part of my education as both my degrees from SU.

As for the more recent hatred and bigotry that have been spun up against trans people, I think Judith Butler, longtime queer theorist—whom I first read in class at Syracuse University, actually—puts it best: "Circulating the phantasm of 'gender' is also one way for existing powers—states, churches, political movements—to frighten people to come back into their ranks, to accept censorship, and to externalize their fear and hatred onto vulnerable communities."

Something I know deeply, from the experience of growing up in one, is that the female body is all about flexibility. This reflects biology—female bodies have greater variation between them than male bodies in countless ways. Just one: A 2022 review study from the University of Vienna is titled, "Female Genital Variation Far Exceeds That of Male Genitalia."[3] Female bodies grow and change in so many ways—and have evolved to do just that, as part of their fundamental function. They can accommodate a fetus, during which time the joints become more flexible, skin expands, weight is redistributed, and more. Then they can revert back to previous form. Breasts can expand and contract and also change internal structure to make breast milk. Female bodies go through menarche, the start of the menstrual period, and menopause. Hormonal and structural changes in the body accompany each of these phases. The ovaries and uterus change every month and then change again.

Given the inherent flexibility and incredible variety of forms of even the most narrow definition of a female body, there's plenty of room for those who feel like women but were born with male body parts to be women too. Women contain multitudes. It's a big red tent.

While gender is socially constructed, biological sex is often defined as the physical characteristics that determine whether an organism produces sperm or eggs. Sounds like a binary, but while it would be awfully convenient if there were a very clear line between male and female, there just isn't. Intersex people exist, and they aren't some tiny minority—they are up to 5 percent of the population.[4] "Intersex" is an umbrella term that describes a variety of natural physical variations, and while some of those are visible at birth, others don't show up until puberty, and some aren't physically obvious at all. This means until we test every baby's DNA when it's born, subject it to a full-body scan, and do that again at puberty, there is no way of knowing how many intersex people really exist. "There are over twelve hundred intersex characteristics that have been identified, most of them invisible, because they have to do with the endocrine system or other systems. We don't recognize when somebody has them, because they're only checked for if there's dysfunction," said Herrick.

Many of us walking around right now have significant physical aspects that vary from a hypothetical norm. That's typical—life on earth is incredibly diverse, always evolving, and fluid; it both responds to environmental cues and randomly mutates. Those of us with white skin or blue eyes, lactose tolerance, resistance to HIV, or denser bones all have genetic mutations to thank for those features. Adaptation and change are keys to survival on a dynamic planet earth, and our collective history is full of variety. Diversity in fetal development, including in sex characteristics, happens commonly and easily during various phases of gestation. It is an ordinary thing, and so are the humans that grow from them.

Understanding the normality and commonality of intersex conditions has led many scientists to question the sex binary altogether. When I visited Sharon DeWitte, an evolutionary anthropologist at the University of Colorado, Boulder, we mostly discussed bones and what we can learn from them about ancient people. But the sex binary comes up in her work too, because, as she pointed out, a binary is also a bias. It predetermines how you understand new information and colors what questions you ask. "If you come into your work thinking sex is binary, that's what you're gonna see," she said.[5]

Considering all this, how can I write a book where I use gendered terms like "man" and "woman"? Partly because I must when I refer to historical information or any scientific study that has used those terms. To use words other than what the original study authors did would be incorrect. Our recent history has relied on a gender binary, and so I must as well. In many places, and where I could, without misconstruing others' work, I have used biological sex as a dividing line. That's both because the work that has been done thus far has divided humans up this way and also because it centers hormonal and physiological differences, not gender.

At this point, almost all of us have been assigned a sex at birth, and for now, that's the definition of biological sex and what scientists are using. There are people for whom this does not work. Many of us alive today have grown up with significant overlap between the sex we were assigned at birth and our gender identity. And there are those who haven't, who are trans or nonbinary. All of us deserve to see ourselves reflected in both culture and science. It's my hope that in the future, all bodies will be studied, which will benefit all humanity.

At the same time, it's also true that we are still living in a culture where some bodies, those labeled female, are extremely understudied and poorly understood. There has been a dearth of reporting and science on the bodies assigned female at birth, and that's frustrating and wrong.

I believe in the capacity for all of us to hold more than one idea in our heads at once, and while I endeavored to include gender-diverse perspectives in *The Stronger Sex*, the truth is that there's little science that includes them. I'm stuck with both the existing and historical biases of the binary in research. There's no way to cover that without reiterating it. I have done this imperfectly, and any mistakes herein are my own, not those of my advisors.

Part One
MUSCLES: POWER

Chapter One

THE CULTURE OF WOMEN'S MUSCLES

The thing they don't tell you is that weightlifting gets you high. It's different from the sweet endorphins gained from other kinds of heart-pumping activities. Lifting heavy weights combines regular feel-good exercise chemicals with a sense of physical power. It's an intoxicating combination.

Weightlifting also helps us grow muscles, of course, which is something many women are still afraid of. Yes, I know about Michelle Obama's arms, and I've seen the "Strong Is Sexy" T-shirts, but plenty of people still push the ideal that when it comes to women's bodies, sleek is sexy and muscular is manly. When I mentioned that I had begun weightlifting classes, my Pilates-teacher friend immediately asked, "Aren't you afraid of getting bulky?"

No. I wasn't anymore. But I had once felt that way too.

I've always been naturally muscular, with sizable calf and thigh muscles. But in the culture I grew up in, muscles were for men. I've seen the online comments for decades now: A woman with muscles is "masculine" and therefore gross. It's one of the reasons trans men and nonbinary transmasculine individuals get so much flak. People assigned female at birth are supposed to be fit, but not too strong. In shape, but not muscular. When you get down to it, it's a very, very specific aesthetic. Like many women, every exercise I've ever done prior to weightlifting was concentrated on minimizing muscle gain and simply becoming "toned."

I signed up for Olympic weightlifting classes at a strength and conditioning gym in my town in November 2023. Following the required one-on-one sessions, where I learned some basics, I joined an all-male group for classes twice a week. It took over a month to really get the hang of both the lingo and the moves, like the clean and jerk, the snatch, and my very favorite, the split jerk, which involves lifting the weighted barbell overhead from a starting position at shoulder height, while at the same time jumping into a lunge. I started working with an empty twenty-two-pound bar. It took several months to feel like I really knew what I was doing in terms of the foreign-to-me movements. Ten months later, I was deadlifting two hundred pounds and split-jerking ninety pounds over my head.

Even before the guys cheered me after that split jerk, I felt fucking awesome. Physical strength is a potent drug of its own.

A woman split jerking is now a lot more common—my newest weightlifting instructor is a woman, the classes aren't always all guys and me, and my gym is owned by a woman. Many young girls today are learning different lessons than I did. I grew up in the 1980s, when men's physical power was very in my face. Between the ubiquitous *Conan the Barbarian* posters, the *He-Man* cartoons, and the Incredible Hulk's, well, hulk, I learned early and often that men can build big muscles, which allow them to lift dainty-girl models, one on each arm, to the sky.

These depictions of men's strength were delivered alongside the rules about what my soon-to-be-a-woman's body could and couldn't do. Those rules came from TV shows, school, and the way-above-grade-level pulpy horror books I read. I was told not only how weak I was—and all females were—but also that if I defied those social norms, I would be a weirdo. I was a girl, so I was less strong in all the ways—but especially physically.

My own body taught me something different. I learned other, contradictory lessons when I listened to my muscles. As an only child I spent hours playing in the woods, dragging and lifting large branches to build forts, shimmying down the steep sides of narrow gorges, and piloting my bike at top speed over bumpy roots. Failure meant dropping a branch on my foot or falling off my bike. I've always known exactly how far I can jump, since underestimating my stride meant losing my shoe in sucking swamp mud. I wasn't taught how to do any of these things.

When I was instructed in how to use my body, it wasn't always helpful. In gym class I was told to use my arms to climb, but after trial and error that left me dangling from tree branches, I learned how to quickly propel myself up the trunk of a giant Norway spruce using my legs to push me from below instead of my arms to pull. I remember feeling frustrated that nobody had told me this.

From the age of four, my grandma raised me to ignore all kinds of cultural norms, and she did the same herself. She was especially dismissive of any that suggested she couldn't do something she wanted to do, whether that was piloting small planes in the 1950s, traveling around the world three times solo, beginning a career in emergency medicine in her fifties, building her own stone walls, or raising her granddaughter. She gave me a running list of chores, and she assumed I was strong and capable enough to do what she'd detailed on the clipboard that hung on the kitchen wall.

As kids do, I pushed my limits. I stacked too much wood into a giant metal wheelbarrow so I could make fewer trips to the woodpile, and sometimes it fell over. I forced the almost too heavy lawnmower up the hill, knowing that if it slid backward I could slice my foot on the blades. I figured out how much easier it was to carry fifty-pound bags of dog food on your back than with your arms in front. I knew I was strong because I could do all these tasks when I was ten years old—tasks that adult men often performed in other homes and on TV. Were some of these things moderately dangerous? Sure. But my grandma was less concerned with safety—despite being an emergency medical technician—than by what she thought more important than a sliced toe or a broken bone: the ability to figure out problems and tricky situations for myself.

At school, I was taught the opposite of what my grandmother expected and what my woods playtime taught me I was capable of. Because I was a girl—as sturdy and tallish as I was—I was expected to ask a boy or a man for help with almost anything that required a modicum of strength. At summer camp I was told to ask a boy for help (not just another person) to haul stuff half the weight of those dog food bags I put on my shoulders at home. I was supposed to ask the elderly man on staff at the library to change the bottle in the water dispenser when I could easily do it with nary a spill at age fourteen. I'd make sure nobody was around and just flip the bottle over and into its snug home without saying anything. Even at that age I knew that showing my strength wasn't acceptable.

Over the years, these gendered messages that ignored my abilities made me feel invisible. I was seen first for my sex, not my strength—for my long, blonde, curly hair, not my muscles. I received the lesson loud and clear: My whole self wasn't acceptable. As a straight, cis young woman, I found this difficult to accept, and it sometimes made me angry. For a transgender or gender-nonconforming person, I can only imagine that the challenges (and the painful dismissal of who they know they are) are even more extreme.

I hid my strength in high school. I was embarrassed by my big calves, though I also relied on them when I began trail running at fifteen. I poked at my thigh muscles as my professor detailed their attachment to bones in my anatomy and physiology class during my years as a premed major at Syracuse University. I wondered if the guys who hired me for a job assembling bespoke computers at an engineering company in Berkeley assumed I could lug the bulky CPUs around, which I could. (They told me I was hired because I was the only person who applied who had experience using hand tools.) I worried that my muscles would turn to fat through years of sitting on my butt all day at office jobs in Manhattan in my twenties, but they never did. I wrote about my complicated feelings about having bigger, stronger legs than the men I dated when I was a fine arts graduate student at Columbia University. At thirty-two, when I reunited with my beauty-queen Australian grandmother, whose once-curvaceous legs were by then skin and bones, seriously weakened due to osteoporosis, I wanted to know why some bodies end up like hers. My other grandma had been cremated with plenty of muscle on her old bones. As I spied middle age in the near distance, I wondered what would happen to me. Would my sometimes-despised, mostly enjoyed muscles disappear one day? There wasn't much information about muscles out there for women, unless you were a bodybuilder. I wanted to know more.

I started talking to exercise physiologists, doctors, and athletes, and writing about aspects of human physiology. I found that while there were half truths embedded in what I had been taught about muscles in the female body, there was also quite a lot of ignorance, exaggeration, hyperbole, fact twisting, and even some outright lies, all of which served to drive home the idea of women's inherent weakness—and to celebrate men's strength.

In short, the lessons I learned about men's and women's muscles had an agenda, and it was a disempowering one for women.

SEPARATE AND EQUAL? YEAH NO.

The lessons about who is strong and who is weak start early (even now), despite a lack of grounding in science and medicine. Research shows (and any pediatrician will tell you) that pre puberty, there is no inherent difference in kids' muscular power, coordination, or physical ability, whatever their gender. For every girl who has outrun most of the kids in her third-grade class in an impromptu footrace (as I did), this is not a surprise. According to a foundational 1998 muscular strength test by researchers at the University of Puerto Rico, girls are as strong, fast, and able to play any sport as boys are up until about age thirteen or fourteen.[1] There are, of course, differences between kids who vary in size, and individual kids go through puberty at different times. For a time, due to earlier puberty, the young female body even has a strength advantage, due to larger size—I remember towering over the boys in sixth grade.

For the first decade-plus of a child's life, separate-gender sports are purely a cultural construction. Sex-separate soccer teams for first and second graders is theater—a false division that's more about parental comfort and expectations than strength or sports ability. This gendered system teaches all kinds of lessons, overtly and covertly. Boys learn that they have the (physical) upper hand, even when it doesn't yet exist, and that girls can't and shouldn't compete with them. Girls learn that it's wrong for them to challenge boys when it comes to physical pursuits, and that they should only play sports with other girls. And all kids learn that the gender binary separates and excludes—and if you don't fit into the category of "girl" or "boy," there's no room for you at all.

These early messages have lifelong repercussions, with plenty of obviously negative ones—girls incorrectly learn that they aren't as strong as boys at an age when they literally have the same abilities, and they are discouraged from aggressive play despite their personal interests.

That almost no little girls play American football with Pop Warner (the largest youth football league in the US) results in very few women playing the sport in any capacity.[2] Which could be OK if football weren't the most popular sport in many places in the US—and if so many other popular sports didn't also have low numbers of girls playing them. Almost a million boys played football—the most popular sport for that sex—in US high schools in 2022, according to the National Federation of State High School Associations.[3] Just under half a million girls participated in their most popular high school sport, track and field. The great news is that girls' participation in sports has grown quickly since Title IX was passed in the 1970s. About 43 percent of high school athletes in 2022 were young women.

Messages that we hear over and over have their impacts, but so does plain-old physical experience. If a parent notices a difference in certain sports skills in opposite-sexed kids, this is part of the reason why. From the earliest ages, boys and girls are taught different games and how to move their bodies in space differently. While this variance can be mitigated by family in the early years, cultural messages and expectations can quickly undo what is taught at home.

Joanne Parsons, a physical therapist and an associate professor at the University of Manitoba in Canada, found that by grade four, there's already a significant gender gap between what boys and girls are good at. Not relative strength, as there is no difference there, but in what skills each gender succeeds at. "Boys throw and kick well, girls skip," Parsons said in a *British Medical Journal* podcast in 2021. "This is not a biological but a societal expectation. Those things have to have some influence throughout that person's life—the way they move is different, and that's determined by environment."[4] Practice makes perfect, and even young boys are taught the physical skills needed to succeed in sports more often and at younger ages than girls. Extrapolate that over a large population, and by five or six years old, girls and boys have learned and practiced different physical skills.

If you're skeptical that early exposure wouldn't make a difference by the time athletes advance to elite levels of sports, new research is showing that it's likely what gives male players at least some of their muscular advantage a decade or more later.

I'd never considered that my years of playing in the woods and doing chores for my grandma set my body up for a lifetime of muscular strength until I met with Sophia Nimphius on a beautiful summer's morning in Sydney, Australia's busy central business district. Under a bright blue sky, we navigated past a marina and into a bustling cafe, each of us ordering a coffee and healthy breakfast bowl topped with fresh fruit—hers oatmeal, mine chia. Nimphius is the pro-vice-chancellor for sport and a professor of human performance at Edith Cowan University in Perth, Australia, and she was in town to speak at the Women in Sports Conference, which brought together experts from around the country. She'd just returned from speaking at the FIFA Medical Conference in Boston, Massachusetts, where she'd presented on popular myths in girl's and women's football (soccer). Earlier in her career, Nimphius worked directly with high-level surfers and with elite rugby, basketball, soccer, and softball players. Now she is dedicated to researching sex-based differences in performance and injuries in athletes.

Nimphius said one underacknowledged reason that the strength of young female athletes seems to differ from that of young male athletes is simply hours of exposure and experiences to develop it—not just as trained athletes, but over their lifetimes.[5] Girls aren't pushed to do physical labor; in fact, it is actively taken away from them and given to boys and men. This has lifelong repercussions. "Tonnage, or training volume, as measured in sports science, is how much weight you have lifted over the course of your training history. It's a crude measure, but it's a pretty good indicator—if you've lifted tens of thousands of kilograms, you're probably quite strong," said Nimphius. "If we go back and look at the difference in tonnage for young women and men over a longer time, you're going to see some big differences. If every time you go to the grocery store, your dad was like, 'Sophia, bring the groceries inside,' that move is the same as a farmer's carry in the gym, right? So then you go to college, and now you're in a weight room. The tonnage you can pick up will be higher because of these small, seemingly innocuous jobs." Women start weight training later than men, and girls start most sports later than boys. That early experience changes bodies, priming them for the muscular power needed in many sports. And girls don't get it—or they get it in a discounted form.

If biological sex were the only reason behind male/female athletic differences, said Nimphius, it wouldn't be such an advantage to be a farm kid, and researchers wouldn't have found that the majority of elite athletes had older siblings pulling them into tougher play. She sees this in the talent-development space all the time, in that individuals with older siblings are more likely to be elite athletes. It's such a well-known factor in sports that FiveThirtyEight ran a feature article analyzing the question "Why Are Great Athletes More Likely to Be Younger Siblings?" in 2020.[6] Nimphius said it was simple: If kids play above their level, it pushes them to be better players. "You either didn't play or you had to compete." A 2014 study of athletes trying for a place on the US women's national soccer team found that three-quarters had an elder-sibling advantage.[7] This all builds the case that girls aren't physically challenged as much as they could be when they are younger, and that impacts their bodies for the rest of their lives. Nimphius pointed out that Abby Wambach, former captain of the US women's national soccer team, was the youngest of seven kids, with two older sisters and four older brothers. She started playing soccer at age four. "It's that kind of training age, and mentality, that is irreplaceable."

By middle school, these social, not physical, factors mean that many girls start their athletics journey with a significant deficit in strength and athleticism compared with the boys they know. If girls don't have the same physical experiences and aren't pushed in the same ways, how can they get as strong as those who have? Or even just as strong as they can be?

Besides resulting in fewer female players with the physical life experience equal to that of males, preventing girls from doing tough physical work could also result in more injuries. A recurring news story in recent years is that injuries to the anterior cruciate ligament (ACL) tend to occur more frequently in female athletes than male athletes.[8] Dozens of headlines from Down Under pointed out that in the first two seasons of the women's professional league of Australian Rules Football (commonly known as "footy"), ACL injury rates were 6.2 to 9.2 times higher than in the men's competition.[9]

Nimphius didn't buy initial hypotheses for the ACL injury rate, which blamed everything from women's "wide hips" and stretchier ligaments to particular hormones for the problem. She said these explanations assumed equal training and disregarded the biased systems that precede elite competition. "We fundamentally don't train equally, much less equitably, in quantity, intensity, or length of time," she said. The ACL injury rate has dropped since 2019 and in the years since, but for a time it was the subject of much concern over what was seen as innate physical weakness in female bodies.

Nimphius pointed to her presentation at the FIFA conference, where she highlighted the average age that soccer academies—training grounds for the elite athletes who go on to form the World Cup soccer teams—start training boys and girls. All kids start at thirteen years old, plus or minus two years. Since girls go through puberty a year and a half to two years earlier than boys, that means they're beginning this critical physical preparation too late, when they've commenced puberty. Most boys start before or at the beginning of puberty. This matters because "what you do through your pubertal years changes your musculoskeletal system. Those changes are lifelong, and there is some evidence to say that you have an increased ability to adapt in that stage," said Nimphius. If you hit that window, it's a lifelong benefit. "In people who are very active during that period, in the bone and hard-tissue research, you can see remnants of that training, even well into adulthood." Once the puberty window has been missed, it's gone forever. Training in those key years around puberty is highly impactful, and "girls aren't getting that advantage, but boys are, and that's baked into the system."

Dance shares similar high-risk movements and knee-joint strain as Aussie Rules Football, but sex disparities in ACL injury rates aren't seen in dance. This is likely because both sexes begin dancing at very young ages (well before puberty) and receive similar training—a fact that points away from physical differences in male and female bodies and toward training and experience as the risk factor for injury. In her paper on the subject, Nimphius wrote that the combination of reduced motor-skills development that comes through practice and shorter preseason training creates a "perfect storm" of female athletes with a higher risk of ACL injury in footy.[10]

Knowing this, she said, can help coaches train better or differently for the current players who never got the advantages of early preparation—and it should also change how soccer academies think about recruiting, for the future of the sport as well as athletes' health.

Looking at training and making modifications—not blaming biological factors—has already resulted in lower ACL injury rates for female footy players, said Nimphius.

Practice clearly shapes our bones and muscles—and other parts too. One oft-discussed area is the difference in spatial ability between men and women, which has been found over and over again through decades of studies.[11] Most researchers have attributed greater male abilities in this area to testosterone's influence on the brain, and this connection has been used as a proxy for male superiority in biological and evolutionary discussions.[12] It's a pat story—men are good at certain things, and they have more testosterone, so it's a clear, hormonal cause and effect.

Probably not. Among those with autism spectrum disorder (ASD), no sex difference was found in rotational tasks.[13] Recent research has determined that much of the sex difference in spatial ability is likely attributable to confidence—which comes from experience. When both sexes were given time to practice a novel task (moving an object in space), they got better at it, and testing ability evened out.[14] Experience, and the confidence it brings, improved results. Self-belief matters: One small Italian study showed, in a group of seventy-one women and eighty-one men, that "women improved performance and reached men's scores in the MRT [mental rotation test] when they were led to believe they were better than men."[15] Simply telling a group of women that they could best the men in the test improved their scores. What does a lifetime of being told the opposite do?

A 2018 study by researchers at Northeastern University looked at throwing ability, another task where testosterone has been credited for boys' and men's better performance.[16] Here, too, they found that practice had an impact: Gender differences in throwing ability for most age groups disappeared when those who were less experienced got time to do the prescribed activity.

Most of the reason that men are generally better at spatial tasks and throwing is because they've probably practiced them during childhood and adolescence, not because their brains or bodies are so different from women's. Structural differences have an impact, but they're not everything. Generally, larger shoulder muscles in male bodies allow stronger overhand throwing, while female bodies' lower center of gravity lends them greater stability and balance. More practice at something over a lifetime—and especially what we do in childhood—makes us better at that thing, whether it's navigating an unfamiliar neighborhood, crocheting, speaking a second or third language, or playing soccer.

In light of this science, the fact that for hundreds of years, women's sports haven't been taken as seriously has translated into less-prepared female athletes—and it still does. When I heard Maggie Mertens speak at the party for the release of her book on women's running, *Better, Faster, Farther: How Running Changed Everything We Know About Women*, she mentioned a fact that gave me hope while also making me realize that true gender equality in athletics is just beginning. After all, the first generation of female athletes born after the 1996 founding of the Women's National Basketball Association is just now coming of age as adult players. "That's how you get a Caitlin Clark—that and a lot of practice," said Mertens.[17]

Girls and women are still trained as "smaller men" due to a lack of exercise science covering female bodies. Combine that with the fact that girls largely don't engage in certain types of sports altogether—like football and wrestling (until very recently, see Chapter Two)—due to assumptions about what girls can and should do physically, and we arrive at a sports landscape with less competition, lower stakes, poorer training, and less financial and moral support.

According to *Best Practice for Youth Sport*, an exercise physiology textbook, the most impactful time for kids to learn fundamental motor skills is between two and eight years old.[18] Girls' repeated exposure to skipping, jumping, and movements that test fine motor control and flexibility, and boys' exposure to kicking, throwing, and catching impact the physical skills they bring to school sports and into adulthood. These experiences are stored in our muscles. "It's mostly culture that causes gender differences. Boys are encouraged to be rough and tumble, so they are encouraged to produce muscle from their earliest activities. Girls are pushed [or] push themselves toward movement that doesn't produce muscles. So we get less strong—it's a cultural phenomenon from a young age," said Kate Clancy, a biological anthropologist at the University of Illinois.[19]

If you weren't encouraged to demonstrate your strength as a kid, even as an athlete, you might not focus on how beefy your biceps are, or how a stronger gluteus maximus can improve your overall athleticism. It happens even among professional athletes. "I worked for an NRLW [National Women's Rugby League] team, and we had a four-week preseason. Some girls had never lifted a weight before—that's going into premiership rugby," Joanna Parsonage, a sports scientist and strength and conditioning coach based on the Gold Coast of Australia, told me. Weight training's growing popularity generally means this isn't as likely to happen in the 2025 team recruits, but change is never as fast as you might expect.

In 2021, a viral TikTok showed the NCAA basketball tournament's weight-training facility for the male players, a giant room containing many sets of colorful weights of varied sizes. The women's weight room was less well stocked than a hotel fitness room, with just a bench and a set of six small dumbbells. As Sedona Prince, a player for the University of Oregon Ducks who made the video, said, "If you're not upset about this problem, then you're a part of it."[20] The internet was set ablaze with women, and some men, annoyed by the fact that at the famous and well-funded college basketball championships, blatant disparity still existed.

When girls and women grow up in a culture that discourages them from testing the limits of their physical strength (and one where boys and men routinely are encouraged to), it's no wonder they believe they're weaker. Women's work of all kinds has historically been seen as less important than men's. Those biases, including from women themselves, cross over into sports and mean less money for female athletes. And so, for years members of the US women's Olympic soccer team were paid significantly less than members of the men's team, despite winning far more games and bringing in the same amount of money—a disparity that was only rectified in 2022.[21]

All this cultural training and baggage also messes with female athletes' self-perception—and it can take years for women to believe that they are capable of competing at the same level as men, even when pursuing solo sports. When I interviewed Jennifer Pharr Davis, a longtime record holder of the fastest known time (FKT) for hiking the Appalachian Trail, she told me that she had first competed for the women's FKT, which she handily won, before finally going for the FKT for all people. She made the switch when she realized she had "limited myself based off gender. My body showed that I could compete with the men in a way that my mind had not expected," she said.[22]

Luckily for all of us, there are some women, like Pharr Davis, who push past what they've been taught about the capabilities of the female body.

DOING IT HER WAY: SMALL PACKAGE, BIG STRENGTH

Mara Jameson (not her real name) fights fires and saves lives in upstate New York. Like many women in careers where they work around mostly men, Jameson has had to be flexible, adaptive, and creative to fit in. That started with the Candidate Physical Ability Test (CPAT), which is "basically like an obstacle course." The CPAT isn't just a fitness test; it's centered around the actual skills a firefighter needs to put out fires and rescue people and pets.

Jameson told me that prior to the CPAT's creation in 1996, the qualification test was just a random collection of tough physical challenges, like running a certain distance in a certain time and doing so many push-ups or pull-ups. Each department had different standards, and there was a lot of subjectivity over what a passing grade was and which wannabe-firefighters would make the cut. That kind of subjectivity in a test can result in a team that "looks the part" rather than one that functions best.

Now it's a standardized test. "They spent a long time and a lot of resources developing a new test, so you can look up each requirement and the specifics of exactly what's involved. It's actually very job specific," said Jameson.[23]

Challenges include dragging a 165-pound dummy while wearing a fifty-pound weight vest. "It's a greater percentage of the body weight of a smaller person—I only weigh about 115 pounds—than for a big guy." She trained for it by dragging a weighted tire back and forth across her driveway. A CPAT challenge that has a particularly high failure rate is walking up a machine that has a conveyor belt of stairs (at one step per second) for three minutes and twenty seconds—while wearing seventy-five pounds of weight. "You're not allowed to touch the railing or get off or stop. If you do, you fail." Jameson said people assume that if they are in shape they can climb some steps, but it really requires specific training. Jameson worked up to walking for seven minutes while carrying ninety pounds. Here her smaller size was useful: "The stairs test is all legs and butt and stability. I have an advantage in that with a lower center of gravity."

The sledgehammer skill test mimics the forcible entry firefighters must perform to access a burning building that's locked. Sledgehammering a door down in a short time requires plenty of upper-body strength—or so you'd think. Jameson assumed as much, and although she practiced, she actually failed her first CPAT because she couldn't pass the sledgehammer test in the time allowed. "That one in particular I was terrible at. My sledgehammer hit was very weak." She was already lifting weights—what else could she do?

Turns out that her problem wasn't strength; it was technique. Jameson worked with a friend who showed her how to use the tool, and she was able to triple the amount of force (measured in the test by a machine) simply by modifying her swing. "It wasn't that I was weak. I just didn't really know how to hit with a sledgehammer in that particular way," she said. The answer wasn't more muscles, it was torque, like the windup and release of a baseball pitch. Jameson passed her test and became a firefighter.

Jameson said that now that she's been fighting fires, the rigors of doing the job have proven the usefulness of the test requirements, meaning it's a fair test. Her fellow firefighters also have confidence that she can do what needs doing because she passed the same challenges they did. Only about 50 percent of people who attempt the CPAT, of any gender, pass it, according to Jameson's experience; other research data shows pass rates of 40–80 percent, depending on population and location. But not many women apply, she told me, estimating that about six women and about eighty men attempted the exam during the time she's been with her department.

She loves the work. When the alarm goes off, "you drop whatever you're doing, you get into your gear, and you get out the door in a matter of seconds," she said. She either drives the fire truck and operates the pump or rides on the back step. "Everybody wants to ride the back step except the guys who are close to retirement. My job then as nozzle man is to put out the fire. If you are the first arriving, you get out, you see what's going on, and you have to estimate how much hose you need." Then the firefighters drag the empty hose toward the fire, getting as close as they can before the pump operator charges the line. When the water starts flowing, the nozzle reaction kickback force is a good percentage of Jameson's weight, with the potential to knock her over. It's simply harder for a smaller, lighter person to handle the nozzle, but her air pack and gear weigh her down a bit, and she can often find a brace to mitigate the force of the nozzle. Coming up with those kinds of in situ solutions is part of being a firefighter.

There are lots of other jobs at a fire. "If you're not the nozzle man, you have a pike pole or some other tool, and you're breaking walls to find hidden fire," Jameson said. In an emergency situation, with a lot going on, it's an advantage to have a variety of skill sets and abilities. "There's a big recognition in the fire service that everybody has different strengths and weaknesses physically and intellectually, and in terms of life experience. But everybody has something to offer." Not every firefighting guy is tall and burly; there are short and slight male firefighters too. She said that although she's had issues with the equipment fitting, some of the guys are so big—due to either muscularity or height—that the equipment doesn't fit them either. Her size is an advantage in confined-space rescues, where a larger-bodied person wouldn't fit. "There's so many different kinds of rescues, and there's so many different kinds of situations. We're all a team."

She said one of the reasons that you don't see many female firefighters isn't just the physical test, which clearly a female body is able to complete. It's the life experience that leads people to firefighting. A lot of the men she works with have done construction, roofing, landscaping, auto mechanics, or truck driving—which gives them specific skills that are practical as firefighters. Knowing home construction can be useful when putting out a fire, for example, because you know what's behind the wall. It's a feedback loop: Women don't do those types of jobs, and so they never learn certain skills that make for a stronger firefighter.

But Jameson wanted to do the job, so she went for it. "I just think things are interesting to do, and I want to do them. The idea that a woman couldn't pass the CPAT, or that someone of my size and dimensions couldn't pass the CPAT, was like, well—maybe. But you know, I think I can do it."

I was kind of amazed at her confidence, and I asked her where that faith in her body came from. She had good role models. Her mother was a high-rise iron worker. Her grandmother had a PhD in psychology (and was told at a job interview that she was qualified but they were going to hire a man who didn't have her family responsibilities). Jameson had been an athlete and a rock climber in her younger years.

"My superpower is a combination of being curious about things, and also not caring what people think," she told me. That easy confidence, that willingness to give something new and physically hard a shot, even in midlife, isn't so common. I couldn't remember when I had heard a woman talk so comfortably about her body's capabilities. Jameson knew that her body was strong in the ways it was strong, and that her strength could be used to complete tough tasks—maybe in the same ways a guy might do them, maybe in different ways, using her own unique talents.

Jameson's words felt like a missive from the hopeful future of living in a woman's body. There wasn't a lot of discussion about overcoming self-doubt, or men undermining her, or unfair tests or standards. There was just pure confidence in her body and an acknowledgment that she could work hard to do what she was interested in doing. That she could give her best as part of a team and it would be enough. That her leaders and teammates would be tough but fair.

Doing the work of firefighting certainly tests anyone's body, but it's not about looking a certain way or having certain muscles in specific places. It's about getting the job done. Unlike athletics, which test specific sets of skills, the work of a human body in the world is more complex and less prescriptive. It's more like Jameson's work—physical challenges that come with a variety of possible solutions—and that's been true throughout most of human history.

Jameson's achievements came from a place of genuine curiosity. Regardless of what she'd been told about what someone of her size and gender could do, she really wanted to find out what would happen if she worked hard and did her best. That's much easier said than done, but her spirit of open-hearted inquiry and her questioning mind really inspired me.

I now embody this question in my ongoing weightlifting journey. Like the many women who have gotten into the sport, I want to see how much I can eventually lift. This will do wonders for my bone health in old age, as I report in Chapter Fourteen. But I'm also doing it for the sweet, sweet high of clacking weights onto my barbell and throwing them over my head in a split jerk, my favorite move. How strong can I get?

Chapter Two

CENTERING WOMEN'S MUSCLES

Audrey Jimenez is super excited to be part of what the Associated Press calls "the fastest-growing high school sport in the country"—girls' wrestling.

Some stats that might surprise you if you didn't realize wrestling is a Thing now: The number of girls' high school wrestling teams has quadrupled in the last decade. The number of girls wrestling was up 60 percent in 2023 alone. As recently as 2017, only six states sanctioned girls' wrestling, but by mid-2024, forty-six states had.[1] The Associated Press reports that in Pennsylvania, the number of female wrestlers almost doubled in 2024, and the number of high school teams in the state shot to 180 that year from exactly none in 2020.[2] That's just one state. Hundreds competed in the first all-female state tournament in Pennsylvania, where in previous years, there weren't enough girls competing, so they sometimes had to wrestle with boys.

Jimenez has been there, done that. After years of playing softball, she was immediately curious when her tío—her uncle—opened up a jiu jitsu gym. ("He's actually my second cousin, but Mexican families, if they're relatives older than us, we'll just call them our uncles," she told me.)[3] Her tío had invited people to come try jiu jitsu training as part of his business opening, and Jimenez enjoyed "figuring it out." She was also considering her future. "I had pretty big goals with softball. I was making regional teams and traveling a lot. I wanted to go to college, and I had dreams to go to the Olympics." But she found herself more interested in jiu jitsu at her tío's gym, where she was also learning wrestling moves. There wasn't as much opportunity with jiu jitsu as there was in wrestling, so she put her focus there to align her interests with her goals. Though COVID chaos affected her freshman year at Sunnyside High School, in Tucson, Arizona, delaying state championships, she won state in the girls' division that year—and in her sophomore and junior years.

After she had won all the competitions she could against other girls, she did what she needed to so she could keep moving up in her sport. "There were maybe, like, one or two other girls in the local clubs, so I was wrestling boys," she told me. In February 2024, as a senior she became the first girl to win an Arizona state title wrestling against boys.[4] She had to get special permission from the Arizona Interscholastic Association to do it, but she said everyone was pretty cool about the situation. The wrestling community was always "very welcoming of me." The roar from the crowd following her historic win was a rich and wonderful sound.

Jimenez has a theory as to why thousands of young women are so into the sport. "As individuals, girls maybe learn a lot quicker—I think we pick things up. So we're jumping levels noticeably. You see girls wrestling just as good as some of these guys, and that brings attention to the sport," she said. A greater push for equality in high school sports from coaches, parents, and school administrators helped, sometimes in combination with advocacy by nonprofit groups like Wrestle Like a Girl. Recent wins in women's wrestling at the Rio, Tokyo, and Paris Olympics, combined with a greater societal acceptance of combat sports for girls and women, have helped. It's no coincidence that arm and shoulder muscles on girls and young women are cool now.

Jimenez's own muscular arms are enviable, and she regularly displays their strength, as anyone who had such muscles would in the 2020s, whatever their gender.

There are a lot of reasons Jimenez personally loves wrestling, from the techniques that vary "whether you're on your feet, or on the bottom, or on top," to how she's always "learning and evolving in the sport, learning more about positioning, or tactics, or strategy." Wrestling is "very chess-like" because you're in a one-on-one competition with a single other person. Jimenez loves being part of a team that cheers her on, but "once you step on the mat, you don't have to depend on anyone else, win or lose, and I really enjoy that."

One of the basic rules of wrestling is that people are divided by weight class, which essentially divides them by size, so it's a sport focused from the get-go on equalizing strength and testing technique. Since women are already wrestling men and winning, could it be a mixed-gender sport?

At the college level, it becomes clear why wrestling, like so many other sports, will likely be gender-segregated into the future. When I checked in with my own graduate alma mater, Columbia University, I discovered that they had a powerful women's wrestling team, captained by Maya Letona (who also wrestled boys in high school). But it's still a club team, which means limited finances and only one coach (varsity teams at Columbia get three to four coaches). That's likely to change soon, because the organization that governs college sports in the US, the NCAA, considers women's wrestling an "emerging sport." It's slated to become the NCAA's ninety-first championship sport, with the first bouts scheduled for 2026.[5] That designation means Columbia and many other schools will now have the framework to organize a woman's team—including the funding that comes along with it. But there's no room in that framework for a nongendered or mixed-gendered team.

College sports are set up based on men's teams and women's teams, from the NCAA to club sports. It's understandable that Jimenez is glad to see women's wrestling become its own category, because more girls will have the opportunity to wrestle by fitting into that existing system. This is incredibly important for athletes and the opportunities that athletics allow for young people. Jimenez described where she comes from as a low-income community, and said wrestling has been a "big outlet" for her. "I have the best family—they're super supportive and everything—but we haven't always been through the easiest of times," she told me. "So wrestling has always been there, and it's been something where I could see myself having a great future."

That future is only possible if women's wrestling is incorporated into the existing college sports system; there's no current way for female wrestlers to use their sport as leverage for a college degree, as so many do in every other sport, unless it is part of the ecosystem of gendered sports. That means its status as a gendered sport is unlikely to change anytime soon. Still, Jimenez is on her way, leveraging her strength and physical savvy to secure a spot as a student-athlete at Lehigh University.

WHAT IS MUSCULAR STRENGTH?

What's better when it comes to physical strength: faster or longer? "It depends" is a totally reasonable answer. Is this a sprint or a marathon? Are we getting a job done that pays by the hour or the project? Are we penalized for mistakes made by speeding along, or is arriving first the only criteria for winning? As humans, we make these kinds of calculations all the time. Sometimes it's worth it in life, or work, to take extra time to do something, and other times, quickness is the name of the game. This is also true in athletics, with the balance of speed to endurance depending on the sport and the event.

But when we talk about muscles and the physical strength they represent, it's often a less nuanced conversation. There isn't a singular definition agreed upon by everyone, but muscular strength is usually a variation on the idea of lifting, pushing, pulling, lowering, or flinging the heaviest weight one can manage in a single effort or contraction. Lifting a lighter weight over many reps (endurance) or moving weight using less energy (efficiency) aren't how most people understand strength, and endurance is often considered a different category than strength.[6] Generally, muscular strength equals heavy plus fast.

Strength thus far has not been about long plus consistent (even though Merriam-Webster defines "strength" as "the state of being strong: capacity for exertion or endurance") or efficient use of power.[7]

Guess which gender has defined muscular strength by what their bodies do a bit better?

Important note: The generalizations here and throughout the data provided below are just that. There are many cisgender women who have more of certain types of strength that I attribute to male bodies, due to different life experiences and/or training in athletics or through their work. People of different genetic backgrounds can have widely varying abilities to add or lose muscle and in where that muscle is located. We've all seen runners with bigger butts and thighs competing against others with very slim legs, because humans are diverse! Trans men and women may have muscles that have been modified by hormone supplementation or may have worked to build up or reduce certain muscles in order to comport with physical ideals around the gender they identify with. Scientists have barely begun to consider how hormone differentials caused by being intersex and diseases that can defy the "norms" of the gender binary, including polycystic ovary syndrome and Turner syndrome, may impact athletic performance. So keep all of this in mind as you read about muscles and sex—a topic that, as you'll see, is pretty complicated even if you ignore this side note (which researchers have thus far mostly been doing).

That being said, let's be clear: Yes, in general, male bodies are better at the particular ability to lift a heavy object using arms and shoulders—due to significantly more practice at doing it, starting as kids, and more muscle fibers in their upper and lower arms, driven by hormones, both of which result in greater muscle mass. This is why, for those quick, heavy lifts, male arm muscles can be up to twice as strong.[8] (There's less of a sex differential in the legs, as women carry more lean muscle in their lower limbs and have a lower center of mass.[9]) The idea here isn't to deny that cisgender guys are strong in certain ways—but that by defining muscular strength in this one way, other types of strength are ignored.

Looking outside the gym and taking into consideration the arc of human history, muscles need to do much more than just lifting or chucking something weighty a few times. Muscles in areas other than the upper body do many other jobs. Core muscles help the human body to maintain balance and to move quickly and accurately, and they are key for moving long distances while carrying or dragging weight, as our gatherer-hunter ancestors routinely did. More flexibility means muscles have a greater range of motion and are less likely to be injured. And how muscles throughout the body use energy determines how long a person can persist in a given task, whether that's hiking through snow, carrying water over a parched landscape on her head, or performing lighter-weight but repetitive motions like tanning a hide or gathering firewood.

Most survival tasks on planet earth depend more on this longer-burn type of muscular power. While quick bursts of strength can be useful in battles and hunting particular species of large animals, muscular endurance and shooting accuracy are more important in finding and killing smaller animals and fish, and for efficiently gathering plants, nuts, mushrooms, and fuel for fires. In short: Brute force has important but limited utility, and lower-impact but longer-lasting muscular power is imperative for a wide variety of necessary tasks.

Generally, the muscles in female bodies excel at longer-lasting power. We know this not only because when you look around the world, you see women doing this work—the water- and baby-carrying, the long-distance foraging trips, the cleaning, the food processing—which requires hours of lower-weight muscular work. (Girls are also socialized to do this type of labor, so their muscles "learn" it.) We know it because scientists have seen it in the lab. Both animal and human studies have shown that across various muscle types, although muscles in male bodies move faster and have a higher maximum power output, at lower weights, muscles in female bodies are able to persist doing a task for longer—sometimes a lot longer.[10]

Experts on how the human body moves—like Sandra Hunter, a professor of exercise physiology at the University of Michigan—use the term "fatigability" to describe how quickly muscles tire. Hunter has been studying muscle fatigability since she began her PhD studies in the early 1990s, and the work came to her unexpectedly. Early in her career, she was conducting what she thought was a "very boring" study that was supposed to simply examine the reliability of a test that looked at how long people of both sexes could perform an isometric (sustained) muscle contraction—think standing and holding a bag of groceries, or a child in one arm.

Hunter told me that, naturally, she tested herself first. She took the place of a subject in the special chair in her lab that's designed for this kind of research. The chair's armrest, equipped with straps, holds the subject's bent arm at a right-angle position. In this configuration, the sensors in the chair measure how hard the person can push upward at a certain force and for how long. Hunter figured out the maximum weight she could lift from that position, then set the machine at 20 percent of that weight. Next, she tried the test she would give others for the study: How long could she hold that weight? Hunter found she could hold the weight in position for twenty to thirty minutes before her muscles failed. She started testing college-age recruits to her lab. One female student bested Hunter and lifted upward for an hour before her arm muscles gave out.

Then a big, burly guy sat down. His maximum weightlifting power was about seven times Hunter's. She set his test weight at 20 percent of that and was shocked at what happened next. "After about a minute," Hunter told me, "I realized that he was going to fail really quickly. I thought he was putting it on—I thought it was a joke. But he lasted just about two minutes. That was astounding to me, because I thought, 'This is still just 20 percent of his maximum.'"[11] He was not the only man to have this issue.

In the published study, Hunter reported that overall, the women were able to sustain the isometric muscular contractions twice as long as the men, even at the same relative intensity.[12] That this seemingly basic fact was a surprise to a young physiologist in 2001 speaks to the lack of research into women's bodies that existed at the time. But after her initial surprise, Hunter started to wonder about the mechanism for the difference between men's and women's muscles that was uncovered in her study: "How come he only lasts two minutes?"

To find out, Hunter created a new study with ten men and ten women of equal strength—they were all capable of the same max lift.[13] Using 20 percent of that weight, Hunter repeated the sustained contraction with them. When the subjects were matched for muscular strength, she found that the sex differences disappeared: Both sexes' muscles failed at around the same time. Hunter thought this was good evidence that the initial study's sex difference might have been due to bodily mechanics—in larger muscles, blood flow can become blocked by the movement of the muscle itself. When muscles don't get enough oxygen, they fail. So the burlier a person is, regardless of gender, the more likely their blood flow might be blocked when holding a muscle in a contracted position.

Hunter didn't stop there. Since she already had these strength-matched subjects in the lab, she thought about her prior study and wondered: Was there another way to test muscle fatigability that wouldn't be impacted by the mechanical effect of larger muscles cutting off blood flow?

She changed the movement performed by the test subjects. Instead of doing a sustained isometric exercise, they contracted their arm muscle, held for six seconds, and then released for four seconds. The weight was set at 50 percent of their maximum, and the contract-hold-release movement was repeated until muscle failure. The idea was that during the release, the muscle would again have access to oxygenated blood, removing the issue of larger muscles blocking blood flow. In this study, she found that the strength-matched women lasted twice as long as the men.[14] So strength for either sex depends on the task, but in terms of sustained muscular work, the finding was (and remains today) that "women on average have more fatigue-resistant muscles than men," as Hunter puts it.

Significant additional research since Hunter's has backed up that statement. In a large narrative review published in 2023 in the *Journal of*

Strength and Conditioning Research, physiologist James Nuzzo gathered and compared a variety of studies on a number of different exercises—from bench press to biceps curl to leg press to trunk extensor—in men and women at different weights. Looking specifically at lower weights, half or more of the studies in the dozens Nuzzo gathered showed that women were able to perform more repetitions of exercises, including lat pull-down, bench press, biceps curl, and leg press. At anywhere from 40 to 80 percent of the maximum weight that the person could lift, press, or extend, many studies found that women could do more reps. The lower the percentage of maximum, generally the better the women did compared to the men, and they did particularly well at the 40–50 percent range. At around 70 percent of their maximum weight, women were breaking even, and men began exceeding them at over 80 percent. This is more evidence that "women are less fatigable than men in tests of isometric muscle endurance," as Nuzzo wrote.[15]

Why would it be easier for muscles in female bodies to persist at lower weights when those in men's burn out? The answer could be oxygen uptake. It was long assumed that male bodies are best at oxygen uptake, since they have higher VO_2 max rates—which is how much oxygen a body can use during intense exercise. But what about oxygen uptake at lower exertion levels? Researchers from the University of Waterloo in Canada had eighteen male and female college students of similar fitness levels walk quickly on a treadmill—but only up to 80 percent VO_2 (not max). They measured their subjects' oxygen uptake and oxygen extraction. The female subjects had about 30 percent faster oxygen uptake while exercising at a moderate pace, which led the researchers to conclude that "oxygen extraction dynamics were remarkably faster in women."[16]

This finding was a surprise to the scientists who ran the study. In a press release, one of the researchers said the fact that women's muscles take up oxygen faster, "scientifically speaking, indicates a superior aerobic system." It also means that women's muscles are less likely to be affected by muscle fatigue—and may be more resilient. "While we don't know why women have faster oxygen uptake, this study shakes up conventional wisdom," said the study's lead author. "The findings are contrary to the popular assumption that men's bodies are more naturally athletic," he continued. "It could change the way we approach assessment and athletic training down the road."[17]

Sandra Hunter thought another possible answer to the fatigability question lay in sex-based differences in the central nervous system. She looked at whether the brain and spinal cord were more effective at transmitting information to and driving the muscles—and found no sex difference.[18] So if the discrepancy between the two sexes didn't arise from the muscle structure itself, or from how the muscles communicated with the nervous system, could it be in how the muscles were powered?

Hunter's colleagues sent her studies that showed just this reason for why muscles in female bodies might be built for endurance work. When a male and a female run or cycle at the same relative intensity for an hour or longer, the muscles in the female body burn more fat and fewer carbohydrates.[19] This has been well established over the last couple of decades in numerous tests, so that even the premarathon nutritional advice for male and female bodies is different.[20] In the short term, burning carbs gives muscles in male bodies a boost, but in longer-duration exercise those same muscles working at an intensity similar to women's muscles will burn out faster (although specific supplementation can help). Muscles in male bodies burn though the carbs that give them a boost, which may be one reason men drop out of marathons more, said Hunter. "Males tend to have a lot more catastrophic endings than females, and they slow a lot more [over the course of the race], especially if they don't fuel appropriately during the race."

Muscles in female bodies burn carbs too, of course, but in conjunction with more fat-burning and greater preservation of the carbs. Fat keeps women going when the carb supply drops off. That's obviously useful in any activity where muscles need power over the long haul. And it may explain why, although men outperform women by about 10 percent in short-duration activities, that gap drops to just 4 percent in long-distance events like cycling, running, hiking, and especially swimming, where women regularly outperform men. (See Chapter Six, on endurance, for much more detail on this topic.)

Muscle fibers are yet another area of difference that may confer a consistency and endurance advantage to female bodies. They tend to have more of the type-1, "slow-twitch" muscle fibers that are especially good for this kind of effort. Those type-1 fibers also give female bodies a significant health advantage. Some research suggests they may also be key to the mystery of why women are less likely to develop type-2 diabetes, despite the fact that women generally carry more body fat and have lower muscle mass, factors that should mean the opposite. Along with estrogen's effects, those extra type-1 muscle fibers in female bodies increase insulin sensitivity, which reduces the chance of getting diabetes, according to a 2014 study that was part of a wide-ranging look at how sex differences affect glucose and lipid metabolism.[21]

But base physical differences aren't the whole story. All humans have a mix of fast-twitch and slow-twitch muscle fibers. The proportion any individual has depends strongly on genetics, but life experience and age also impact it. Resistance training increases fast-twitch muscles; endurance exercise increases slow-twitch muscles.[22] And everyone loses fast-twitch muscle fibers as they age. Again, life experience matters. My own childhood and teen years spent hauling wood, dog food, and cat litter, as well as moving sizable stones to build or repair walls on our property—and these days, my weightlifting regime—mean that I will have more fast-twitch muscles than I would have without all that work. It's just a bonus that I also have the gene type that helps me build them quickly. If I were pulled into a lab study, I might skew the results based on my atypical life activities for someone of my gender. But that's not my biology, just my life.

It gets murkier. I have been a trail runner for thirty years now, but as an independent athlete, I create my own routines. I've found that I tend toward a long warmup and then enjoy moderating between intense sprints and slow jogs or even walking. That's just what feels "fun" to me, and in my dance practice, I have a similar cadence. Have I chosen dance, trail running, ocean swimming, and weightlifting due to my genetic strengths, childhood chores, or something else? How much do the sports we choose reflect what we are naturally good at? As Sophia Nimphius showed in the previous chapter, what we practice shapes our bodies.

I can't help but wonder—despite my relatively active lifestyle—how different my strengths might be if I were a lifelong member of a hunter-gatherer culture, walking ten miles on a regular day, training to shoot, trap, or chase game? Or what if I, with the same body I was born with, had grown up in an early agricultural community, where I ground grain with heavy stones, hauled water and firewood, and did muscular work from dawn until dusk? The fact that I was born when I was means I got to *choose* the physical activities I engaged in. Throughout most of human history, very few people chose how they used their bodies.

When it comes to muscular endurance, there's also more to consider than purely physical abilities. Sandra Hunter, who has continued to study female muscles over the course of her career, came across the work of Robert Deaner, a psychologist who thought that sex-related performance differences in athletic pursuits were attributable to women being less competitive and assertive than men. Among nonprofessional athletes, women run marathons at a more even pace due to their less assertive nature, posited Deaner. (If this sounds like a high-level rehash of the familiar "it's all in their heads" argument that many of us might be frustratingly familiar with when women's bodies are the topic, that's how it sounded to me too.)

Hunter explained the same pacing phenomenon that Deaner attributed to attitude by arguing for her glycogen research. The fact that men's and women's bodies fuel differently would lead to men running faster at the beginning of a race, when they are fueled, and then dropping out at higher numbers when they run out of gas (glycogen). Hunter thinks that happens not because men are more competitive, but because they burn through their carb stores. "Rather than argue against [Deaner], we joined forces," said Hunter. Their collaborative paper looked at results from fourteen marathons and showed that, indeed, women paced more evenly than men did.[23] Each researcher saw it as proof of their own hypothesis, compromising in the conclusion section of the paper that pacing differences "may reflect sex differences in physiology, decision making, or both."

This doesn't mean that psychology or experience has no place when we think about muscles and endurance exercise. Repetitive, onerous tasks are common in many types of work traditionally done by women. That foreknowledge and plain old experience could translate into women running a marathon at a more even pace on average than men. To counter Deaner's hypothesis, it's not women's lack of competitiveness; it's a lifetime of practice in the art of patience (and the experience to know that it's an effective way to reach your goal).

In top long-haul athletes of any gender, pacing is a key skill for winning, but for women, it may come more naturally based on life experience. Maybe in this skill, male athletes want to follow women's lead.

Lastly, hormones also assist the muscles in female bodies with strength and endurance. A 2024 study in the journal *Muscles* gives a solid overview. The authors reported that estrogen reduces muscle breakdown and repairs damage to muscle through interaction with specific receptors on muscle cells. "Estrogen also modulates inflammation by decreasing the expression of pro-inflammatory cytokines, which can otherwise lead to muscle degradation," the authors wrote. Estrogen promotes the development of those fatigue-resistant type-1 (slow-twitch) muscle fibers. These specific examples of the myriad ways estrogen benefits muscles also explain why studies have shown that hormone replacement therapy can increase muscle mass in female bodies.[24]

THERE'S MORE TO MUSCLES THAN PHYSIOLOGY

As you can see from the volume of information, theories, knowns, and unknowns, researchers don't yet know exactly why men's muscles are stronger in some tasks and women's are stronger in others—though all kinds of science is currently looking at a variety of causes, and we'll continue to learn more. But the physiology of any body still grows up and lives in a gendered world. Even toddler girls and boys are pushed in different directions when it comes to taking physical risks and what games are deemed suitable for which gender. We all grow up both using and thinking about our bodies differently due to how we are taught to view ourselves and the expectations our caregivers have. There's what we are allowed to do, what we are encouraged or taught to do, and what we are taught to fear.

With that in mind, remember that most people who take part in scientific studies are those who have the time to do an experiment *and* who are accessible to researchers. That means the pool of research subjects is skewed young, healthy, and white. The number of studies performed on college students is understandable; that is who cash-strapped researchers have outside their lab doors. Yet it is not representative of all of humanity. Unless there's a specific push for diversity as part of a study, bodies raised in poverty, Black and brown bodies, disabled bodies, trans bodies, and older bodies are usually underrepresented compared with their prevalence in the population. So the biases that the studied bodies carry with them into the lab, combined with the exclusion of many other bodies, ultimately cloud the research results.

Including everyone means we get better data and conduct better science, but diversity encompasses more than just the categories I mentioned above. For instance, consider how much and what kind of exercise study subjects regularly do. That's only sometimes noted in research on exercise science. If participants in scientific studies are taken from the general population, the biases they grew up with—such as girls being less likely to haul heavy loads, play sports (and even if they do, receive proper training), or engage in muscle-building activities—follow them into the lab.

"Traditionally [researchers] sample with no understanding of the training background of the people they're sampling, meaning you get societal bias. Men tend to be more exposed to resistance training, or have been exposed to life where we asked men to pick things up for us," Nimphius pointed out. How many and what kind of workouts someone does (if they exercise at all) will obviously impact their performance on a test of muscular ability. "You don't know if it's sex underpinning [the results] or the lived experience," Nimphius said.

She thinks a better way to nail down the differences between men's and women's muscular capabilities would be to conduct research on bodies that are more equally able. In this case, testing people who are more similar to each other can answer a specific question by reducing the noise from cultural effects. In a joint study led by Nimphius with researchers from Appalachian State University in North Carolina, they did just that. It took more time and effort than a traditional study since the researchers had to first test their subjects to find a group that had sixteen men and sixteen women matched to ability. Once they had their cohort, Nimphius and her team attached a battery of sensors to their subjects that measured muscle activity in four different areas of the quadriceps and the hamstrings during a squat. They found that the difference in muscle activity was not "statistically significant between men and women."[25] In other studies she's conducted (see Chapter Fifteen), Nimphius has found similar results in performance. When skiers had equal strength and training, the differences between their abilities were individual, not sex-based.

To truly determine what sex-based muscle differences might or might not exist, Nimphius wrote in a passionate commentary that sports-science research needs to design studies that "consider a multidisciplinary approach to describing or matching male and female participant groups by including basic physiological measures of strength and better descriptions of sporting age or skill (e.g., years at highest competition level or sport-specific skill) and athletic training age or skill (e.g., years of resistance training or measure of movement skill). This is a much stronger approach to elucidating whether observed differences are a function of a modifiable trait versus that attributable to gender."[26]

Humans have a mix of kinds of muscular power, depending on life experience, and a range of genetics that dictate what types of muscles they can grow. There are men who are better at endurance and women who are better sprinters. Nonbinary, trans, or intersex individuals are also going to be better at one or the other—not due to their sex or gender, but because humans are different.

Our culture generally centers men's bodies when it comes to athletic pursuits, and muscle-based activities especially. You think muscles, you think men. But if each sex has muscular advantages and disadvantages, or if they just vary among all people depending on gendered life experience, either way that means we are missing out on half of what humanity is capable of—half of what's interesting about muscles, half of what there is to learn about how muscles work in conjunction with the rest of our bodies, and half of our capabilities as human beings.

The science is both catching up with female bodies and beginning to consider the variety of genders and how they may impact athletic performance (or not), at the same time that athletic opportunities are expanding (which is great!)—but these are recent phenomena. The world is still far from equal, especially in lower-income countries that are dependent on women's unpaid labor (and where women who perform paid work are paid less). There are still many places in the world that have robust soccer teams for boys, creating an athletic pipeline for them, while girls are caring for younger siblings, helping their moms with household work, and lugging water and firewood (typical jobs for young women). Furthermore, it's illegal to be trans in more than a dozen countries, and it's de facto disallowed in dozens more.[27]

Female endurance advantage could be more closely tied to these kinds of work experiences. It could also be purely physiological, attributable to muscle fiber type, the greater quantity of fat-burning muscle in female bodies, better oxygen uptake, or hormonal influences. Most likely, it's an interconnecting system of the physical, experiential, and the psychological.

We are far from knowing the full extent of what female bodies are capable of. It's exciting to think that the future holds some of the answers—but the past does as well, and can enlighten us to the possibilities, because we didn't always live as we do now.

PREHISTORIC PEOPLE POWER

Our ancestors' lifestyles were radically different than ours. Whenever I imagine what it would be like to depend on my body to feed, shelter, and transport me—rather than as a meat sack I have to take out to exercise like a fleshy pet with cute hair—I get a tiny glimpse of how different my relationship with it could be.

During prehistoric times, infant mortality was higher than it is today, but not as grim as previous estimates have made it out to be, new research has found.[28] If we survived childhood, we could reasonably expect to live until age seventy. (See Chapter Fourteen, on longevity, for more on that topic.) We would have been working and playing seasonally, doing different types of physical labor depending on the time of year. We know from anthropologists who look at present-day hunter-gatherers, as well as evidence from ancient settlements, that the physical work humans did in the hundred thousand years of gathering and hunting is different from that of the agricultural work that almost everyone did later on (and some still do). In turn, agricultural work of course varies more than factory labor, which is again very different than the sedentary life of an office worker who goes to the gym three days a week.

The human body is very flexible, so we can live all these different ways and still experience relatively good health. Whatever activities (or lack thereof) we engage in over the years of growth and through puberty change our bodies for a lifetime, as Nimphius's research shows. And what we *keep* doing as we grow keeps changing our bodies too. Outside the physiological impacts, how much or little we depend on our bodies in day-to-day life profoundly impacts how we see and understand our own bodies and the bodies of others.

Our genes haven't significantly changed in the past hundred thousand years, so we are capable of what our ancestors were and vice versa; our prehistoric ancestors would look and sound like us. Yet due to their childhoods, communities, and the ways they grew up, they were also very different—but scientists can learn about the limits of the human body from even very old bones.

Anthropologists who measure the minute marks that muscles make on bones can estimate how much our ancestors walked: an average of about 7.5 miles a day (12 kilometers).[29] And we know via DNA and teeth, which can show a record of geographical location in their enamel, that hunter-gatherer women traveled just as far and fast as men did, moving camps and communities hundreds of miles.[30]

Even after the agricultural revolution, though, women did a tremendous amount of physical labor, according to the bones they've left behind. In a 2017 study at the University of Cambridge, researchers performed CT scans of arm and leg bones from skeletons of central

European women dated 7,300 to 1,150 years ago (early Neolithic era and into the medieval period). They compared them to the bones of modern-day women, including both sedentary women and athletes (for example, soccer players and women on Cambridge's championship rowing team).[31] The researchers found that the ancient women's arms were up to 16 percent stronger for their size than those of the rowers (who trained twice a day and rowed seventy-five miles a week), and 30 percent stronger than those of nonathletes. Interestingly, their legs were about the same strength.

Alison Murray, now at the University of Victoria, and first author of the study, calls this the "hidden history of women's work." It is likely, according to Murray, that this arm strength came from the many hours spent tilling soil, harvesting crops, and, importantly, grinding grain to make flour. For thousands of years, the grinding of grain was accomplished by hand using two large stones. In places where anthropologists have observed this work going on today, the women might spend five hours a day doing it. That definitely makes an hour-long gym session focusing on arms and core seem a lot less impressive.

Of course, men were doing some of this work too (the grain-grinding seems women-specific, but plowing and tilling soil was done by both sexes)—but the idea that women weren't strong enough to do heavy labor wasn't a concept that would have made sense to people at the time. Women were indisputably strong enough to do the hard work of prehistory and early agriculture, and they did it, as their remains show. And both men and women would have known that if they skipped work, everyone would have less to eat.

We can't know how women of the prehistoric past, or even those of a thousand years ago, thought about their bodies (though some observers have speculated about this based on sculptures and art). But we can be sure of one thing: Because their lives were so different from ours and their bodies were more intimately tied to their own and their tribe's survival, women's relationship with their bodies—and with others'—was much different than our own. Those women who were stronger than today's rowers from grinding grain? They would have been physically formidable, which really challenges ideas about women's supposed frailty and men's "natural" brawniness. There's no reason to think that some of our ancestors thought women's muscles were anything other than valuable, desirable, and even beautiful. Without them, we might not be here today.

WHEN WOMEN'S MUSCLES WERE BEAUTIFUL

Until quite recently in human history, the ability to opt out of physical labor was something only the rich could do. It became a sign of wealth and power to be delicate and weak. It showed you could pay—or enslave—someone to do hard work for you.

When women's power was reduced (for example, following the Norman invasion of England in 1066, when women lost the ability to divorce and own land they had previously held), their value as objects owned and managed by men grew. The less power they had, the more women's bodies became status symbols for rich men (and for the poorer men who imitated them). The female body was most useful—most "beautiful"—in a male-controlled culture when it was strong enough for reproduction, but not any stronger. Its purpose was decorative and reproductive.

Unless, of course, you weren't rich, or were a slave—then of course you had to work in ways that depended on strength, and you'd have the muscles that came with that labor. So women's muscles became "ugly." Of course, it wasn't just muscles; racist notions about bodies also influenced ideas of beauty, and those began in the decades after the slave trade really got going in Europe. As Sabrina Strings wrote in her book *Fearing the Black Body*, Black women who came to Europe as part of the slave trade directly affected representations of beauty during the late-fifteenth and sixteenth centuries. At first that was a positive thing. Strings wrote that the most famous artists of the era (think Leonardo da Vinci, Michelangelo, and Raphael) were from northern and western Italy and the low countries (modern-day Belgium, the Netherlands, Luxembourg, and parts of northern France and western Germany). The major cities in those regions were hotbeds of the Renaissance but also "served as key ports in the expanding slave trade," Strings wrote. "Consequently, black women often appear in meditations on beauty by the era's most important artists."[32]

It wasn't long before social distinctions between low-status slaves and higher-status white women became critiques of African women's faces and bodies, rendering the features associated with them undesirable. Cementing the cultural distance between the races was also a way to justify racism. You can't really enslave people you admire, after all. These ideas traveled to the United States along with the slave trade. And so features associated with poor people and slaves (visible muscle, darker skin) were considered "ugly," and their opposites (small, fine, delicate, and white) were considered "beautiful." We still see many of those influences in beauty culture today and in critiques of women's bodies—even those of athletes—deemed "too muscular." The critiques get especially cruel when talking about the bodies of women of color.

Strong, muscular women have been with us all along. They just haven't often been celebrated or acknowledged. Sometimes they've been hidden in plain sight. Following Strings's examination of the art of Peter Paul Rubens (1577–1640) in her book, I took another look at the Antwerp-based painter and found something surprising. Rubens was so well-known for depicting plus-sized women in his paintings that the word "Rubenesque" became yet another adjective to describe the body of a larger woman. Viewing his artwork online, I noticed that it wasn't really fat that gave Rubens's women thick thighs and shoulders, but muscle. I looked closer, zooming in until the pixels exploded and my screen was a mess of colors. Could I have been viewing strong women's bodies this whole time, been told they were fat, and just accepted what I had been told?

While looking for higher-resolution photos, I learned that Rubens—being a savvy businessman as well as an artist—was well known not just for his huge oil-painted canvases but also for his prints. In a genius move, he instructed the artisans in his workshops to make black-and-white printed versions of many of his most famous paintings to sell to art lovers who couldn't afford his giant masterpieces. The Royal Museum of Fine Arts Antwerp (KMSKA) has many of these prints, and I was curious to view them in the most hi-res version of all: real life. After I caught a korfball game in Delft (see Chapter Fifteen), I took the train to Antwerp to see the prints by special appointment.

I appeared at ten a.m. on a Monday in the all-white, super-minimalist entrance foyer of the KMSKA and met up with curator Lizet Klaassen. She badged us through a plain, unmarked door that opened into a long hallway. Then we went through another secured door and down a hallway—and then through another door and hallway. And another. Deep in the belly of the museum, there were paintings everywhere, many of which were displayed in giant frames like posters in an art store. We wound our way up a steep set of metal stairs to the print room.

Surrounded by brick archways that curved overhead, and a deep hush, Klaassen told me that the space we were in had been built during the Cold War as a nuclear-proof bunker. She pointed to the edge of the wall where it met the floor. Hinged flaps ran almost the length of the space. "The only way to get the big paintings out of the galleries above is to lower them through here, and then down below." I looked up and saw openings in the floor above, to the gallery. It was a legit secret passage—for paintings.

Klaassen opened a special folder on a large desk under a bright light, and I looked closely at the Rubens prints. As I examined them, I realized that my online observation had been accurate. In print after print (for example, *Susanna and the Elders*), I saw long flowing hair, billowing robes and fabrics, pretty faces—and thick thighs and calves with muscular definition. In *The Prodigal Son* two women are working in a barn. One shows the meaty forearms she's earned from her work hauling a bucket for the pigs she's watering. Another female figure in the background moves hay from an overhead storage area to the hungry animals below, using a large, heavy pitchfork. These are working women, and they look like it. (Their muscular definition is easier to see in the two-toned prints, and less distinct but still visible in the paintings the prints were designed from by Ruben's assistants.)

As the authors of *Rubens & Women*, Ben van Beneden and Amy Orrock, put it, in Rubens's paintings, "women are not simply passive erotic objects to be observed, but enduring and active agents of their destiny." His ancient goddesses and nymphs have "strong" bodies and display "a powerful generative force." Rubens's point of view was influenced both by his mother, who kept the family together when his father was imprisoned, and by the powerful women he worked with as he gained influence as an adult. Rubens the artist and businessman was also a diplomatic agent; for over a decade he traveled to Holland, Spain, and England to negotiate peace treaties on behalf of Archduchess Isabella Clara Eugenia, who governed the Spanish Netherlands from 1621 to 1633.[33]

The archduchess was an admirer of Rubens's strong women. She displayed his paintings in her hunting castle, and she wasn't the only powerful fan of the Flemish artist. Philip IV's wife, Isabella of Bourbon, commissioned *Diana and Her Nymphs Setting Out for the Hunt* especially for her rooms. Diana, the Roman goddess of the hunt, is depicted with well-muscled arms and shoulders, thighs and calves. She's surrounded by doting hunting dogs and flanked by her nymphs, who are also strong, and one of whom is pushing away a muscled satyr. Diana is wearing a chiton, "a one-shouldered antique garment that was also worn by the Amazons, a fierce race of female warriors."[34] Rubens's gory and fabulously violent painting *The Battle of the Amazons* portrays a lot of powerful female arms and shoulders (and yes, bared breasts) as the warriors stab and fight and fall off bridges alongside the men they are fighting. (See Chapter Fifteen for much more about the real Amazon warriors.) In another of Rubens's huntress paintings, *Diana Returning from the Hunt*, the goddess's deltoids and biceps literally ripple with muscle.

Looking at Rubens's work anew made me think about what I had been told about strong female bodies my whole life. I was told that even when I was looking directly at a clearly muscular woman—I had seen some of these paintings before—I was seeing something else, something considered unattractive in my culture: fatness. Understanding the racial and cultural context around why I was taught that it was more "beautiful" for women to be toned, not muscular, helped me realize that those ideas are about disempowering women. I had been taught to be blind to women's muscles, but I wouldn't be anymore. Now I knew what I was looking at.

Part Two
THE CYCLE: FLEXIBILITY

Chapter Three

THE POWER OF PERIODS

There's a lot more period acceptance and positivity these days than when I was a kid. But as I was writing this chapter, I came across a Reddit AITA (Am I the Asshole?) post that so perfectly encapsulated where we (still) are about periods that I feel compelled to share it.[1]

The post was written by a mom who wanted to know if her thirty-eight-year-old brother-in-law was out of line when he complained about seeing her young daughter's used tampon in the daughter's bathroom trash can. He told the mom that what he saw was "disgusting," and that her daughter should be more discreet. The mom explained to him that menstruation was "a natural part of life for women and nothing to be ashamed of," and then she reminded him that the bathroom was her daughter's "private space." The brother-in-law got angry, said he shouldn't be "confronted with such things" at her house, and told the mom she should clean out the cans before his visits. Then the AITA writer got angry back, told him to grow up, and used some other undisclosed choice language. The post-writer's husband, the parents-in-law, and even friends said she was out of line and should have "been more understanding of his discomfort," which is why she wrote in to ask: AITA?

The fact that a grown man was not only "grossed out" by seeing a tampon in the trash but then proceeded to bring it up, get in a fight with his sister-in-law about it, and then get backed up by other family members—I was not the only Redditor to feel outraged. It's not the initial disgust that bothered me; we all have pet peeves, and people can feel about period blood however they want to. The issue is that the man felt he had a leg to stand on when he sought to shame the young girl and her mother about it. Fundamentally, he thought he should be protected from this specific substance. Remember, his words were: "confronted with such things."

Disgust over menstruation has a long history and is very common among men and some women. What's important here isn't policing individual ideas of what's gross and what's not, but the fact that this is normalized—which has repercussions that hurt those who menstruate.

Let's do a quick thought experiment around the idea of what's normalized and what's not. How many movies have you seen where someone is bleeding? And how many of those occasions has it been menstrual blood? If aliens were to prepare to meet humanity by watching all our movies and TV shows, they would absolutely know how to cause and help a bloody nose, a broken leg, or a head wound. But would they even be aware that half of people have periods every month for decades of their lives?

GENERATIONS OF MYTHS ABOUT FEMALE BODIES

How did we get here? Beginning with the "father of medicine," Hippocrates of Kos, who got female bodies very wrong. (Yes, the same guy who inspired the oath that is still recited by medical students: "First, do no harm," and so on—aka the Hippocratic oath.) To be fair, Hippocrates got many things right, from refuting the idea that diseases were punishments from the gods (as if being ill wasn't bad enough, you've also managed to piss off the all powerful!), to inventing the concept of the case study for an individual patient, so protocols could be followed rather than just left up to doctors' memory or intuition. He made vows to treat all people with all illnesses, whether they were enslaved or free, and to never abuse his patients' bodies. He even went so far as to recognize that the origins of diseases could be different depending on a person's sex, and that it was important to speak directly to women to "learn correctly from the patient the origin of her disease." Good stuff—for 400 BC.

However, Hippocrates was also living and practicing medicine amidst the patriarchal culture of ancient Greece, where girls were the property of their fathers, and women their husbands. Medicine followed culture in that women were considered "weaker, slower, smaller versions of the male human ideal, deficient and defective precisely because of their difference to men," wrote Elinor Cleghorn in her scholarly but super-readable book about the history of illness in female bodies, *Unwell Women: Misdiagnosis and Myth in a Man-Made World*.[2]

Since, as Cleghorn detailed, women's power in ancient Greek society was limited to their "mysterious" ability to bear and raise children, that idea was mapped onto their bodies: "Women were under the dominion of male authority, and medical discourse legitimized this by making women's bodies subordinate to the whims of the very organ that defined their social purpose." From the foundational texts of scientific discourse, according to Cleghorn, "women emerged as a mass of pathological wombs."

The idea that the uterus (and its associated menstrual periods) were to blame for women's physical and mental health complaints continued into the Dark Ages and the rise of Christianity, which borrowed many ideas from ancient Greece and Rome, especially when it came to medicine. Medical information was censored by the early churches. The men who taught that Eve was made from Adam's rib, responsible for the ejection of humanity from the Garden of Eden, and that women were to be "submissive" and "quiet" were the same ones in charge of what medical info was disseminated. Luckily, because doctors at the time were forbidden by the church from examining the female body, most women were treated by female healers and midwives, who probably provided care that was a little less demeaning to women. Until many of these women were accused of witchcraft. During the sixteenth and seventeenth centuries, almost fifty thousand people were executed for the crime of witchcraft, and 80 percent of them were women. (Cleghorn notes that most of them were over the age of forty.) This was a simple and effective way to eliminate the most powerful women from the population—and literally kill off a significant source of knowledge about female bodies, leaving a huge information gap to be filled with misinformation.

Men's bodies were seen very differently. During the early modern period in Dutch history, when ideas of what constituted proper women's work were narrowed and domestic labor was epitomized, the authority of men was touted as "natural."[3] When Antonie van Leeuwenhoek, who invented the modern microscope, took a first look at his own semen in 1677, he concluded that each contained a mini human inside it. All it did, he posited, was unfold and grow in the woman's body, who nurtured it (he rejected the notion that women produced eggs).[4]

So the connection between female and male physiology and what a woman or man "should be" was cemented. Well into the twentieth century (and even the twenty-first), cultural and medical narratives depicted the menstrual cycle the same way women often were: as wasteful (of bodily energy and resources), mysterious, and irrational. Eggs were described as placid ladies-in-waiting, beneficent, and nurturing, which mirrors the stories told about how women in Western culture were supposed to act. These stereotypes informed the narratives told in scientific journals and medical texts, which in turn influenced how doctors and researchers thought about male and female reproductive organs.

Attaching negative, passive attributes to ovaries, periods, and the uterus (and positive, active qualities to male reproductive organs) continued well into the era when many doctors practicing today were being trained. In the classic text *Medical Physiology*, edited by Vernon Mountcastle and still in use in the 1980s, he wrote, "Whereas the female sheds only a single gamete each month, the seminiferous tubules produce hundreds of millions of sperm each day."[5] The female "sheds" (passive) a single item—when actually it's quite a battle between the eggs—and the male "produces" (active) volumes of useful material.

In 1991, Emily Martin, an anthropologist, took a survey of textbooks being used by medical students at Johns Hopkins University. Her essay "The Egg and the Sperm: How Science Has Constructed a Romance Based on Stereotypical Male-Female Roles" fluctuates between hilarious and disturbing.[6] It's also a total takedown of how cultural fairy tales get written into science. Martin noted that in text after text, the language used for eggs described them as queenly, princess-like, and regal, wishing and hoping for a sperm (like a woman pines for a man, of course). The egg is always an object that waits, passively, for its rescuer sperm. This was in material intended for med students!

According to a description from Martin's essay (with phrases from the textbook in question appearing in quotation marks), in one book the authors

> liken the egg's role to that of Sleeping Beauty, "a dormant bride awaiting her mate's magic kiss, which instills the spirit that brings her to life." Sperm, by contrast, have a "mission," which is to "move through the female genital tract in quest of the ovum." One popular account has it that the sperm carry out a "perilous journey" into the "warm darkness," where some fall away "exhausted." "Survivors" "assault" the egg, the successful candidates "surrounding the prize." Part of the urgency of this journey, in more scientific terms, is that "once released from the supportive environment of the ovary, an egg will die within hours unless rescued by a sperm." The wording stresses the fragility and dependency of the egg, even though the same text acknowledges elsewhere that sperm also live for only a few hours.

Those are the nicer categorizations. Ovaries are also described as "scarred, battered" organs that are inherently wasteful. Menstruation is depicted as a failure to get pregnant, and so when the lining makes its way out as a period, it is described as waste, scrap material, or debris. Although sperm production could also be seen as wasteful, with all those extra sperm so much debris, this extraneous material is now a wonder. One text marvels at the seminiferous tubes within the testicles, which "span almost one-third of a mile!" and are able to produce a million sperm cells a day, leading the author to wonder, "How is this feat accomplished!" While the egg is transported, drifts, or "is swept" along, streamlined sperm "deliver" their genes to the egg, using efficient tails, so that with a "whiplash like motion and strong lurches" they can "burrow through the egg coat" and "penetrate" it.

Martin provides five pages of examples. This is just some of the cultural background informing what doctors practicing today have learned about women's bodies. Those who went to medical school in the 1990s would be reaching their professional apex in the 2020s, while those trained in the 1980s are now emeritus advisors, board members, and mentors. It's important to keep this in mind when we examine what has changed and what hasn't.[7]

A line can be drawn between the history of demeaning language to describe menstruation and the female reproductive system and the lack of interest in studying female bodies more broadly and the menstrual cycle in particular. That's been changing and continues to, thanks to the work of mostly female researchers. Historically, where there is more attention and detailed data, it's usually focused on pregnancy and fertility, not on women's health in general. But the menstrual cycle affects the female body in many ways outside reproductive ability. (To illustrate: The National Institutes of Health [NIH] spends over $200 million dollars a year on infertility research, but just $29 million on endometriosis. That's up from just $6 million in 2017 due to recent outcry about the disease, which affects around 10–15 percent of all female bodies, compared to 11 percent of those affected by infertility—and that percentage comes from only those trying to conceive, not all female bodies.[8]) People like me who are child-free feel this lack of care and information especially keenly, since health, not reproduction, is our primary goal.

It can be shocking to learn about the many unknowns that surround menstruation, which has wide-reaching health effects, can begin as young as eight or nine years old, and continues until menopause. That's about half the lifetime of half the people in the world. Researchers are now working to fill in the significant data holes on this subject, primarily in adolescent and perimenopausal people. For example, the most current and commonly used research on first periods is a 2008 study of just ten girls' hormone levels.[9] Natalie Shaw, a pediatric endocrinologist at the NIH, initiated the first long-term, large-scale research into this subject in 2019, titled "A Girl's First Period." According to the National Institute of Environmental Health Sciences, "Despite its undeniable importance, surprisingly little is known about the biological underpinnings of a girl's first few menstrual cycles."[10] The earliest results of Shaw's study on menstrual pain in girls was shared at an Endocrine Society meeting in 2024. Thanks to Shaw's work, plenty more information will come to light about this understudied area in the coming years.

A FLEXIBLE CYCLE BY DESIGN

If you are someone who has a period, you may have noticed it's not always the same from month to month. Sometimes the full cycle is a bit (or a lot) shorter or longer than other times. You might get cramps occasionally or always or never. Sometimes you bleed for three to four days, other times longer. One month you get grumpy right before you bleed, and another you realize you have more energy. You might feel yourself ovulate, or you might not realize that some people experience that sensation because you never have. You might be able to do everything you normally do while you have your period, from work to working out, or you might have intense migraines or digestive effects or depression that makes it hard to get out of bed.

We are taught that our cycles are supposed to fit within pretty narrow parameters, repeated every month like clockwork until pregnancy or menopause. Recent research shows that this is far from the case. The menstrual cycle is highly adaptable and reactive to a number of inputs—by design.

That "perfect," regular cycle—the kind I saw as a grainy projection on a pull-down screen in high school, or that you might have seen in a three-dimensional YouTube illustration if you're younger—is a simplified cartoon of the reality. Those representations are like tourist placemat-maps compared to the smartphone maps that include topography and traffic conditions. They're 2D illustrations versus the 3D reality.[11]

The menstrual cycle isn't like those simple depictions because it is meant to change, fluctuate, and adapt to current conditions. It can, and does, react to illness, immune strain, nutrition, and stress. This variability is imperative because the menstrual cycle requires an energy commitment. It simply doesn't make sense for the body to use resources it lacks to menstruate and ovulate. This is a smart hedge. If resources are scarce, having a period isn't just a waste of energy in that moment. It's dangerous over the longer term because it could lead to the greater resource strain of nine-plus months of fetal growth, combined with the stresses of childbirth, the postpartum phase, and breastfeeding. The sensitivity of the menstrual cycle to environmental conditions is a check on the system, a way to preserve the health of the individual by preventing the menstruating body from making the mistake of committing to a much larger energy investment in pregnancy.

Our bodies have evolved this incredible sensitivity to protect themselves. Importantly, menstrual variabilities don't only reflect a temporary lack of food or illness; they can indicate all kinds of health issues. In 2015, the American College of Obstetrics and Gynecology (ACOG), supported by the American Academy of Pediatrics, advised that health care professionals consider the period a fifth vital sign. The menstrual cycle can be utilized as a health monitor—as useful an indicator of overall health as body temperature, pulse rate, breathing rate, and blood pressure.[12] The list of diseases and disorders that are linked to abnormal menstrual bleeding on the ACOG site is long. It includes thyroid issues, various tumors, sexually transmitted infections, pituitary disease, platelet function, bleeding disorders, and more. This vital sign is possessed only by those who have uteruses that shed a lining each month.

Rather than seeing menstrual adaptability for what it is—a way to keep the body strong and efficient—medical history includes myriad takes that interpret the cycle's flexibility as instability. Under the negative light that has been thrown on the female body for hundreds of years, the menstrual system isn't resilient or a window into health concerns. It's unreliable. Untrustworthy. Weak. Those who start with the idea that female bodies are inferior to male bodies then see menstruation as proof of their beliefs.

This is how you twist a physical strength into a story of weakness, one of the many ways women have been disempowered. Cycle variability is just one example.

THE EPIC BATTLE OF THE EGGS

I'm sure you've heard the variation on the phrase "big balls," given a feminist twist to become "big ovaries." I thought it was a bit silly, but after I did a deep dive into the newest research into ovarian egg production, I'm a convert. While both sperm- and egg-producing structures are important for manufacturing the stuff that makes new humans, the ovaries do something way more interesting, more complex, and more powerful than balls do. And for the record, eggs, the largest cells in the human body, have a volume about ten thousand times greater than that of sperm cells.[13] Much bigger. Just saying.

Let's get into some biology to understand the powerful egg. Male testes produce sperm packets, which are genetic packages with their own motors. They are one of the few cells in the human body that are designed to function in another person's body. Which is cool and, obviously, necessary for sexual reproduction. Sperm competition occurs on a short timescale, as the cells only last about a month and a half inside a male's body (and up to five days in a female's body). On the whole, sperm production is pretty straightforward and happens much like the creation of other types of specialized cells in the body (all of which have their interesting and unique aspects, for sure).

For the most part, sperm is the way it is because it's responding to the biology of the egg. "When we talk about sperm evolution and behavior, what we often fail to recognize is that the egg and its surroundings are involved in every step of the process," Rachel Gross wrote in her book *Vagina Obscura*. She quotes Syracuse University biologist Scott Pitnick, who studies sperm evolution: "If I had to wager, it's that female reproductive-tract traits are evolving independently of ejaculate traits," he told Gross. "Males are just struggling to keep up."[14]

The manufacture of motile sperm, while it has complexities all its own (like producing a flagellum), is a less complicated process than what happens in the body of someone who is ovulating. Eggs are formed in the body of a female fetus *before* birth, and they mature and compete over a lifetime. The egg (or, rarely, eggs) released during each cycle comes from a vast supply produced while the person was in the womb. (But note that it's totally normal to have eggless cycles, especially just following menarche and prior to menopause.) Yes, this means the egg that created you was first made when your mother was a fetus inside your maternal grandmother's body.[15] Babies with ovaries are born with all the eggs they will ever have—about one million to two million oocytes (immature eggs).

Throughout life, regular, ongoing culling events destroy unworthy eggs and redistribute extra, valuable parts of those cells for use elsewhere. Throughout the process, only the very best eggs get to the stage where they can compete to be Egg of the Month.[16]

All these numbers depend on how many eggs you start out with, but by puberty, a kid's body contains three hundred thousand to five hundred thousand eggs.[17] By age thirty, about fifty thousand to sixty thousand are left, or around 15 percent of the original supply (that number varies a lot by the individual). By menopause, estimates suggest there are anywhere from one hundred to one thousand eggs remaining in the ovaries. If this seems like way more than even the Duggars could use, that's because so many are lost in the truly great battle each month among eggs to be The One.

Ovarian follicles are tiny sacs that contain the immature eggs a person is born with. There are thousands of them, one for each oocyte, holding them snugly inside, protected. Follicles don't just keep the egg safe; they also produce estrogen, and they include cells that can secrete other hormones or form connective tissue. The follicles are the building blocks of female egg selection. Follicle development is turned on and off by hormones and "tightly regulated by crosstalk between cell death and survival signals," as Chinese researchers at Sun Yat-Sen University wrote in a 2013 paper.[18] They added, "The ovary is an extremely dynamic organ."

Here's why: Since the 1960s, animal science researchers have known that most domesticated animals ovulate in a staggered process that's more like several waves than a single event.[19] But it wasn't until twenty years ago that research determined that humans undergo follicular waves too. This is important because it means there's not just one battle of the eggs each month, but multiple rounds, gladiator-style, to determine a winner. This isn't a one-and-done competition, but an ongoing fight to the death—for life—every month.

We might not know this if, in 1997, Angela Baerwald hadn't been determined to find out if the follicular waves that occur in animals happened in humans too.[20] Part of the reason for the almost forty-year delay between animal and human tests is down to researchers' long-standing disinterest in female bodies. But it was also about waiting for the right technology to come along. Transvaginal ultrasounds, which only became widely available in the mid-1990s, allowed Baerwald to safely and fairly easily examine research participants' ovaries. (Previous work in animals was done by removing and comparing cow ovaries after slaughter. By the 1980s, live mares' ovaries could be scanned using the first ultrasound machines for this purpose—which were much too large to use on human bodies.)

In her tests, Baerwald saw fifty research subjects every day over the course of one of their menstrual cycles. She scanned their ovaries to count how many unreleased eggs, nestled in their follicles, were present. Yes, that was a lot of scanning and counting, and it went on seven days a week for months. The results upended the previous story of how humans ovulate.

Baerwald's groundbreaking research, published in 2003, "challenges the traditional theory that a single cohort of antral follicles grows only during the follicular phase of the menstrual cycle."[21] That's restrained scientist writing for "This radically changes what we know about ovulation."

Baerwald's research points to busy, dynamic ovaries, not part-time workers. It shows that, like other mammals, human test subjects underwent two or three waves of follicle development, with major and minor waves. "Major waves are those in which a dominant follicle develops; dominant follicles either regress or ovulate," Baerwald and her colleagues wrote.[22] Minor waves are defined by their lack of a dominant follicle; the eggs are "competing," but none are winners. Ovulation is spurred by estrogen: Whichever egg makes the most hormone in the shortest amount of time seems to be the winner, but all the specifics aren't yet understood. What is clear is that ongoing, complex chemical actions and reactions occur in the ovaries with varying waves of intensity. Baerwald's later work found that the number of waves a person has each cycle, and how many are major or minor, depends on the individual.

Baerwald's discovery was called "profound" by those who work in the field of assisted reproductive technology because it upended the idea of a single, simple cycle.[23] This knowledge allows more egg retrievals for those dealing with infertility and may help explain why some people can ovulate (and get pregnant) at unexpected times. It's also a win for basic research into how the female reproductive system works.

Since the body dedicates resources to these repeated egg battles, they must be important. During a follicular wave, the eggs' mitochondria (remember the powerhouses of the cell from biology lessons?) replicate, increasing in number from about two hundred to about six thousand per oocyte, a process that uses energy to create more power in the cells for their battle. This is just in the early competition of a single follicular wave. Dozens of eggs grow their mitochondria in anticipation of being Egg of the Month.

To recap: During each twenty-eight(ish)-day cycle, pre-eggs grow into semi-eggs, and they do battle by growing their internal power plants and hormone production, a type of chemical warfare, to become the dominant egg. Then the cycle repeats in another follicular wave. Depending on the person, it may repeat again, until the Egg of the Month is chosen. Every human alive today, save those created with in vitro fertilization (IVF), are made from that powerful Final Egg (yes, I'm referencing the horror-movie cliché).[24] The Final Egg will grow to have more than three hundred thousand mitochondria, and to fully develop its powers and complete its total takeover of the ovary, it uses the hormones built up by the unsuccessful eggs. Ultimately, the failing eggs must be destroyed.

If that's not an epic battle, I don't know what is.

EGGS ARE CHOOSY (BUT WE KEEP FORGETTING THAT)

Making all your eggs at once, stress-testing and dumping most of them, and having one (or, rarely, two, three, or more, as is needed for twins, triplets, etc.) available at a time for fertilization is a mammalian adaptation. It represents a shift in reproductive strategy, according to Lynnette Sievert, a biological anthropologist at the University of Massachusetts, Amherst.[25] That shift was away from a more ancient method of reproduction, which fish, amphibians, and most reptiles still employ to great success. They make both eggs and sperm in great quantities throughout their lifetimes. Female fish and frogs expel masses of eggs into the water, and the males shoot, deposit, or generally aim their sperm in the eggs' direction. The eggs that get fertilized then develop—or don't, due to environmental conditions—or they are eaten by predators. Sea turtles have sex, and they lay hundreds of fertilized eggs at a time, repeating this cycle until they are elderly. So do oviparous snakes (viviparous snakes give birth to live young). For all these animals, reproduction is a numbers game. Lots of eggs, lots of sperm, plenty of fertilized eggs and hatchlings, with just a few offspring surviving to adulthood. In many cases the newly hatched turtles, tadpoles, and wee snake babies provide an important food source for other animals who live in their ecosystem, like a biological offering to the greater community.

This more-reproductive-stuff-is-better design is still employed by male humans, but not by females. "Human males still follow the fish pattern. They're still putting out a million sperm," Sievert said. "They're not cleaning the sperm, they're not putting out the best sperm, they're just putting out all the sperm like a fish." She wonders why female mammals made a significant shift away from that model. "Why was there never a selection on male sperm and mammals to be like eggs? Something shifted that separated the sexes," she said. It's an unanswered biological question, but there is at least one possible answer: control.

Female mammals house inside their bodies the mechanisms through which eggs (and sperm) are used for reproduction, while amphibians, reptiles, and fish let outside ecological conditions like temperature, predators, salinity, and pollutants decide who lives and dies. Both strategies are clearly effective, but why would mammals have shifted away from a successful model? It could be that longer-lived mammals are able to store epigenetic information about local conditions as they grow, which could influence when and which eggs and sperm are utilized. The choices about which eggs or fertilized eggs continue to develop and which don't are made before, during, or even after conception, resulting in offspring that are best suited to current conditions.

Why all this trouble to "turn your body into an eggshell"—as Cat Bohannon put it in her book *Eve: How the Female Body Drove 200 Million Years of Evolution*—when the real eggshell or other reproductive strategies work so well? It could be explained by a combination of energetics and fine-tuning. By bringing both fertilization and the growth of their young inside the female body, mammals can use their lived experience (not just conditions at the moment of conception) to affect which traits are selected for. At the very earliest part of the process, female mammals' bodies do this by controlling both which egg and which sperm are preferred.

The fact that eggs choose sperm is a basic biological fact that has been "discovered" quite a few times over the years. As Emily Martin detailed in her memorable paper, the prevailing narrative once painted sperm as the active party in fertilization, with all the speedy, tough sperm racing to be the first to penetrate the egg's outer membrane to deposit their DNA. Back in the mid-1980s, it was first discovered that the egg is actually the active decider in fertilization. The egg does this by using its zona pellucida—a thick, protective protein coat—to chemically grab on to sperm.[26] A couple of years later, we got an even more nuanced picture of the zona pellucida; turns out the egg can screen an incoming sperm cell, test it, and then either reject or admit its DNA. While the sperm wiggles back and forth, it can't break a single chemical bond, but the egg can.[27] Research in the 1990s and since has further supported this model, and it's now widely accepted.

Still, over the last twenty years, this information has been treated as a "new discovery" over and over again, proving the stickiness of the sperm-as-hero story. In 2017, *Quanta Magazine* published an article about a researcher whose work was "challenging . . . dogma," asserting that "the egg is not the submissive, docile cell that scientists long thought it was."[28] Two years later, a University of Virginia magazine article, "Fertilization Discovery Reveals New Role for the Egg" stated, "The old notion of the egg as a passive partner for sperm entry is out. Instead, the researchers found, there are molecular players on the surface of the egg that bind with a corresponding substance on the sperm to facilitate the fusion of the two."[29]

An explanation for this continual "rediscovery" of previously known scientific information about the interaction between egg and sperm was featured in a *Ms.* magazine article in 2024 about Evelyn Fox Keller, a pioneer in the field of feminist philosophy of science.[30] "One of Fox Keller's key findings was that seemingly neutral assumptions in biology can in fact be gendered. Keller's informed social analysis of the sciences paved the way to approach science as a cultural phenomenon," wrote Kalini Vora. The fact that some researchers and the science press continue to repeat the same "discoveries" for decades illustrates how gendered ideas stick in our culture and can hold science back.

The newest evidence indicates that not only does an egg decide which sperm to admit; the egg may attract or repel different sperm even before they reach the egg. In 2020, scientists at Stockholm University, in collaboration with colleagues at the University of Manchester, discovered that eggs release a chemical that can attract sperm as they travel.[31] They also found that different eggs attract different varieties of sperm. In some cases, the eggs attracted sperm that did not come from the woman's partner. They figured this out by obtaining reproductive material from couples who consented at an IVF clinic in Manchester, UK. "Each experimental block comprised the follicular fluid and sperm samples from a unique set of two couples, exposing sperm from each male to follicular fluid from their partner and a non-partner," the researchers wrote of their methods.

Chemosensory communication between eggs and sperm allows "female choice and bias[es] fertilizations toward specific males," the authors of the 2020 paper wrote. What are the egg's criteria? It's unknown at this point. It could be selecting higher-quality sperm or sperm that's more genetically compatible in some way. "This shows that interactions between human eggs and sperm depend on the specific identity of the women and men involved," one of the researchers, John Fitzpatrick, told the science news platform Labroots.[32] Fitzpatrick thought that the choice of sperm was entirely up to the egg.

The science shows that, contrary to some cultural stories, the menstrual cycle is highly sensitive in order to conserve energy; eggs go to war each month and only the strongest survive; the winner egg sends out come-hither signals to the sperm it prefers; and then it chooses which sperm to unite with to make a potential new human being.

So much for the inherent weakness of women's bodies and the passive female reproductive system.

A DORIAN GRAY UTERUS

The eggs and ovaries aren't the only female reproductive parts with some long-ignored powers. As Gross put it so well in *Vagina Obscura*, "Dynamic, resilient, and prone to reinvention, [the uterus] offers a window into some of biology's greatest secrets: tissue regeneration, scarless wound healing, and immune function."[33]

The uterus, as we all know, grows and discards a lining every month or so for decades, unless interrupted by pregnancy, when it serves as an embryo's first source of nutrition and protection. But that lining is way more than just sometimes-inconvenient blood.

The process of rebuilding the lining starts even as the old one is being shed. It begins with a thin layer of epithelial cells, arising from the remnants of the discarded tissue, which rapidly cover and repair the raw surface. Out of this grows the new uterine lining, sending spiraling blood vessels out to nourish and support the growing glands, which sense an embryo if fertilization occurs. This entire process is driven by estrogen. In just a week or so, a new lining, measuring 1/3 to 1/2 inch thick (8–15 millimeters), is built.[34] Ovulation occurs (or not, since not every cycle drops an egg), and then the tissue is shed, causing some wounding to the uterus's interior walls. Immune cells assist in healing the uterus, but you'd never know there was any damage to look at it a week later: It heals scarlessly, every time, up to five hundred times in a menstruating person's life.

When Australian researchers wrote about this process, they compared the "unique 'rapid-repair' endometrial environment" to the "slower repair" skin environment.[35] They also identified some remarkable substances in menstrual blood. Nearly two hundred proteins were found in higher quantities in that fluid compared with those found in the other blood circulating throughout the body. These include enzymes and enzyme inhibitors (which help break down the lining and facilitate repair), antimicrobials, and antioxidants.

If this information gets you thinking that the uterine lining might be a treasure trove of material for health research and innovation, you're entirely correct. In fact, the truth is bigger than that. In addition to the components of menstrual fluid listed above, period blood also contains specialized cells called mesenchymal stem cells or MSCs, first identified there in 2007 by Caroline Gargett. Her work has revolutionized our understanding of where stem cells can be found, leading to an entirely new field of research in reproductive biology.

I spoke with Gargett on a cool day in early autumn from my father's home on the coast south of Sydney while she was at her office in Melbourne.[36] The neighborhood kookaburra interrupted our conversation a few times. Her discovery underscores the importance of including diverse perspectives in science and demonstrates how, when we allow it, life can lead us in important new directions.

Gargett was forty-five when she earned her PhD in hematology and accepted a position at Monash University in the obstetrics and gynecology department to study endometrial angiogenesis (the mechanism by which the uterine lining's capillaries are formed). It was while doing this work that she truly began to understand the incredible regenerative power of the uterus.

Endometriosis is the growth of uterine-like tissue outside the uterus, most commonly in and around the intestines and other lower organs, though it can migrate throughout the body—even to the neck. It affects one in seven of people with periods, usually takes close to a decade to diagnose, is incredibly painful, and is underresearched for the reasons outlined in Chapter Eight, on women's pain.

One of Gargett's early hypotheses was that specialized, remnant cells in the base of the uterine lining—cells that quickly heal the organ and regenerate the endometrium each month—didn't stay put. She suspected that these cells might be stem cells and were also present in period blood. This could be problematic because in most menstruating bodies, some menstrual fluid travels from the uterus into the peritoneal cavity (the large space in the torso below the heart and down to the pelvis) through the fallopian tubes. Her hypothesis was that if this blood contained stem cells that were functioning outside the uterus, it could be the cause of endometriosis.

Stem cells are rare, comprising about one cell in five hundred. First, Gargett proved that the migrating cells were unique cells, and then she proved they behaved like stem cells. Because this was new territory, she had to publish several studies to establish that these menstrual-derived cells could both replace themselves and produce more mature cells, continuing this process over time. Then Gargett demonstrated that they were specifically mesenchymal stem cells—cells capable of differentiating into fat cells, bone cells, and cartilage—per the standard criteria for definition as MSCs accepted by the International Society for Cellular Therapy.

This early work was necessary but took years. "It was slow going for us at the beginning, because we just had to get to these basic experiments, you know, and we did all the basic studies," Gargett said. It was well worth the effort. While some scientists initially ignored her research, she is now recognized as a trailblazer in the field and is working on a number of applications for these stem cells, with a primary focus on female-specific health care. One of her team's key projects involves incorporating these cells into a bio-printed, degradable mesh designed to assist with recovery from pelvic organ prolapse, making the implant more compatible with the body. Prolapse occurs when the vaginal walls weaken, often due to overstretching and damage, usually from carrying a fetus.

Gargett explained that the specialized stem cells can do two important things for prolapse. First, they "definitely alter the inflammatory response and dampen it right down," making the mesh more tolerated by the body. The cells might also implant and reproduce where they are placed, which could help strengthen the vaginal walls and repair other damage to pelvic floor muscles. This approach has been proven effective in animals as large as sheep, and human trials are next.

But what if prolapse could be avoided in the first place? Gargett thinks a woman's own stem cells could be used for this purpose: "We wanted to actually give them to women postpartum. Vaginal birth can be quite traumatic. A lot of women have trauma; it's been very underrecognized. About 60 percent of women do—it could even be higher—and we think perhaps they don't really fully recover," Gargett said. These women may develop prolapse later in life. "We want to give them their own cells back, just inject them into that area, say on their six-week checkup."

That got me wondering. I had heard of cord-blood banks where parents store their baby's blood for potential future use in treating disease. Could people who menstruate one day bank their own MSCs from their periods to improve women's health? "Yes," Gargett said. Thinking like a scientist, she envisions that such a bank would first be a boon to researchers, since it's a "very noninvasive way of getting these cells, [and] there's quite a lot of them available." She imagines both public and private banks, but hopes public ones will take the lead.

Outside reproductive health, MSCs are currently being researched for their healing potential in a number of areas, including wound healing, liver disease, Alzheimer's, stroke recovery, heart disease, and more.[37] Research published in the *World Journal of Stem Cells* states that MSCs derived from menstrual blood "are expected to become promising seeding cells for diabetes treatment because of their noninvasive collection procedure, high proliferation rate and high immunomodulation capacity."[38]

In addition to their myriad possible applications, menstrual-blood-derived stem cells, unlike stem cells from most other sources, are both versatile and easily accessible. "The endometrium is the only tissue from which adult stem cells can be retrieved via a routine office-based procedure without anesthetic," a release from the Yale School of Medicine states.[39] Otherwise, bone marrow is a common source of stem cells, and they can also be found in placentas, amniotic fluid, and fetuses.

For decades controversy has surrounded the use of stem cells in medicine and research due to their extraction from aborted fetuses. Stem cells have been used to develop vaccines against polio, measles, mumps, rubella, chickenpox, whooping cough, tetanus, and rabies, and they are currently being used to develop vaccines against Ebola, HIV, and dengue fever. Additionally, these cells are used for research into Down syndrome, degenerative eye disease, and other conditions.[40] Still, the scientific use of stem cells has been banned several times in the United

States. Most recently, in 2019 the first Trump administration abruptly banned NIH research using fetal tissue, a move that was reversed by the Biden administration in 2021.[41] In 1988, President Ronald Reagan had banned the use of fetal tissue for research, an action that was upheld by President George H. W. Bush when he took office.[42]

The existence of a source of stem cells that is produced by half the population on a monthly basis for much of their reproductive lives and is able to be acquired in a noninvasive manner is a game changer. Researchers at the School of Medicine at Zhejiang University in China wrote that MSCs derived from menstrual blood "have no moral dilemma and show some unique features of known adult-derived stem cells, which provide an alternative source for the research and application in regenerative medicine."[43]

This is a perfect example of why the study of women's bodies is important for all of humanity, and the fact that menstruation has been stigmatized, or even just ignored, has hurt us all. As Gargett pointed out, "I think [menstrual blood] has been a very overlooked biofluid. Because we do poo, we cough up stuff from our lungs, which is pretty gross. Menstrual fluid isn't that bad; it's only a bit messy," she said. It does need to be handled quickly as there are enzymes that break down the cells she wants to use. Collecting a sample via a menstrual cup on day two of the cycle appears to be the best method.

In the 1920s, there was still debate about whether menstrual blood could harm plants and animals, with papers discussing it as a possible "menotoxin" until the mid-twentieth century. In 1936, research was published deliberating whether menotoxin from a pregnant mother caused her child's asthma.[44] Can you imagine a world in which semen had ever been seriously considered toxic?

Not only is menstrual blood not toxic or harmful, but the opposite is likely true. This much-maligned blood could hold the key to healing—for all human beings, not just those who menstruate. What other discoveries are waiting to be uncovered deep in the less-studied female body?

Chapter Four

HACKING THE FEMALE BODY TO WIN AT SPORTS

The first ticker tape parade in New York City was held in 1886 to mark the dedication of the Statue of Liberty. Since then, the Manhattan tradition, which travels along the "Canyon of Heroes" from The Battery up to City Hall, has celebrated local notables, politicians (Eisenhower, twice), adventurers (Amelia Earhart, also twice), royalty (King Hassan of Morocco), and war heroes (the women of the armed forces following World War II). In 2019, the confetti parade was thrown for the US women's national soccer team, who'd won the World Cup for a second time. Forward/midfielder and all-around badass Megan Rapinoe captained the team to victory in what was called a "history-making performance" with sixteen million Americans looking on.

Often, postwin sports coverage is a festive yet boring exercise in ego stroking—self-congratulatory but bereft of new, interesting information. That year was different. When asked about what had helped bring her squad the World Cup victory, head coach Dawn Scott started talking about periods. Scott explained that the team members had been not only tracking their menstrual cycles, but also using the information to inform their training regimen. Part of the team's win was attributed to Aunt Flo.

Unexpected! Periods only cause problems and get in the way, right? Scott's quotes to the UK's *Telegraph* newspaper went viral, earning worldwide coverage and opening the door to a conversation about the menstrual cycle in athletics. The discussion even made its way to lots of people who don't watch soccer, via a segment on *Good Morning America*. "For a few players, I always noticed that just before they started their cycle, their recovery fatigue was increased and their sleep was less," Scott told the *GMA* hosts. "I was noticing it for three or four players and thought, 'We're six months out from the World Cup, how [can we] help that?'"[1]

THERE'S AN APP FOR THAT!

Was it possible that the monthly menses, long thought to be a wasteful energy drain, could be harnessed for athletic training, elevating and adding to the strength of female athletes? Scott, like many women, had known about menstrual cycle tracking for years, but she wasn't sure how to use the information to improve her team's training. Then she partnered with Georgie Bruinvels, who spearheaded the female athlete program at Orreco, the company where she went on to develop a period-tracking app, FitrWoman.

Bruinvels had spent years researching menstrual cycle phase and its impact on athletic performance; she had written her doctorate on the subject for University College London. Bruinvels's research upheld the common knowledge many people with periods would agree with: The hormone shifts associated with the menstrual cycle can make athletics feel more challenging.

Stacy Sims is also an advocate of period tracking for athletics. Sims, an exercise physiologist and nutrition scientist who has directed research at Stanford University, in California, and the University of

Waikato, in New Zealand, wrote a book, *Roar*, on the subject of women's unique training needs, and has made her career studying sex differences in sports medicine.[2] If you've heard that "women are not small men," it's not just because that's the title of one of Sims's TED Talks. She's trademarked the phrase. She also helped design the Nike Sync app (tagline: "Harness the Power of Your Menstrual Cycle").

Sims told me from her office in New Zealand that there's a good reason for considering the menstrual cycle during training, while also pointing out that it's mostly irrelevant in terms of game-day strategy: "If we look at mindset, and the positive self-talk that athletes have, and all the things that go into peaking for that one point in time [game or competition day], the menstrual cycle effect is negligible," said Sims. "That's because the psychological component and some of the other physiological responses . . . can supersede that."[3]

Sims and others say that adrenaline, screaming fans, and professional athletes' training override hormone impacts. The proof is in the pudding: Rose Lavelle scored the winning goal at the 2019 World Cup, the day before her period began.[4] At the same time, no athlete can succeed without day-in, day-out training. That is where, Sims said, "using a phase-based method gets results." Knowing whether an athlete is in the luteal or follicular phase can provide a hormone-backed boost to different types of training. But aside from whether you are running or lifting, tracking menstrual cycles can help in other ways too. Coach Scott and Bruinvels have developed strategies to mitigate the negative aspects of periods on the soccer players by changing diet, sleep, and other lifestyle factors to offset hormonal impacts. They found this positively impacted training days, helping athletes stay stronger throughout the month—and stronger overall.

Those who work with athletes report a small but real connection between training athletes with their cycles in mind and gaining greater strength. But how about Sims's assertion that game-day impact is negligible? The answer is, like many things having to do with our individual human bodies, some athletes are probably more impacted by their periods than others. A 2022 study showed that female soccer players ran shorter distances at their matches when they were in the early follicular phase—that is, when they were menstruating, and estrogen and progesterone were low.[5] This was a small study of only eight players that relied on estimations about hormone levels at various stages of the cycle rather than on blood measurement of hormones, but it looked at three seasons of data, and there seemed to be a real effect.

At the 2022 European Championships, the fastest British woman on record, sprinter Dina Asher-Smith, had to quit a race due to a severe calf cramp. Prostaglandins, chemicals similar to hormones, cause the uterus to contract during the menstrual period to shed its lining, but can also cause cramps in other body parts, particularly calves. When Asher-Smith spoke to the BBC, she was frustrated by the situation and called for more research into the menstrual cycle: "More people need to actually research it from a sports science perspective because it's absolutely huge. I feel like if it was a men's issue, we'd have a million different ways to combat things, but with women, there just needs to be more funding in that area."[6]

The state of women's sports science backs Asher-Smith's statement. Historically, just 6 percent of sports research has looked at female bodies exclusively. Researchers of the 2021 study who uncovered that stat wrote simply, "At present most conclusions made from sport and exercise science research might only be applicable to one sex."[7] So what we know about athletics and the menstrual cycle is very limited. When reflecting on why he hadn't focused more on female athletes in his work, Iñigo Mujika, a swimming coach and sports physiologist at the University of the Basque Country in Spain, was brutally honest. He admitted that, in hindsight, women were likely excluded because "female hormones were too complicated."[8] That's changing, and quickly, because the dearth of information appeals to graduate students in exercise physiology who are hungry to stake their claim on new ground. Sandra Hunter, the professor of exercise physiology we met in Chapter Two—and one of the few researchers who has studied women's bodies and sports performance for decades—said it's now a hot topic among her grad students. The same is true for Sims, who supervises PhD students studying topics like sex differences in concussions and ACL rehab based on menstrual cycle phases.

Grad students aren't the only ones interested in this understudied subject. Women's health research in sports has recently been acknowledged as fundamental. "The effect of the menstrual cycle on physical performance is being increasingly recognised as a key consideration for women's sport and a critical field," wrote the authors of a 2021 review paper from the University of Adelaide in Australia.[9]

Shruthi Mahalingaiah is one of the principal investigators in the Apple Women's Health Study team at the Harvard T. H. Chan School of Public Health. The Apple study uses crowdsourced data to answer pressing questions about topics like the menstrual cycle and sports, and activity patterns among nonathletes. "There's just an explosion of different kinds of tracking that put us on the cusp of learning a lot about ourselves as individuals and kind of population level trends," Mahalingaiah told *New York* magazine.[10]

The app FitrWoman is used by coaches and athletes, including the US women's soccer team and the US women's swimming team. A statement on the front page of the app's site reads, "What makes female athletes different should be celebrated and embraced, not avoided or marginalised."[11]

On a puffy-clouds-in-a-blue-sky summer day in New York City, I met with Jess Freemas, a female-focused exercise physiologist who works on the FitrWoman app.[12] I was running late because I had stopped to hear a cellist practicing al fresco under a pretty pedestrian bridge, and as I jogged through Central Park on my way to meet her, I found myself in good company. Runners of all ages, body conditions, genders, and states of undress zipped past me, some headed to and others away from the Shuman Running Track that encircles the Central Park Reservoir. Dr. Jess, as she's known by patients and colleagues, met me atop Bridge 27, which I had chosen for nostalgic reasons: My father, born and raised in Manhattan, had painted canvases of the park's bridges in the late 1950s. A painting of the lacy, ornate Bridge 27 hangs above me as I write this book.

Dressed in the NYC uniform of head-to-toe black—athleisure style with leggings and fab sneakers—and crowned with long, shiny chestnut hair, Freemas was eager to show me the latest update to the FitrWoman app. We plopped down on the edge of the Great Lawn in the sunshine, and she pulled out her phone.

The newest version of the app (which was in beta when I saw it) is intended for the individual athlete. The home page, thoughtfully designed and blessedly free of pink, featured boxes in greens, blues, and purples, displaying all kinds of information tailored to the user. Dr. Jess had already been entering her cycle info, including how heavy or light her bleeding was and other physical effects, throughout the month. She stressed that this also includes positive aspects of hormonal changes, like elevated mood, better coordination, and increased energy, alongside symptoms like cramps, bloating, and fatigue. The app offers options for mental health–related feelings as well, including anxiety and joy. The latest update also incorporates the type of contraception an athlete is using, if any.

All this info is factored into the predictions and suggestions the app makes. On the home page, it's clear where the user is in their cycle. The app displays the current day and its position within the four phases—bleeding, follicular, ovulation, or luteal—each represented by a different color. Typical hormone levels for each stage are also shown, allowing the user to see when progesterone is increasing or estrogen is decreasing. Each box is clickable for more details. "Essentially, it goes through what's happening at this time of your cycle and the physiological and psychological things that might be affected by those hormones," Freemas told me.

The more information the user provides over time, the better the app theoretically learns, making it easier for an athlete to track her body's changes and notice patterns. For example, by recognizing that you are likely to ovulate soon, you can anticipate a higher body temperature (which may be important when training in the heat) and understand that an increased heart rate is normal. Users can view their cycle data in easy-to-read graphs and charts, along with any physical or mental symptoms—both negative and positive—they've logged, to identify trends over time.

After tracking a user's cycle for a few months, the app can indicate when they are approaching the time of the month when they might experience mood fluctuations, allowing them to consider preemptive measures. "They can ask themselves, 'What do I need to do to be proactive about this?'" said Freemas, which permits better-informed decision-making. The app provides food recommendations, such as boosting protein and fat intake during certain phases, or increased antioxidant intake to help with certain symptoms at other times. When we met, Freemas was in phase three of her cycle, making it a good time for her to eat extra protein. "It's possible you may need more protein during this phase compared to others because progesterone is catabolic to amino acids and so easily breaks them down," she explained. She made the point a few times that all these recommendations are research based, and part of her job is to keep on top of the studies and adjust the app to reflect significant findings.

Freemas's enthusiasm for her work comes from her own lived experience. She found great solace in tracking her cycle after she was diagnosed with premenstrual dysphoric disorder (PMDD), which is characterized by severe and chronic mood and physical effects the week before and during the early days of one's period.[13] What she learned still informs the mission behind her work. "One of the big things when I first started really trying to manage my PMDD was tracking, because my mood would just switch, and I could realize, 'God, I'm feeling really, really off today and I'm feeling really bad about myself,'" she said. "There would be a week where I would just feel bad about myself for feeling bad." It was an upsetting and recurring experience. But when she started cycle tracking, she noticed something important: "Oh, this hits at the same time every month. So I feel more calm and comfort knowing [my mood] is related to my hormones, and I'm going to be OK." She said her mindset has totally switched. "Now I'm like,

I'm just going to take care of myself and do what I need to do to feel better, and not be so hard on myself." If she can help athletes—who often face high-pressure situations and stressful, complicated lives—achieve a sense of calm and control similar to her own, Freemas feels she's accomplished something important.

Since FitrWoman is used by world-class sports teams, the athletes' side is only half the equation. The other is the coaches' site, where they can quickly view information from all team members who are logging their data into the app. A coach might be able to see, as Freemas puts it, that "a player is logging fatigue, and they don't normally log fatigue, right? So they can find out if something is going on with that player through open communication." A coach could also know that a player is experiencing cramps or is feeling extra energetic, which could influence their training and recovery.

After leaving the animated Freemas and walking through the park back to the West Side, I couldn't help but reflect on both the utility of the app and also what it says about period shame. Freemas exhibited no embarrassment or reticence when she talked about her PMDD, her bleeds, or her irregular cycle. It was like discussing an ankle injury or my friend's Crohn's disease—simply a fact of physical life. For athletes, it's important to gain a deeper understanding of these experiences. Knowing the effects of one's menstrual cycle is simply more useful information for an athlete and a coach to consider in a training plan. I was struck by the absolute normality of the conversation, and I thought, "This is what it would be like if men had periods." (Of course, I'm talking about cis men here, as trans men can have periods.) It's just a matter of fact, something to utilize when useful and mitigate when problematic—whatever it takes to play the game and win.

I smiled to myself as I wandered over the Great Lawn, music drifting from teenagers playing in a brass band for a small crowd, the new skinny towers looming above the edge of the park in the distance. New York City is always innovating and evolving, and finally, so is the culture around menstruation. Change is good.

DIY TRACKING AND HACKING

What if you aren't a world-class athlete? Weekend warriors as well as casual runners, cyclists, swimmers, and hikers are striking out on their own with some Wild West–style biohacking, learning as they go. If you're interested, below is a summary of Sims's suggestions as a starting point for tracking your workouts to your cycle. (A similar guide can also be found at the Nike Sync web page.[14])

The beginning of the cycle probably has the most variability between people. Starting at day one (first day of menstrual bleeding), estrogen and progesterone are low, which might make it easier to work out hard. Other individuals may need more energy dedicated to menstruation itself. Taking this time to rest versus pushing it depends on how your period feels to you. This is the time of the month when it's most important for you to listen to your body. (I am one of those people who have their best runs just before and at the start of my period, so if having more energy at that time of the month sounds impossible to you, I can attest that it's true for some.)

Once the period is over and hormones are still low, energy, speed, and recovery are at their best. During this follicular phase, estrogen rises slowly and steadily from day six or seven through day fourteen (or the middle of the cycle), potentially providing more strength, endurance, and focus for sprints or heavier weight training.

During ovulation—around the middle of the cycle, days thirteen to sixteenish—it can be a bit easier to overheat due to higher average body temperature. In addition, blood plasma volume is about 8 percent lower, which means the blood is less efficient at delivering oxygen and clearing out lactic acid, so you might want some extra cool-down time.[15]

After ovulation, there's a big drop in estrogen levels going into the early luteal phase. (Right after ovulation, I always notice a significant drop in energy, and this is pretty common. The same run I did a few days earlier feels like more of a chore.) Think moderate-intensity resistance and weight-training workouts plus endurance aerobic exercise, fueled by extra carbs, rather than sprinting and HIIT (high-intensity interval training) workouts.

During the last week of the cycle, the late luteal phase, estrogen keeps dropping, and progesterone declines too. This is a good time for stretching, flexibility, and recovery exercises, like yoga and Pilates, long walks, and lighter weightlifting loads.

As we learned in the previous chapter, different people have different cycles, there is no "normal" cycle, and cycles can vary over a lifetime. So although the above is a good general guide, the point of training according to your body's cycle is to discover how to align with *your* body's cycle, not a generic one. That's where the power lies.

SLOW DOWN: CYCLE TRACKING'S DETRACTORS

Speaking of power, aside from learning about the amazing competitive and healing abilities of the ovaries and uterus—which fundamentally changed how I saw the female body—I wanted to know: Could understanding the menstrual cycle help female athletes reach a place of parity with male athletes?

The truth is, it's too early to tell if this aspect of training is a key to greater performance, or if adjusting other impacts, like those related to culture or type of training, would make a more significant difference. Despite the many apps now available to help you "train to your cycle"—and the experts who claim to have seen positive results—there are also those who question the concept. In 2021, Australian researchers published a narrative review that looked at all available research on the subject of the menstrual cycle and performance, including two meta-analyses, one that included seventy-eight studies and another that included twenty-one studies.[16] Some studies focused on sedentary women and others on elite athletes, and they incorporated a variety of tests, including anaerobic, aerobic, and strength assessments. After reviewing the data from all the studies, researchers did not find "clear, consistent effects of the impact of menstrual cycle phase on physical performance."

Some caveats: Authors of that paper (and the authors of the papers they reviewed) acknowledged that many of the individual studies included in the big-picture analyses weren't what scientists called "high quality." Either a given study had a small number of participants (like the one on soccer players I cited earlier), or it had high heterogeneity (meaning there was a lot of variability in the data, which usually indicates an issue with the study design). Maybe the study needed to narrow its lens or adjust the parameters of the information collected to make better sense of the results. Studies with high heterogeneity are often considered flawed, and it's an indication to researchers to look for those flaws. Additionally, low-quality studies can raise questions about the research methods used.

Brianna Larsen, a senior lecturer of sports and exercise science at the University of Southern Queensland, specializes in female athletes and the menstrual cycle. She knows these studies well. She explained that while having a better understanding of one's physiology is important, and she's seen some evidence for phase-based training, the research is not yet strong enough to make blanket recommendations for all menstruating athletes.[17] She said training for strength during the follicular phase might offer advantages over training the same way throughout the whole cycle. Furthermore, there's some evidence that higher estrogen and lower progesterone during the follicular phase may stimulate more growth hormone and reduce inflammation compared with the luteal phase.

"There is starting to be some promising stuff," she told me, but added, "What we've seen so far is so highly variable that I think some of these apps recommending all people change their training according to different times of the month are a little bit premature." Larsen also pointed out the complexities of applying this knowledge in real-world situations, because research is usually conducted outside high-performance environments. "It seems at least plausible that having all athletes in a women's team train differently (i.e., according to their own cycle) could have other impacts on team cohesion/dynamics that negate the benefits of phase-based training. Given [that] phase-based training is a lot harder to program for than team programming, I think it's too early to recommend this approach to coaches until more research is done."

Disagreement is typical in newer fields of study, and this subject certainly qualifies. It doesn't mean the scientists lack expertise or that there's nothing to be learned. Rather, studying a new topic can make it challenging to determine the best scientific approach. Testing methods, data collection, tracking, and follow-up all need to be refined and standards established. For example, previous standards for measuring heat tolerance, carb burning, and VO_2 max levels were based on men's bodies, so these scientific building blocks need to be re-established for female bodies. Larsen is one researcher who is actively working on setting those standards.

Updating sports-medicine research to include female bodies presents unique challenges. First, several hormones fluctuate throughout the menstrual cycle. Although fitness apps typically track against a standard twenty-eight-day cycle, many people have more variable cycles. It's also perfectly normal for a woman's cycle to change month to month based on factors like stress levels, nutrition, sleep, and time changes, as discussed in the last chapter. This means that while these apps can serve as useful guides for athletic training and understanding one's body better, their information isn't specific enough for rigorous research use.

To accurately understand how menstrual phase affects performance, researchers can't assume that a subject follows an average cycle—such as presuming that ovulation occurs fourteen days after the first day of their period, or that by day eighteen they are well into their luteal phase. Estimating or assuming hormone levels is now considered poor practice in research. Knowing exactly where a person is in their cycle is key for researchers (though costly and time-consuming), but it ultimately leads to much higher-quality data.

The only reliable way to determine hormone levels in a given body is to test subjects' blood each time they enter the lab. This added level of data collection—on top of whatever researchers are already examining—is "just logistically hard," said Larsen. "If you're testing a group of men, you say, 'Hey, all of you be here on Saturday, and we'll test you.'" However, if she wants to assess women's ability to, for example, thermoregulate during intense exercise in the luteal phase versus follicular phase, every female participant could be at a different point in her cycle on a given day. This means both that the phase needs to be determined by a blood test before the research can commence, and that testing might require multiple days.

"And that's before you even think about menstrual cycle dysfunction, which throws up different hormonal profiles—and also before you throw in the use of hormonal contraception, which completely changes the hormonal profile again," Larsen said. "So you're working with a big soup of different hormonal profiles on any given day, and to try and tidy that up neatly and test a homogenous group of athletes and compare them—it's just really, really difficult." Good thing she is determined. "It's not to say that because it's too hard, we shouldn't do it. We absolutely should."

These complexities are not just time-consuming for both researcher and subject; they're also expensive. Securing enough funding to support large studies that can accommodate varying schedules, multiple tests, and coordination is a challenge. As Larsen notes, this is why we keep seeing "little studies that add little pieces to the puzzle, but they don't ultimately answer the question." Researchers are navigating the financial and logistical limitations that are their reality.

Testing hormone levels is key to getting high-quality, scientific answers about menstrual cycle phases for another reason too. Athletic performance can be affected by more than biology, since, like all of us, female athletes have received a lot of messages over time about their bodies and menstruation, many of them negative. It matters that female athletes consistently report that they perceive their performance to be "relatively worse" when they had their periods.[18] Expectation impacts performance.

Meredith Reiches, a founding member of the GenderSci Lab at Harvard University and now an associate professor of anthropology at the University of Massachusetts Boston, has researched exactly this. Reiches wrote that when it comes to perceptions of menstruation, research suggests that cultural baggage is influential, with respondents in one study remembering feeling worse during their periods than they reported at the time.[19] "Thinking back to what their premenstrual and menstrual phases were like, people may unconsciously shape their narratives to meet the expectation that they had negative experiences," Reiches wrote. She explained that in another research study designed to test beliefs, women reported symptoms based on the cycle phase they were told they were in, rather than the phase they were actually experiencing. "That is, when told that they were premenstrual, they reported having premenstrual symptoms although they were not in fact premenstrual," Reiches wrote. So if an athlete hears that a certain cycle phase will inhibit specific abilities, their performance might be affected simply because they believe it will be.

This puts researchers in a tough spot, given the history of women's health complaints being ignored or downplayed by the medical establishment. On one hand, it's important to take seriously the negative experiences individuals report about their menstrual cycle. "Failing to do so can lead to the under-diagnosis of medical conditions like endometriosis, particularly for women of color, whose pain and suffering tend to get taken less seriously than the medical complaints of white women," Reiches wrote.[20] On the other hand, there has historically been so much cultural storytelling about periods that it can distort our personal experiences. I've been there—friends of mine simply don't believe that I feel energetic just before and on the first day of my period. Feeling differently than I was "supposed" to feel led me to doubt my own experience—and to shut up about it for a long time due to that doubt. The long shadow of menstrual expectations is yet another reason why direct hormone measurements are so essential.

Menstrual cycle tracking could turn out to be a big nothingburger or a breakthrough for female athletes. Right now, though, it's still an open question. Other factors discussed in this book—such as sex differences in responses to environmental stimuli; or female athletes having less lifetime experience, receiving less training, underfueling (see Chapter Seven), or being paid less—might have a far greater effect on an athlete than any hormonal variabilities.

PERSONALIZED PERIOD TECH

New technology could make hormone tracking much simpler, both for regular people curious about our cycles and for scientists. What if, instead of a blood test for exact daily hormone readings, you could just pee on a stick—like a pregnancy test—scan it with your phone, and find out how much estrogen, progesterone, and luteinizing hormone are in your body that day? What if you could compile those daily tests over the course of the month to accurately track when your body really ovulated, or how long your luteal phase really was?

Amy Divaraniya, the founder and CEO of Oova, did exactly that.[21] I met with her in a busy café in the midtown Manhattan building where her company is based. Oova grew out of Divaraniya's frustration with her own experience: She dealt with an irregular menstrual cycle from the time she got her period through college. At first her concerns were brushed off by health care professionals, but after college, she saw a doctor who diagnosed her with polycystic ovary syndrome (PCOS). A chronic condition, PCOS can cause hormonal imbalances (e.g., excess androgen), irregular periods, and ovarian cysts. It's incurable but can be managed.[22] "They took one look at me—I was studying for finals and had gained weight, and hadn't had my eyebrows done in awhile—and were like, 'You have irregular periods, you have PCOS.' They didn't do bloodwork, no ultrasound, nothing like that," she told me with an edge of anger in her voice.[23]

She dutifully took the medications she was prescribed, and got "every side effect in the book." She suffered from anemia due to constant bleeding and endured depression for several years. Then she saw a new ob-gyn who prescribed the proper bloodwork. She told Divaraniya she didn't need medications because her lab work showed she didn't have PCOS. "I was like, 'Are you fucking kidding me? Do you know what I just went through for three years?'"

This experience set Divaraniya on the path to creating Oova, which produces hormone-testing kits and an accompanying app for consumers. The first application was designed to help those dealing with infertility gain a better understanding of their cycles in (hopefully useful) detail. Divaraniya explained that it's common for people to assume they ovulate in the middle of their cycle. For someone with a thirty-two-day cycle, they might think the best time to try to conceive is around day sixteen. "But it may not be the case. Ovulation could be as early as day ten or eleven. And then you'd completely miss your window for getting pregnant, right?" she said. For those dealing with infertility, having this information could make the difference between becoming pregnant and not.

Knowing hormonal details at this level can foster a deeper connection with and understanding of one's body, said Divaraniya. "Our users are able to notice nuances in the hormone data month over month. Like noticing, 'Oh, when I ovulate from my left side, my cycle is thirty-one days, but when I ovulate on my right side, it's twenty-seven days.'" She considers the ability to recognize patterns at this level "really meaningful."

As a child-free woman in my forties, I was intrigued by the idea of using Oova to determine whether I was perimenopausal. I'd heard that perimenopause—the years leading up to menopause, which is defined as the first full twelve months without a period—could begin in the late thirties. Although I didn't feel any different, and my periods were the same as ever, I wondered what my hormones were saying. Did the pinch I felt in my side each month during the middle of my cycle mean I was still ovulating? I wanted to know!

Divaraniya was way ahead of me. Oova had launched its perimenopause program a few months before we met. As soon as it was available, I gave it a try and logged data for a month. I learned that I was still ovulating, and my hormones confirmed that I wasn't perimenopausal yet. This kind of insight can help someone figure out if symptoms they're experiencing are due to hormonal changes or something more serious. For many issues, doctors are quick to dismiss female patients around my age with the diagnosis "it must be perimenopause." These assumptions can lead to delays in treatment or even missed diagnoses. Since everyone who menstruates will eventually go through menopause, knowing where you are in that process is incredibly empowering.

Oova's tests could also simplify hormone tracking by scientists, offering an alternative to blood tests. Understanding individual differences in monthly cycles might benefit many: regular people wondering if their health issues are hormone related; those curious if they're nearing the end of their period years (like me!); people dealing with infertility; scientists who want to directly test their subjects; and, of course, athletes.

"The most powerful thing you can do is track your own cycle," Brianna Larsen told me. Doing so offers the most personalized approach to training for athletes—and the rest of us. In her research, she's noticed that some athletes show clear monthly patterns in their strength, while others don't—because, again, we all have different bodies. One person might find cycle tracking—or even specific hormone tracking with a tool like Oova—incredibly useful, while another might not.

Larsen, like Divaraniya, thinks it's powerful to tap into the specifics of one's own cycle, and it's a practice even nonathletes can benefit from, for overall wellness reasons. If there are days during your cycle when you know you generally feel more powerful, keep them in mind while you plan your training. On days when your energy dips, consider what kind of nutritional support might help. By tracking your cycle or your hormones, you can anticipate those times and take action before feeling off. If you tend to overheat at certain times of the month, is there a way to do some precooling before training? "I do think there's absolutely some merit in menstruating athletes knowing their own cycle and using that to their advantage," Larsen said.

There's a lot to keep in mind—the shortage of high-quality studies; the emerging research into menstrual cycles and performance; athletes' self-perception; the challenges of identifying exactly how hormones impact muscle strength, recovery, and fueling. But one thing is clear. As the authors of the 2021 metastudy concluded, "There is a need for further research to quantify the impact of menstrual cycle phase on perceived and physical performance outcomes and to identify factors affecting variability in objective performance outcomes between studies."[24]

STAY SAFE

Now for the elephant in the room. While the science behind all this is fascinating, I've been writing this chapter in a world where period tracking isn't monitored by corporations or the government. But the reality, especially in the United States, is that tracking periods carries significant risks for anyone who menstruates. As *New Yorker* writer Jia Tolentino pointed out, all of us who use smartphones, computers, apps, Amazon, email, and texting are part of a system that generates "vast, matrixed fatbergs of personal data assembled by unseen corporations to pinpoint our consumer and political identities."[25]

Government agencies are looking. Tolentino wrote, "In August, 2022, Mozilla reviewed twenty pregnancy and period-tracking apps and found that fifteen of them made a 'buffet' of personal data available to third parties, including addresses, I.P. numbers, sexual histories, and medical details. In most cases, the apps used vague language about when and how this data could be shared with law enforcement." She goes on to detail how the Department of Homeland Security "purchased access to location data for millions of people in order to track them without a warrant"—which was only revealed because the American Civil Liberties Union looked into the incident for a 2020 lawsuit.[26]

With abortion now banned or severely restricted in many US states, there are very real concerns about how data from period-tracking apps could be misused. This goes way beyond app developers selling your data to Google, Meta, or Amazon for targeted ads (annoying) and into the possibility of data being used to prosecute individuals for actions deemed illegal in some states. Although some apps claim to prioritize privacy, a BBC report indicates that when faced with a subpoena from law enforcement, companies comply and hand over data 80 percent of the time.[27]

Jessica Burgess was prosecuted in Nebraska for helping her daughter obtain medication abortion pills, with evidence obtained from Facebook messages after police subpoenaed Meta.[28] Both mother and daughter were found guilty and sentenced to jail for violating state law based on their online conversation. While this case doesn't involve a period tracker, it's an example of how information we consider personal can be used in court, leading to serious legal consequences.

Concerns about these issues have persisted in the United States since *Roe v. Wade* was overturned. In a 2024 study, researchers investigated the privacy practices of twenty popular menstrual-tracking apps and identified "problematic practices." The study revealed "inconsistencies across privacy policy content and privacy-related app features, flawed consent and data deletion mechanisms, and covert gathering of sensitive data."[29] Ruba Abu-Salma, an assistant professor of computer science at King's College London, told the BBC, "While female health apps are vital to the management of women's health worldwide, their benefits are currently being undermined by privacy and safety issues."[30]

If you live in a state or country where abortion is illegal, it's worth thinking carefully about what apps you use and what information is stored on them. A Reddit user summed it up when they wrote, "If the data being accessed could result in jail time, think really really hard before making any digital record. This is one instance to use paper. Easier to hide, and quicker to completely destroy."[31]

Part Three
PERSISTENCE: DURABILITY

Chapter Five

FEMALE FAT IS FUNDAMENTAL

Sometime after Latoya Shauntay Snell had passed the twenty-two-mile marker in the New York City Marathon, she heard a man shout, "It's gonna take your fat ass forever, huh?" Her home-stretch high, which she'd been enjoying until that moment, quickly turned to anger. She stopped in her tracks and yelled at the tall, balding white guy, losing time. She shook it off and finished the race. When she got home, she posted about the incident on Facebook, and other runners said they'd seen what had happened and told Snell, "He wasn't worth it."[1]

That was back in 2017, and it wasn't the first, or the last, time Snell heard from people who have opinions about her body size and abilities. Snell describes herself as a Black, plus-size, queer, and chronically ill athlete, and so she hears often about how she doesn't belong.[2]

Snell loves running marathons. I'd found her on Instagram, where the algorithm decided to do some good in this world and introduce me to the Running Fat Chef, aka Snell. First her videos made me smile, because she has powerful energy that telegraphs out though the tiny phone screen. I couldn't help but notice her fabulous style (currently: fierce loc knots in a bob with blue-green tie-dyed ends and killer black lipstick). But a few videos into getting to know Snell, I found myself crying. As many online activists and influencers do these days, she educates by bringing the most heinous, dumb, ridiculous comments to the fore and responding to them. And there it was—some rando on Instagram suggesting that Snell should start with shorter runs of 5K or 10K so she could "lose some weight" before she tackled the NYC marathon in the fall. Snell, in a bright pink T-shirt with the word "unfollow" printed across the front, replied to this comment by simply listing off her experience. It includes finishing over three hundred races, including three 50Ks (thirty-one miles each), two 60Ks (thirty-seven miles each), and a 100K (sixty-two miles). She has run the NYC Marathon seven times. The next slide on Snell's reply to this commenter includes her cycling feats, her deadlift of 425 pounds, and her bench press of 225 pounds. She ends the list with some humor, mentioning that she's also a "profanity expert, professional neck snapper, side-eye Olympian," and, "I pole dance stiffly for fun."

So very many people can't seem to wrap their brains around the idea that "My body size and disabilities don't make me inexperienced," as a post on her @iamlshauntay Insta account reads.[3] Some people refuse to believe she's an athlete at all. Snell is very tell-it-like-it-is and brooks no shit (like a born and bred New Yorker), but she's also patient—to an extent—and deeply interested in changing people's minds. She refuses to apologize for who she is or how she runs.

Snell didn't plan on becoming a running influencer sponsored by a shoe company. She grew up in East New York, one of the toughest parts of Brooklyn, and dealt with a trifecta of challenges: Her charismatic, outgoing dad was also a drug addict, there wasn't always enough food in the house, and her neighborhood was dangerous for a kid. "What's typically offered to Black and brown communities is that we have hopes and dreams and aspirations to be a basketball player, to be a football player. Girls are encouraged to do the cheerleading team.

You might have a dance squad," Snell told me on a long call that got deep and personal for both of us.[4] The stories she related about her father echoed those of my mother, also an addict. When she said that his unpredictability was difficult, and that sometimes his "humanity would really shine" (but then the drugs would kick in), my belly clenched in recognition. Thankfully, both Snell and I have done a lot of therapy—and have found some solace in exercise.

Snell started running when she faced a health crisis. Eleven years ago, while working as a chef, she experienced terrible back pain. A doctor misdiagnosed her with disc degeneration. The pain was actually caused by uterine fibroids, which were only recently diagnosed and addressed through surgery. She said she now moves around "with little to no pain" for the first time in her adult life.

As is the case for so many plus-sized women, both Snell's doctor and her physical therapists assumed her back problem was caused by her weight. This is probably part of the reason why they didn't look further into her pain. Snell took their advice seriously, concerned that if she didn't reduce her weight, her condition would get worse. "If I'm going to lose my mobility, why not use it the best way that I can for as long as I have?" Still hurting, she started walking and doing yoga, and she threw herself into cycling. An online friend convinced her to sign up for a half marathon. She was at the track when she (literally) ran into the group Black Girls Run. "They were so nice and just welcomed me. Before I knew it, I was making friends," she said. When she told them about her half-marathon aspirations, they started teaching her the ropes—how to breathe, how to open up her stride, and what training meant. Her first race was a 10K in January 2014.

She also changed how she ate. Snell said her eating started off healthy, but the pressure to lose weight was coming from every direction, and she quickly developed an eating disorder. "People were just selling me on this idea of what they perceived about fat people—that they have to be lazy. They're slow, they're slovenly. And I was just like, 'But I'm plus size, and I'm not lazy.'" She thought losing weight would make her a better version of herself, and she lost a hundred pounds. She was featured in *Redbook* magazine, and she was running a sub-ten-minute mile. She was part of the Nike Training Club, riding her bike fifty to one hundred miles per week, running forty miles a week, and strength training. Sometimes she ran at one or two a.m. to get her miles in before work. She was still a full-time chef, preparing elaborate meals for friends and family but not eating them herself. She looked tired, her skin was breaking out, and her mood was rotten.

On her way to work one night, she collapsed. Helped by a stranger on the NYC subway and by her coworkers, she got to the hospital. She explained her "psychotic routine" to the emergency room doctor—how much she was exercising, working, taking care of her son, and how little she was eating—and was diagnosed with an eating disorder. She was doing all of this while in significant physical pain from the "back problem" that losing weight was supposed to fix but hadn't. She had grown used to living with the pain.

She went into recovery for her eating disorder, gained some of the weight back, and felt a lot better. She finally received a proper diagnosis for her pain after cycling through doctors. She fell off the recovery wagon and got back on again. She started seeing a good therapist. Still, the message was the same: Despite how awful she had felt at a lower weight and how much better she felt weighing more, the "happily ever after" had to mean weight loss.

As she began to earn media attention, she said it was like an anomaly to people when she said she wasn't trying to lose weight. "They're like, 'What do mean you're a runner that's not trying to lose weight when you're fat?'" She was scouted by HOKA in 2019 and started posting on social media about running as who she is, not who anyone else thinks she should be.

Today Snell is still running and still not skinny. She wears lots of color when she's out training, and she films her ups and downs as an athlete. "Bright colors mean that you're going to be seen. It's something that you 'don't do' as a plus-size person, as a woman, as a Black person," she told me. She rejects those ideas because color reminds her of an older man who inspired her on the track many years ago—he wore colors, ran, danced, and was clearly having fun.

"When I got into the athletic space, it was really mostly to be able to preserve my life," Snell said. To some extent her voyage as an athlete has accomplished that, but in unintended ways. It turned out that losing weight wasn't a cure, and she's done most of her lifesaving herself, with the help of her husband, some evidence-based medical care from doctors who listened, and a couple of good coaches. One of those was Megan Roche (featured in the next two chapters), who told Snell she had to eat well and that she needed to "debunk everything out of my head."

Part of that debunking was discarding the notion that she had to look a certain way or be a certain weight to be a successful runner—or even a fast one. Snell now espouses "slow marathoning" and said she's in it for the joy. "I was able to enjoy the sport a little bit more when I started slowing down. I was able to focus on my breath. I was able to prioritize my form over trying to keep up with everybody else." Looking a certain way has given over to feeling a certain way, which has resulted in less injury and much more satisfaction. "I'm able to focus on what's happening around me versus trying to rush through it." Snell is running her way, and she's in it for the long haul.

Like a true marathoner, Snell keeps going. And like a true revolutionary, she turns the bitterness of hateful comments into sweet and sour learning moments that linger long after the next Instagram video has loaded. She's using her 122,000-strong (and quickly growing) following on social media to change what we think about who runs and who doesn't. "I get to inspire the people to live in their own bodies, to be honest about what they're going through, and not have to use somebody else as a measurement tool," she said.

FEMALE METABOLISMS ARE WILY AND COMPLICATED

Anthropologists have discovered that even while gathering and hunting most of the calories for their communities, and carrying, feeding, and caring for offspring, the average human female historically required 25 percent fewer calories than the average human male. (Yes, women hunted too; see Chapter Fifteen.) That remains true today. In the first trimester of pregnancy, a woman doesn't need any extra calories, and through the second trimester, her recommended calorie consumption is still below what a same-sized male would need. It's not until the final trimester of pregnancy that the female body needs more than a male's, by a little over one hundred calories per day.

This variability isn't only due to size differences; sometimes males and females are the same size, or women are larger. Human beings aren't very dimorphic (different physically between the sexes) compared to other primates. The body size differences between males and females in human populations overall amount to only about 10–15 percent depending on the group, making us minimally dimorphic compared to gorillas and orangutans, which are about 50 percent dimorphic.[5] Some of the sex variance in energy needs is due to body composition, as determined by the complex interaction between genes, hormones, and lifestyle, and some of it is due to deeper metabolic differences.

The majority of energy any body uses in a day—about 65 percent—goes to keeping the heart, liver, kidneys, and brain operating, even though they account for only about 5 percent of body weight. The rest of the energy a body uses is based on two things: (1) how much muscle and fat it's carrying, and (2) activity level. So a man and a woman with similar height, weight, and amounts of muscle and body fat—as well as similar labor and exercise schedules—should need a similar amount of calories. But since female bodies tend to have about twice as much fat and less muscle than male bodies, they generally need fewer calories.[6] Furthermore, female-pattern fat distribution (pear shape) has fewer health impacts.[7]

The idea that most of the difference in caloric needs between male and female bodies is due to body size and composition was backed up by a groundbreaking 2021 study on human metabolism which included eighty coauthors submitting gold-standard data from over sixty-five hundred subjects.[8] It found that men's and women's metabolisms are pretty much the same after accounting for differences in height, weight, and body composition. This finding ran counter to long-held assumptions: "These are basic fundamental things you'd think would have been answered 100 years ago," the study's principal investigator, Herman Pontzer, an evolutionary anthropologist at Duke University, told *The New York Times*.[9] Pontzer brings up a great point, which is that we definitely should know more than we do about metabolism and metabolic differences in all human bodies. This study was an important step in that direction.

Some caveats to the study: First, the researchers found that metabolic rate can vary by up to 25 percent from person to person—that's a large difference! While they didn't find variability between the sexes, there was quite a lot between individuals, which raises additional research questions. And some studies can be too large, said Stacy Sims, the TED-Talking exercise physiologist and nutrition scientist whom we met in Chapter Four.

Sims said all kinds of metabolic variabilities exist between groups of people, from genetic differences to geographical ones. Sun exposure and temperature can impact individuals' metabolism, and so can age (Pontzer's study reflects that one). Health issues like thyroid diseases, which affect about 12 percent of the US population (the rates are lower worldwide), diabetes (affecting almost 10 percent of the US population), or any disease that affects the liver or pancreas can alter parts of the metabolic system.

In her rapid-fire, high-energy way, Sims ripped apart Pontzer's huge study.[10] Not only did she insist there are sex differences in metabolism (which decades of other research has repeatedly shown), but, "We see there's differences not only between menstrual cycle phases, but also as you get closer to peri- and postmenopause. That has to do with insulin resistance," she said. "So this study was a really huge disservice to women, in particular, because they just got lumped into the same category as men."

She explained: Female bodies have more metabolic flexibility and variability over time, so combining female and male data can obscure important female-specific variances. Typically, male bodies age more gradually since testosterone declines slowly and regularly over time, about 1–2 percent a year after age thirty.[11] Women age more variably as a result of significant changes in hormones on a shorter timeline due to menopause. Putting men and women together in the same data set smooths the spikiness of women's variabilities, rendering them invisible. Sims said she's seen this issue come up time and again in studies. That the Pontzer study reproduced this problem visibly frustrated her. "It's a very good representation of men aging in a linear fashion, but the study did not take into account the changes that happened for women," she said. Sims argues that reading the Pontzer study uncritically could result in women at different life stages getting incorrect advice based on men's bodies, yet again.

FAT, METABOLISM, AND CULTURE, OY

The small yet meaningful differences in metabolism between the sexes are driven by both genetics and estrogen, which help women's bodies preferentially store and use long-burning fat rather than fast-burning carbohydrates. This burning of fat over carbs is likely why female bodies can persist in both lower-weight and isometric muscular work (as covered in Chapter Two) and excel in long-distance running, cycling, swimming, and hiking. The metabolic flexibility of female bodies provides an advantage in most physical activities—except for the short, intense bursts of energy that require what experts call "glycolytic capacity." Male bodies have a very real advantage when it comes to explosive strength due to glycolytic capacity, but at the cost of less flexibility, lower efficiency, and a higher propensity for metabolic disease.

Fat is a foundation of the female body's strengths. Yet not only has it been demonized, most of us (including me) have been taught to detest it. From the time I was a kid, I received consistent messages about the aesthetic horrors of fat. Incredibly rude comments about fat bodies have long been justified by the supposed health harms of fat. As a 1990s teen, sing-screaming along to Hole and hard-underlining a paperback copy of *Women Who Run with the Wolves*, I felt the need to hide my body under giant clothing to avoid judgment. As an adult I've realized that much of what I learned about fat was overstated, oversimplified, or simply incorrect. Of the many legends and half-truths I've learned about my body, those about fat frustrate me the most, because they are the source of so much self-hate and self-harm. Not to mention a terrible waste of time and money.

Learning how important fat is for female bodies, and how it's actually the source of unique physiological abilities, gave me a new perspective. In short, fat is power.

"You can never be too rich or too thin" is a quote attributed variously to Wallis Simpson, Coco Chanel, and Babe Paley, though Truman Capote swore he coined it. Whoever thought up this toxic phrase couldn't have been more wrong. Having too little body fat makes any mammal vulnerable to illness and even death. Too little fat on a human female's body leads to period cessation, triggering production of much less estrogen, which has cascading health implications, but especially bone weakness. Too much fat? More research needs to be done to understand what that means and what the health implications really are, especially for female bodies.

Fat comprises about 25 percent of the female body's total weight, compared to the average male's 15 percent (the amount varies by age for both sexes). Since those born female evolved to carry this extra fat, it's deposited in different places, is affected by different hormones, impacts health differently, and is used in different ways than fat on male bodies. As in other areas of research, what we know about fat often comes from tests and studies conducted on male humans—or male mice. What we know about fat and health for women is, at best, incomplete and, at worst, misleading. Starting with the body mass index (BMI).

Over the years in my work as a health and science writer, I've had a front-row seat to the so-called obesity epidemic. I've seen how the BMI—created in 1832 not by a doctor but by Adolphe Quetelet, a Belgian mathematician and astronomer who was interested in average body sizes—got turned into a diagnostic tool to measure "fatness." It's enlightening to read Quetelet's original research, because his many careful measurements of nine thousand white Belgian people were obviously never meant as a health evaluation.[12] Quetelet was clearly and primarily interested in finding and documenting population averages for height and weight of people at different ages. His was a data-gathering project. He saw his charts as a "solution of the following problem of legal medicine: *To determine the age of an individual after death, from the aggregate of his physical qualities*" (italics are Quetelet's, from *A Treatise on Man and the Development of His Faculties*, which is the basis for the BMI charts in use today).

Once you understand this context, it's clear why BMI doesn't account for racial or individual differences in where fat is carried, how much muscle a person has, frame size, or any other variabilities. So how did it become a key health metric used by doctors and other providers? Nutritional epidemiologist Ancel Keys decided in 1972 that these charts could be used as a measure of body-fat percentage. And the medical establishment followed because it was easy to put people into categories with a simple measurement. Unfortunately, BMI is not useful as a way to measure health. Iliya Gutin, a research assistant professor at Syracuse University, called BMI an "arbitrary, subjective label for categorizing the population."[13]

For me, the harm of BMI is personal. Once I first heard about the metric in my twenties and read the early coverage of how fat harms health, I worried. I've always been in the "overweight" BMI category. Despite my always-present muscles, I thought I must be fat, and I believed being fat would hurt me.[14] Based on that information, like many other women, I engaged in unhealthy behaviors to try to lose weight so I would be in the "normal" BMI weight category.

Determined to "get healthy," I went through a period when I ate so little and exercised so much that I acquired mononucleosis (glandular fever) and several concurrent infections, including pink eye, because my immune system was compromised. At that point, according to the BMI charts, for a few weeks I was finally in the "high normal" weight zone. I've never been as sick as I was during that time, either before or since.

In recent years, I have had patchy health insurance as a resident of the US. As a result, I've changed health care providers often. Each time I've seen a new doctor, without looking at or inquiring about my lifestyle, and seeming to ignore my (excellent) lab reports, they have advised me to lose weight, based on my BMI alone. I know I'm far from the only woman who has been caught up by BMI and the legitimizing force of the medical field's embrace of it. That's why the possible harms of BMI as a health care tool are especially important to consider for female bodies.

The reliance on BMI in health care settings doesn't just push otherwise healthy people toward eating disorders. It also negatively affects care when doctors assume extra fat causes conditions it doesn't (like Snell's "back problem") and thereby miss illness or disease.

Without being checked for other fertility issues first, women who have trouble getting pregnant have been instructed to lose weight. Ranae Lammonby told Australia's ABC News that she underwent gastric sleeve surgery in order to lower her BMI, a step her doctor recommended so she could get pregnant. She lost the weight, but it turned out endometriosis was the cause of her infertility, a condition nobody had checked her for.[15]

Due to BMI, people are regularly turned down for surgeries like hip and knee replacements, reconstructive surgeries after cancer and kidney replacement, and gender-affirming care. The Mayo Clinic reports that people die trying to lose weight to qualify for a kidney transplant, despite evidence showing that those with higher BMIs still clearly benefit from a new kidney.[16]

Doctors or insurance companies who refuse surgeries like these say it's because people with higher BMIs experience more breathing issues under anesthetic and more postsurgical complications. Some do—like those with existing cardiometabolic issues or diabetes. Not everyone with a high BMI has these conditions. If a person is healthy and active, that matters more to surgical outcomes than their weight, say doctors who screen patients individually rather than relying on BMI.

The evidence that higher weight means more complications during or after surgery has been refuted. In a UK study of almost five hundred thousand knee replacements over a decade, researchers found that "normal-weight" people actually had slightly higher mortality rates during the study period. The researchers wrote that their analysis of the data "shows no evidence of poorer outcomes in patients with high BMI."[17]

Yet BMI requirements persist, even for individuals undergoing top surgery, a procedure that removes breast tissue and reshapes the chest for transmasculine and nonbinary people. This is a real irony, because transmasculine people with higher BMIs often have larger chests and are unable to bind (hide) their breasts. The very people who need this lifesaving surgery are most likely to be denied it. Similarly, trans women seeking bottom surgeries often undergo crash diets in the weeks prior to their procedures to meet the prescribed weight. This continues despite a 2021 paper published in the journal *Transgender Health* that states BMI criteria "serve as barriers to essential surgeries and do not have an empirical basis."[18] A 2024 study that examined the relationship between BMI and postoperative complications in over twenty-three hundred trans men who underwent top surgery found no severe complications for patients with higher BMIs.[19]

Considering a person's individual health status when evaluating their suitability for surgery makes sense. Reaching conclusions based solely on their weight doesn't. It isn't scientific.

NOT ALL THE SAME FAT

Female bodies preferentially carry fat between muscles and skin (subcutaneously), most typically on the hips and thighs. Male bodies tend to carry fat around the organs in the abdomen (viscerally). Visceral fat is well known to increase the risk of heart disease, diabetes, and other health problems. When women carry too much visceral fat, it hurts them too; subcutaneous fat doesn't have the same dire impacts. Where on the body a person carries fat can vary, as determined by genetics.

A 2019 study from Uppsala University, in Sweden, involving over 360,000 individuals confirmed the sex-based differences in fat deposition. The researcher—surprised by how definitive the results were—described them as "striking."[20] Part of this difference is explained by hormones. Trans people who have medically transitioned consistently note changes in fat deposition locations after they've been supplementing with testosterone or estrogen.[21] As hormone supplementation for trans people becomes better understood in the coming years, it could lead to health insights for all people. Still, hormones don't fully explain the sex-based differences in fat deposition, and this is an active area of research inquiry.

Can excess fat hurt people's health? In some cases, yes. The proof is in the many examples of those who have improved their health by losing body fat. The opposite is true as well. Health conditions and/or medications can often cause fat gain. The point is that good health care assessments consider more than BMI. "We need to shift the way we're thinking about fat away from sheer quantity towards distribution, which seems more physiologically relevant," Guillaume Paré, a professor of pathology and molecular medicine at McMasters University, told *The Montreal Gazette*.[22]

According to Sonia Anand, a better way to measure the fat that impacts health (the kind directly linked to heart disease and diabetes) is through waist circumference. Anand, a cardiologist and expert in population health at McMaster University in Ontario, Canada, recently finished a large study in which participants' body fat was measured by MRI, the gold-standard measurement for fat tissue. She told me that comparing visceral fat to waist circumference shows a high correlation. "Not everyone needs an MRI. Just measure the waist circumference, and you have a pretty good idea of their risk of future things like diabetes, abnormal lipids, sometimes higher blood pressure, and cardiovascular disease," said Anand.[23]

Another measure, called the body roundness index (BRI), takes into account waist measurement, height, and weight to estimate abdominal fat. In a 2024 study of almost thirty-three thousand people (with an impressive fifty/fifty sex ratio and good racial diversity), body roundness was found to have a strong connection to mortality; the rounder a person was (or, interestingly, if their measurement indicated extreme leanness), beyond a certain ratio, the more likely it was to harm their health.[24] Since I'm heavy due to a lot of muscle, the BRI calculator put me in the "healthy" zone because I'm not very round. If only I'd heard this when I was younger.

The mess of a BMI system weirdly puts many athletic people in the "obese" category because they have a lot of muscle. It's not just bodybuilders who get misclassified. The American rugby player Ilona Maher's BMI measures almost 30 ("obese" is 30 and over), and she was trolled online for it. As an athlete, she knew better than to take BMI seriously, responding to the haters with a burn as strong as she is: "Alas, I'm going to the Olympics and you're not."[25] After she led the US rugby team to its first-ever medal at the Paris Olympics, she did a lengthy photo shoot for *Sports Illustrated* as part of its digital swimsuit issue, showing off in a variety of one-piece swimsuits and bikinis exactly how muscle-packed her "obese" body is.[26]

Heavier people aren't the only ones harmed by a reliance on BMI. It also leads to lighter people's health issues being overlooked. In a study of forty thousand Americans from 2016, about a third of those who were classed as "normal weight" were metabolically unhealthy based on standard criteria, including blood pressure, triglycerides, cholesterol, insulin resistance, and more.[27] The study also found that almost a third of the people labeled "obese" based on the BMI scale (and 16 percent who were in the most obese categories) were found to be metabolically healthy. A doctor or clinician making a determination about someone's health using only BMI would be getting it wrong much of the time.

This makes BMI an "extremely imprecise proxy" for fatness and not a proxy at all for overall health, wrote S. Bryn Austin, of the Harvard T. H. Chan School of Public Health, and Tracy K. Richmond, a professor of pediatrics at Harvard Medical School. "In fact," they continued, "BMI misclassifies a large percentage of individuals as unhealthy when indeed they are metabolically quite healthy."[28]

The problems with BMI multiply when you take into account people of different ethnic backgrounds, as Anand has. She published one of the first studies to show that BMI is correlated with health risks at different weights for different groups of people.[29] When she examined the elevation of cardiovascular risk factors in a large population sample, she found that dangerous blood lipid and glucose levels started at much lower BMIs "for nonwhite ethnic groups, including South Asians, Indigenous people, as well as Chinese people," she told me. Anand's findings, along with other data that backed it up, led to new guidelines for use of BMI in Asian populations by both the American Diabetes Association and the World Health Organization Asia-Pacific Region.[30] But what about other groups? Research has also suggested that BMI isn't predictive of health for Black and Inuit people.[31] Complicating things further, Anand says it's also not all about skin color, since genetic differences are more than just skin deep. Among people with Black skin, "you may have women from Senegal and women from the Caribbean, and they could have very different average BMI."

Years of criticism about issues with BMI led the American Medical Association to release a statement in 2023 saying that while BMI may be useful in looking at population-level statistics, it "loses predictability when applied on the individual level."[32] The statement also acknowledged that the AMA recognizes issues with BMI "due to its historical harm" and "its use for racist exclusion."

If fat were equally bad for all bodies, then you'd expect that male and female bodies in the fattest categories would die at the same rates. In a study of 190,000 people from 1964 to 2015, researchers found that the biggest women (defined as "morbidly obese" based on their BMI) had rates of cardiovascular death of 2.4 percent compared to men's rate of 5.1 percent.[33] Some research suggests that the lower incidence of metabolic disease in female bodies is due to specific mitochondrial genes active in their fat cells.[34] It seems likely that for these reasons, a study of almost four million adults, published in *The Lancet* in 2016, found that obesity is correlated with nearly three times more deaths in men than in women.[35] Perhaps this is because, as is widely known among medical professionals, female bodies are generally more metabolically healthy and have better insulin sensitivity—contrary to what might be expected due to carrying extra fat.[36] Male bodies have 50–100 percent higher fasting blood glucose levels, and they clear glucose about 15 percent slower than female bodies. Glucose uptake by muscle in males can be up to 50 percent slower. This is one reason that men are more than twice as likely to develop diabetes as women.

Fatness affects people differently, and it's perfectly possible to be fat and healthy, especially if regular exercise is part of your life. A September 2024 study in *Nature Metabolism* found that endurance exercise not only improves cardiometabolic health in people who are "overweight" or "obese" according to BMI; it also causes structural changes in even the least healthy type of belly fat.[37] When the belly fat of regular exercisers and sedentary people with the same BMI was compared, the exercisers had more subcutaneous fat than visceral belly fat, lowering their risk for health issues like diabetes. The exercisers also had lower inflammation and more blood vessels, which, according to researchers, meant they were using that fat to fuel their workouts.[38] Even belly fat's negative impact varies.

Tremendous harm has been done by conflating fatness with poor health. From overexercising to fraught relationships with food to full-blown eating disorders that kill people, the effects of demonizing fat, especially in women, have hurt so many. We can hope this is changing, but the evidence doesn't seem to bear it out. A 2024 study showed that body dysmorphia is still six times more common in young women than in young men.[39] Although only slightly more women are categorized as obese according to the BMI chart (and evidence shows that fat negatively affects women's health less than men's), over 80 percent of the people who take semaglutide drugs like Ozempic are women.[40] While many people have taken these drugs and found marked health improvements—demonstrating that the drugs' mechanism helps regulate the body's insulin production—for others, it is prescribed solely for weight loss due to the persistent social perception that fat is unhealthy in and of itself.

THE FAT-BRAIN CONNECTION

In the early 1900s, experts widely considered the consumption of dietary fat unnecessary. Yet when George Burr and his lab-tech wife, Mildred (she had earlier worked in the University of California, Berkeley, lab that discovered vitamin E), removed all fat from the diets of lab rats, the animals got sick, their paws and tails became inflamed, and they died after a few months. Nutritionists at the time insisted that a lack of fat wasn't the problem, but no matter what vitamins and supplements were added to the rats' diets, Mildred and George's rats kept dying.[41] When the Burrs reintroduced fat to the rats' diets, the animals recovered; they published these results in 1929. Determined to uncover the reason behind the rats' reactions to the no-fat diet, the couple conducted further research, eventually demonstrating that essential fatty acids are indeed vital nutrients for mammalian health. Not everyone immediately accepted the importance of dietary fat. The Burrs once received a letter of condolence from a fellow scientist who said they must be in error.[42]

Besides subcutaneous fat and visceral fat—both of which are types of white fat and are good for storing energy—our bodies also contain brown fat, which produces heat to keep us warm in cold temperatures (babies have a lot of this type). The fat in our bodies also assists in recovery from injury or illness, and it plays a significant role in one key feature that sets us apart from our primate cousins: our giant brains.

Human beings are generally very proud of our brains, but we wouldn't have them without fat, which is imperative for both brain development and maintenance. The brain is almost 60 percent fat; it even contains pools of phospholipids (fat molecules with a phosphorous head and fatty tail), which are crucial for both the physical structure of the brain and its ability to transmit information from one part to another.[43] Recent research shows that in people who don't eat enough calories to accumulate sufficient body fat, the brain undergoes a profound change—it literally shrinks.

A joint 2022 study between the Keck School of Medicine at the University of Southern California and the Stevens Neuroimaging and

Informatics Institute analyzed the brains of anorexia nervosa patients (80 percent of whom are women).[44] The researchers found "notable reductions" in three measures: cortical thickness, subcortical volumes, and cortical surface area. They were concerned by this finding because it could indicate the destruction of brain cells and their connections.[45] But there was some good news. According to a media release about the study, "By comparing nearly 2,000 pre-existing brain scans for people with anorexia, people in recovery and healthy controls, we found that for people in recovery from anorexia, reductions in brain structure were less severe. This implies that early treatment and support can help the brain to repair itself."

A lack of bodily fat may also affect those who don't suffer from anorexia. It is the suspected culprit behind other types of brain degeneration. Research out of the UK has found that people who were "underweight" in middle age (with a BMI below 20) had a 34 percent higher risk for dementia in later life.[46]

Fat creates a membrane around each cell, providing structure and protection while allowing hormones, nutrients, and other important compounds to enter and other substances to exit. Fat is an important messenger for the immune system, and it helps with recovery from illness and grueling physical work by reducing inflammation. Studies show that women's higher percentage of bodily fat is one reason why they recover from strenuous exercise better than men.

Like so many other areas, our understanding of how fat works, how much is healthy, and how much is harmful for female bodies will increase in the coming decades. One area where we have a better understanding of fat is its importance to the reproductive system.

FAT: THE "THIRD OVARY"

It wasn't until 1970 that Rose Frisch linked adolescent weight to the start of menstruation. Almost irrespective of height, menstruation begins once a person weighs around one hundred pounds. If this seems like basic information that should have been known long ago—since it doesn't exactly require advanced technology to research—that's true. Not only was this data about growing girls' bodies considered news back then, but Frisch had been discouraged from researching the connection. As Sylvia Tara wrote in *The Secret Life of Fat*, Frisch was told by male doctors that studying weight in women was "not worth investigating."[47] Nevertheless, she persisted. When Frisch presented her findings, she was mostly ignored, even by her colleagues at Harvard University. Thankfully, a few researchers in a group that focused on endocrinology and reproduction took her work seriously.

Once Frisch identified the connection between weight and the onset of menstruation, she sought to determine whether muscle, bone, soft tissue, or fat was the factor that pushed the "start button" for periods. In 1974, she published her foundational study, "Menstrual Cycles: Fatness as a Determinant of Minimum Weight for Height Necessary for Their Maintenance or Onset," in the journal *Science*.[48] She had discovered that fat was the key to menstruation, yet she was again mostly ignored.

Frisch's work gradually made its way into the world, and she began hearing from fertility specialists who realized her research could benefit their patients. One of them was Lawrence Vincent, a radiologist in New York City whose office was near a ballet studio. He sometimes treated dancers and observed that they often appeared undernourished and unhealthy, despite their rigorous exercise regimen. Frisch's work prompted him to consider how the savage weigh-ins that punished weight gain impacted the dancers' health. Vincent and Frisch collaborated on a study involving some of these dancers, establishing that those whose body fat dipped below 20 percent had irregular periods, while those with body fat below 19 percent stopped menstruating altogether.[49] This confirmed the connection between adequate body fat and menstruation—and later, fertility. "No one, it seemed, not even women's health experts, was aware that the reproductive cycle required body fat," wrote Tara. It took years for this information to reach the doctor's office.

Estrogen, the key to many of the female body's strengths, comes from two sources: the ovaries and the subcutaneous fat that female bodies preferentially store on the hips and butt. For this reason, some doctors have called fat a "third ovary." Both sources of estrogen—ovaries and fat—are required to produce enough hormone for reproductive ability.

Having too little body fat—and the consequent drop in estrogen—can cause periods to stop. It also directly impacts bone strength. "Fat and bone are like twins that come from the same birthplace. They can even turn into each other when prompted—it's been shown in a laboratory that after differentiating into a fat cell, that same cell can be provoked to turn into a bone cell," wrote Tara. The stem cell "decides" whether to become fat or bone depending on the needs of the body. Heavier people (or those who do weight-bearing exercise) need stronger bones, and so the stem cell becomes bone. The opposite happens in those without enough fat: The lower estrogen production causes stem cells to convert to fat, resulting in the weaker bones often seen in anorexic women.

Once a female body goes through menopause and the ovaries stop producing estrogen, fat becomes the only source of the hormone. This means that older female bodies need fat to produce estrogen for several important health reasons, including maintaining bone strength. It may also explain why it's natural to gain fat with age: When fat cells aren't able to produce sufficient estrogen, the brain notices and signals the body to retain more fat (to increase estrogen production). Hence, postmenopausal women tend to carry more fat. It's not a perfect system—while those with more fat produce more estrogen, individuals whose fat is primarily stored around the abdomen face a higher risk of heart disease, stroke, and inflammatory diseases.

Like menstruation and pain sensitivity, fat has been considered a female weakness. However, because female bodies preferentially use long-burning fat for physical work (whereas men primarily rely on fast-burning carbohydrates), fat is essential to what makes women's bodies powerful and strong. It plays a crucial role in reproductive capability, but it also contributes to stronger bones, balances hormone levels, supports longevity, and is vital for brain health. It's key to feats of endurance as well.

If fat makes female bodies powerful, is that why our ancestors, who generally lived in more egalitarian societies (especially preagriculture), revered fat—and why male-dominated systems demonize it?

Chapter Six

ENDURANCE, THE FEMALE SUPERPOWER

Below a mountaintop in Tyrol, Austria, rests a glacier called Hintertux. It takes three gondola rides to climb the 10,500 feet to reach an opening in the side of the glacier. A tunnel dug into the ice and compacted snow leads to an ice palace crowded with glittering frozen stalactites. Deeper still, a small Caribbean-blue lake of meltwater floats, a liquid core inside a body of ice.

Jaimie Monahan traveled there with friends to immerse herself. Unafraid, clad in a purple bathing suit, she dove into the zero-degree-Celsius water (thirty-two degrees Fahrenheit), which had been carried by the glacier for a thousand years. As she floated inside the belly of ice, the ancient walls shining back at her, her laughs were muffled by the close confines. In a video of the moment, she flipped over and started backstroking through the cavern just so she could "look up and kind of see the shape of the tunnels and everything."[1]

"The water is so clear and cold and pristine, I'm not exaggerating to say it was magical. It's one of the coolest things I've ever done," Monahan told me, with that far-away look of warm joy that comes over the faces of dedicated travelers, crushing teenagers, and cat lovers when describing the object of their affections.[2]

It means something when Monahan said that the awe-inspiring dip was one of the coolest things she's done. She has a long list of truly gobsmacking feats to show for four-and-change decades of life. A world-renowned cold-water swimmer, ice swimmer, and ultramarathon swimmer, the native New Yorker has swum the 28.5 miles around Manhattan thirty-one times, setting a Guinness World Record. That includes making the loop four times in a row, in a nonstop, solo swim of forty-five hours (setting another Guinness World Record). She's set other swimming firsts all over the world, including the Ice Sevens Challenge—which means she has swum an ice mile on all seven continents. That was her first Guinness World Record.[3] An ice mile is what it sounds like: swimming a mile in near-freezing waters.

Wearing a swimsuit, goggles, and a cap, Monahan is dressed just like the lappers at your community pool when she swims, and that's by preference. Swimming without a layer of wetsuit neoprene between her and the water makes her feel "really vulnerable" but also allows her to experience the local conditions "as fully as possible." Sans wetsuit, she feels it all. "When you get close to an ice formation, you feel the cold coming off of it, and when you breathe, the air actually feels different around it."

Sometimes, as she swims past land dotted with fields of snow and iced-over rocks, she's accompanied by curious animals. During her Antarctic ice-mile swim, penguins dove behind her, catching her wake.[4] You can find a video of this event on YouTube—picture metal-gray waters and stark white glaciers floating in the water behind a solo Monahan, her arms and legs pushing her through the water in a quiet rhythm. In other feats, she has swum with turtles, and in Siberia's Lake Baikal, she was joined by two earless seals (which the

Russians call nerpa). They playfully dove underneath and around her while she competed in a relay race.

Monahan's connection to the natural world crops up in most of the stories she tells me about her swims. As someone who feels a kinship with marine wildlife and enjoys breaststroking through lakes, rivers, sounds, bays, and along ocean coastlines myself, I get it, but an hour or so is usually my limit (then I'm more of a float-on-my-back-staring-at-the-sky type). But how about that cold?[5] What about the hopelessness, the boredom, of that eighth hour, the thirteenth?

How exactly do you refuse to give up when—while swimming across the English Channel in notoriously choppy seas, having almost reached the shore after ten hours—you hear that the tide is against you, so you have to swim for six more hours? I feel like I'm failing the interview when she answers.

She tells me that she lets her mind "zone out" and she gets into a "kind of moving meditation." She said that during the Channel swim setback, she thought she could swim for twenty-four hours if she had to, so sixteen "wasn't so bad." I struggled to understand.

"How do you know that?" I asked. "How does it feel to be in hour ten and know you could swim for twenty-four hours in that chilly, choppy water?"

She considers the question seriously. "It was just a feeling," she said. "It wasn't something that I knew for sure. I was able to, so I guess I wasn't wrong."

Talking to Monahan was amazing and mystifying. It was also the opposite of the self-aggrandizement common in many sports interviews because Monahan is incredibly chill. There is something of a monk-like persona to her that most reminded me of my favorite meditation teacher, Tara Brach. After badgering Monahan with questions and receiving what I thought of as Brachian answers, I was frustrated.

But then I realized, in a big "duh" moment, that it's this very aspect of her personality that allowed Monahan not just to win endurance challenges, but to invent records to be set for the first time. We know there's more than pure physical ability to athletic triumphs, but it's easy to forget that. I did.

Most athletes—especially those who participate in endurance sports—say that mental perseverance is crucial to pursuing and winning long-distance events. In addition to Monahan's abiding faith in her body and her deeply calm personality, the fact that she is biologically female may be an advantage when it comes to the particular records she has set.

As I've noticed in my own wild swimming excursions around Puget Sound, women tend to dominate the cold-water meetups. Of the three different groups I swam with in 2024, two were all women and one included a couple of younger guys who were new Wim Hof devotees. In Sydney, Australia, where I spend part of each year with my elderly dad, I see the same thing. Yes, Aussies of all ages swim all summer—but the vast majority of the people swimming the coastal rock pools filled with seawater as autumn deepens are women, usually middle-aged and up.[6] There seems to be something about long-distance swimming, and cold-water swimming, that female bodies seem especially well equipped for.

There are plenty of theories as to why female bodies might be particularly well suited to this kind of work. The most common assumption is that it's simply a body-fat thing. Since female bodies generally have more fat, they stay warmer and more buoyant. I wasn't able to find much data on this, but many, many experts and websites claim it to be true. Monahan doesn't think it's that simple. She said she's seen amazing things from people with "all sorts of body types," and she's witnessed "very lean people that can race in the ice and do really well because they've acclimatized themselves to it."

While we might not yet know the mechanism behind the phenomenon, the advantage for women in endurance swims is real. In a 2020 narrative review of swimming and gender, sports-science researchers from Europe, North America, and Asia wrote that women overperformed men in "long-distance open-water swimming (distance of about 30 km or 19 miles), especially under extreme weather conditions (water colder than about 20°C or 68°F)."[7]

When it comes to Monahan's specialty, for which she's earned the moniker Queen of Manhattan, she's in good company. In thirty years of recordkeeping for the swim around Manhattan Island, the top ten female finishers were 12–14 percent faster than the top ten male finishers.[8] And the fastest woman to ever swim the Catalina Channel in California was faster than the fastest man by twenty-two minutes, with women on average almost fifty-three minutes faster when data from 1927 to 2014 was analyzed.[9]

Monahan has competed for many years in all kinds of endurance races, including triathlons, marathons, and ultramarathons (defined as any race over marathon length). Ultimately, she finds that endurance swimming is easier for her than running or cycling. "Swimming is not as taxing as running a marathon. It's not the same kind of wear and tear." She just prefers it, and she's great at it. In that way, she's like most of us: attracted to doing the kind of sports her body is good at.

RUNNING LONG

For years, Jennifer Pharr Davis held the fastest known time (for both sexes) for completing the Appalachian Trail, which runs 2,189 miles in the eastern United States. "I think endurance is certainly a form of strength and resiliency," she told me. "There's certainly a uniqueness to being very muscle-bound and being very fast. But I don't think there's any less value in being able to persevere, and in a lot of ways that can serve you better over the long term."[10]

I've been trail running since I was fourteen and got into multiday backpacking trips when I was fifteen. My idea of a great afternoon was hiking fast through the woods and gossiping with my best friend until it got too dark to see the trail. Still, the trail run known as Big Dog's Backyard Ultra, aka Big's, sounds like my idea of a horror story. That's not hyperbole. As they did in Stephen King's early novel *The Long Walk*, participants in Big's run until they can't run anymore, and the last person standing wins. Nice and simple. (In King's book, the racers are shot dead when they can't continue; at Big's, they are cheered on for their chutzpah when they drop out, sometimes after running for literally days, and many return for the next year's race.)

Held in rural Tennessee, Big's is just one of many ultra-endurance races that take place around the world. In the US alone there are over fifteen hundred each year, and thousands more in dozens of other countries, each with its own unique conditions, rules, and location (roads, trails, high altitude, sea level, desert). Big's focuses purely on endurance, not speed, which makes it somewhat unusual. And almost unimaginably grueling. Yet sign-ups are competitive, with more runners than available spots. It says something about the racers that author Leigh Cowart covered Big's in 2019 for her book *Hurts So Good: The Science and Culture of Pain on Purpose*.

According to Cowart's description, the participants in that year's race were mostly white men in their thirties and forties. There were some older folks, some women, and a few people of color. In total, seventy-two runners were at the starting line. Runners have an hour to complete a 4.167-mile loop, and then they must wait for the start of the next hour, when they will run the same loop again (so, yes, they rest between loops). After twenty-four hours of running, they will have covered exactly one hundred miles. In 2011, the first year the race was run, the winner completed eighteen loops or seventy-five miles, but since then, all the finishers have made it well past the hundred-mile mark, and most complete over two hundred miles—more than two days of running.

Why do humans test themselves this way? Maybe it's something primal. Running is part of who we are and how we evolved. We are not and have never been the fastest animal, but we do have endurance that trumps that of many of the creatures we eat (and some of those that want to eat us). As Cowart wrote, the characteristics that differentiated us from earlier humans like *Australopithecus* were also those that enabled us to run long distances. As our bodies evolved, "The shoulders dropped and decoupled from the neck, allowing the body to move separately from the head. The forearms got shorter while the legs got longer, and the connection between the spine and the pelvis got more robust. The bones of the feet shifted and squared up, while the surface area of the ankle, knee and hip joints got wider. A huge ligament showed up from the back of the skull down the spine. We grew butts."[11]

Besides these anatomical changes, humans have little body hair, and, crucially, are able to sweat. This ability to self-cool means we can out-endure running animals, especially in the heat (sometimes even, I was surprised to learn, horses).

Back in 2019, at Big's race, Cowart stayed up until dawn watching the runners, and while a few dropped out, forty-three survived the first night. Some runners pee blood (not because they are dying, but because all that running results in bladder slap, which is what it sounds like, causing abrasions), and others have digestive issues (because the lack of rest doesn't give the gut time to process what they consume). Still, runners need to keep up their strength. "Years ago . . . LA-based endurance athlete and coach Jimmy Dean Freeman described ultra running to me as 'an eating and drinking race with a little exercise and scenery thrown in,'" wrote Cowart.

Two hours after nightfall on day two of running, there were just eight competitors left of the original seventy-two: two women (Katie Wright and Maggie Guterl) and six men. It wasn't until the end of day three that Guterl won, having completed her sixtieth loop, for a total of 250 miles. This was the first time a woman won at Big's but not the last. The next year, Courtney Dauwalter won, besting Guterl by eight more loops, reaching a little over 283 miles.

Both these victories are proof that women can and do win mixed-gender endurance races. As Cowart put it, "Running fast is not the only way to run. What about running long?"

ENDURANCE BOOSTERS IN FEMALE BODIES

When it comes to endurance work, there are significant advantages to using fat as a fuel source (as female bodies do) over carbs. Female bodies have other endurance advantages too. Women are generally better at managing pain, even if they experience more of it, and demonstrably better at pacing, fatigue resistance, and possibly heat and cold tolerance (see Chapters Eight and Nine for a much deeper dive into these topics). For endurance races that take place outdoors, which is almost all of them, these other strengths start to outweigh the advantages male bodies typically have, like larger lungs, hearts, and muscles.

Not everything we know about female endurance abilities comes from athletics. In fact, because women have so long been excluded from sports, whether through the soft bullying of discouragement or overt banning from races, most of what we know about this area of female physical strength comes from medicine, physiology, evolutionary biology, and related sciences.

As the title of a 2018 research paper attests, "Women Live Longer than Men Even During Severe Famines and Epidemics."[12] Female bodies are more likely to survive a wide variety of long-term physical stresses. Females' endurance advantages evolved by meeting the physical demands of the earth's varied environments over time, generation after generation.

A quick refresher from Chapter Two: Muscles in female bodies tend to last longer at repetitive tasks and have better fatigue resistance or durability, whereas muscles in male bodies offer more explosive power. Megan Roche, the lead researcher at Stanford University's Female Athlete and Translational Research Program (FASTR), reminded me that "fat metabolism is one big reason why females are able to do so well in endurance performance compared to men."[13] Any body subjected to the stresses of endurance needs to be good at burning fat, and the female body simply has a natural advantage here.

In addition to these muscle and metabolism advantages, the female body—even the version most of us run around in, which isn't that of an elite athlete—has a fundamental remit to be able to handle significant physical stress. It is literally built for such stress, because growing a fetus from scratch requires a tremendous amount of energy and physical power at levels from the cellular to the organ. More is needed than just successfully utilizing the extra eighty thousand calories a pregnancy requires to maintain bodily functions.[14] That energy has to be used to build a new and complex organism—and keep it alive for almost a year. Making enough milk to feed a growing baby is an even higher metabolic ask.

There are limits for any human body—you can only ingest so many calories and turn them into so much energy in a given period of time. At least part of the reason the most celebrated athletes in the world are able to achieve what mere mortals don't is because they are able to take in and use energy at rates that are double or triple what the rest of us can. Olympic swimmer Michael Phelps ate over ten thousand calories a day during his long periods of training—and he used them. But for most of us, there's a lower ceiling on how much food energy we can convert to physical energy. Surprisingly, research shows that even an average female body can do what it was once thought only athletes could.

Researchers at Duke University quantified this in 2019 when they analyzed the ultimate limit of human endurance.[15] They recruited subjects from the bonkers Race Across America, which goes from the Pacific to the Atlantic Ocean and takes twenty weeks to run. Runners completed about twenty-five miles a day over six days a week. Understandably, not everyone who started the race finished it. From a research perspective, it was both an incredible opportunity to test metabolism over the course of an ultra-endurance event, and a tough task. Each runner's metabolism had to be calibrated, so before the race started, the link between their heart rate and metabolic rate was determined. This measurement is "highly individual," said Cara Ocobock, one of the authors of the study and now the director of the Human Energetics Lab at Notre Dame University.[16]

Ocobock and her team performed a number of other tests too, including having the subjects walk on specialized plates to see how much force they generated, and obtaining mobile ultrasounds of the subjects' hearts. The runners wore activity monitors at the beginning, middle, and end of the race. The battery of tests was repeated once the race was over.

Collecting the data at different locations along the race course was logistically difficult and expensive. "You're asking somebody who is going through a grueling experience to add on all these additional stressors and things to think about, like, 'Oh, did I put my heart rate monitor on?' Or, 'Did I weigh myself before and after the race today?' That's a lot of extra mental strain on an individual who's already under a lot of strain," said Ocobock.

For the three racers who finished the race and were enrolled in the study (all men), a wealth of data was uncovered. Incorporating data from other, shorter endurance competitions (like the Tour de France and ten-day Arctic treks), the researchers found that even among these very athletic individuals, there is a metabolic limit. In the long term (weeks or more), that limit maxes out at about 2.5 times the body's resting metabolic rate. Yes, an athlete can go higher than this in the short-term (the shorter the time period, the higher they can go), but the scale is pretty logarithmic. The shorter the race, the higher the human body can push the metabolism. For endurance races, the limit is 2.5 times.

There's another group whose metabolic limits get close to that 2.5 number for months at a time: pregnant people.

POWERS OF PREGNANCY

During a typical pregnancy, the body's energy use increases to 2.2 times the basal metabolic rate. This makes any pregnant person an endurance expert, living for months close to the limit of what the human body can handle. "The physiological changes that accompany pregnancy mirror quite closely—with only a couple of exceptions—endurance training changes that all bodies experience," Ocobock told me.

Although these changes are not linked to the X chromosome (meaning both male and female bodies undergo alteration during endurance training), they likely have their evolutionary roots in the demands placed on the female body during pregnancy. The adaptations for pregnancy were then passed down through the generations, benefiting individuals of all sexes, Ocobock told me. Depending on genetic variations, all humans have some ability to adapt to endurance exercise over time, thanks to our female ancestors. Estrogen likely gives females an additional bump in endurance capacity, said Ocobock.

She also pointed out that pregnancy isn't the only metabolic stressor for humans; lactation requires even more energy. During gestation, the female body breaks down nutrients, which enter the bloodstream and move through the placenta to the fetus, a somewhat direct energy transfer. But that all changes at birth. "The moment that little sucker comes out, they move, they scream, they cry, they have to digest on their own, they have to maintain their own body temperature," said Ocobock, making me laugh at her double entendre. The baby's caloric needs increase after it is born due to all the work it has to do on its own, and milk is so packed with calories, nutrients, and fat that "it's actually metabolically more expensive to lactate than to be pregnant, because unlike during gestation, in lactation, nutrients flow through the bloodstream but then have to be converted into the milk," said Ocobock. Something to keep in mind next time you see someone nursing or pumping breast milk.

Despite the heavy energetic toll placed on the pregnant and lactating body, it's becoming more common to see weekend warriors and elite athletes engaging in challenging runs, rides, or swims while pregnant. Paula Radcliffe, who has won the London and New York City Marathons three times each (and who held the fastest women's marathon time for a sixteen-year stretch), was heartily criticized in 2007 when she trained—including doing intense hill climbs—until the day before she gave birth.[17] She delivered a healthy baby, Isla, but later said she had made mistakes. Not that she trained while pregnant, which didn't cause her any problems, but that she went *back* to training too hard too quickly after the birth. She injured her sacrum by not allowing herself a long-enough recovery period, she told *The New York Times*. Still, she won the New York Marathon that year—and the next year too. During her second pregnancy, in 2011, she also continued training until the birth, but she took things a little more slowly afterward, to give her body time to recuperate.[18]

That same year, Amber Miller also ran a marathon while nine months pregnant. She began the Chicago Marathon with the idea that she'd see how far she could get, having run thirteen miles the week before. According to her recap of the day in *The Guardian*, she figured she wasn't endangering the baby since she was so close to her due date.[19] So on a beautiful October day, she started the race and, feeling really good, kept going at a nice, easy pace. She finished the marathon in just under six and a half hours. Nineteen minutes after she crossed the finish line, her husband caught up with her. After a short rest, while still in her running gear, she began laboring and went to the hospital. She delivered her child, June, in about four hours. "I couldn't believe that it had all happened on the same day—it was the longest day of my life," Miller told *The Guardian*. "On reflection, giving birth is definitely harder than running a marathon. Give me a marathon any day."

Unsurprisingly, Miller received a lot of flak in the media following her marathon finish, including from doctors. (She also got some supportive comments from other doctors.) Miller's run and the variety of reactions it spurred, even from health professionals, point to the lack of knowledge about what pregnant bodies are capable of. In a short paper looking into Miller's experience, two physiologists examined a number of possible physical hazards to Miller and her fetus. They concluded that her feat was "'no big deal' physiologically speaking."[20]

There's little research on this subject, especially for athletes. When a group of elite female runners from five countries—who had collectively participated in fourteen Olympic games and seventy-two world championships—was asked where they found the best information about their pregnancies and athletics, they cited other distance runners as the most trustworthy sources. They ranked medical professionals second and coaches last. All of them had between one and three children, and although they received plenty of counsel—including unsolicited advice from random people in their communities—they found most of it uninformed or irrelevant to their situations, according to a University of Ottawa study.[21] This leaves female athletes as pioneers, relying on their intuitions about what their bodies can and can't do. In a study of 110 competitive runners, most continued running during the first two trimesters of pregnancy, but only 31 percent continued into the third.[22] There was a very low rate of injury. After birth, 84 percent continued to run while breastfeeding and reported that "running had no effect on their ability to breastfeed."

Radcliffe's and Miller's anecdotes are just that: one-offs. But they point to the fact that at least some pregnant bodies are capable of enduring significant metabolic stresses on top of carrying and nourishing a fetus.

In Western cultures, records show that for at least two hundred years, until about the 1960s, women were instructed to minimize exercise and exertion while pregnant. (Remember, it was in 1967 that Katherine Switzer was chased away from the Boston Marathon for running as a woman.) As kinesiologists at Kansas State University wrote in a 2015 review paper, "Many pregnancy guidelines surrounding exercise and pregnancy during the 1950s (and 1960s) had little scientific basis, and were predominately vague, cautionary, and reinforced the notion that pregnant women were frail."[23] That advice depended on social class; most pregnant working-class women, whether during the medieval period or the 1950s, were expected to continue with their duties. Of course, so were enslaved and indentured pregnant people, though depending on who they worked for, they might have been given lighter duties. (That led to some indentured servants being sold to the local parish when they became pregnant or, under a 1696 revision to laws governing indentured servants in Virginia, "reimbursing" their masters for lost work with additional service. According to *Encyclopedia Virginia*, "A woman servant who had a 'bastard child' would be punished by having an extra year of servitude added to her contract or 'pay one thousand pounds of tobacco to her master or mistress.'"[24] Never mind that all too often, the father of that "bastard child" was the servant's master.)

This class divide for pregancy and work continues. Today in the United States, there's still no mandated paid leave for pregnancy, either before or after birth, even for physically laborious jobs. As a result, many can't afford to take the three months of unpaid leave that are guaranteed by federal law (and even that law applies only to those who meet specific requirements). In the UK, twenty-six weeks of paid leave at 90 percent of salary is the norm. Australians will receive twenty-six weeks of paid family leave starting in 2026. Japan offers fourteen weeks (including six weeks before birth), and Swedes get over a year of paid leave.[25]

For nonathletes, health authorities around the world agree that exercise during pregnancy is healthy unless there's a contraindication due to a health issue. Still, much is unknown. In a 2020 paper, researchers pointed out that "there are still controversies and scant knowledge on the role of regular exercise on delivery outcomes, including mode of delivery and length of active labour." In this study of 105 people, those who exercised during pregnancy had shorter labors (by three hours) and were more likely to have a vaginal birth.[26] The UK's National Health Service advises yoga, dancing, walking, and even running during pregnancy, stating, "There is evidence that active women are less likely to experience problems in later pregnancy and labour."[27]

This makes sense. It's not as if our ancestors were sitting around all day while pregnant. In the recent past, they would be working to keep a farm going or laboring in home-based businesses like brewing, baking, leatherworking, or pottery. Even earlier, prior to the advent of agriculture, women (like men and any other gender their tribe recognized) were busy hunting and gathering, caring for other kids, creating art and crafts, butchering animals, processing food, carrying stuff, and walking over all kinds of terrain. Even though pregnancy and breastfeeding are sources of additional physical and metabolic stress, the female body had to be able to work too. This foundational requirement could be one reason that female bodies are especially good at endurance work of many kinds.

MORE CLUES IN REVEALING OLD BONES

Complex organisms like human beings have evolved fundamental abilities to withstand genetic and environmental stresses. This broader concept is known as canalization. As the authors of a joint

Canadian–US study note, "The developmental and genetic mechanisms that produce this phenomenon are very poorly understood."[28] While the exact processes of canalization remain unclear, it is evident that animals, including humans, survive and thrive because of it.

Amanda Hale, a forensic anthropologist, explained canalization to me, because it relates to an important sex-linked concept she's studied: the female buffering hypothesis. A few years ago, when she was a PhD candidate at North Carolina State University, Hale's research focused on examining lesions and grooves found on ancient human bones and teeth (hypoplasia) to learn more about the lives of the people whose bodies the bones came from. She became an expert bone reader. Hale now works at the technology company SNA International for the Defense POW/MIA Accounting Agency, where she uses her skills doing some of the most difficult work I can think of. She helps identify the remains of US service members who are unaccounted for.

Hale's expertise allows her to estimate how much physical stress—like disease or lack of food—a person has lived through, even using bones that are hundreds or thousands of years old. Remains tell important stories. "What you see generally, regardless of the social structure of the population, is that women live longer than men, though you see more stress markers in women," she told me. "Women are surviving these stresses—what does that mean? That's where the whole theory of female buffering comes from—that there's something about our physiology that allows us to be more resistant to stress."[29]

One possible explanation for female buffering was found recently in cadaver guts. Erin McKenney is an assistant professor of applied ecology, also at NC State. She was celebrating her successful PhD defense over ciders when she realized that one of the people lifting a glass to her from across the bar worked in the school's anatomy lab. This was fortuitous, as McKenney had been wanting to find a source of human cadavers for some time. She wanted to take some new measurements of human guts, but due to the complexity of the human anatomy, there's no way to get these measurements from live subjects. At the bar that evening, McKenney made her way over to Roxanne Larsen, then a researcher and lecturer at Duke University. Over her fizzy beverage, McKenney asked Larsen if she could collaborate with Larsen's students in the anatomy lab. Could she ask them to do some citizen science on bodies that were already being used for medical education? "Roxy was like, 'Yeah! But right now, you should be celebrating!' And I'm like, 'Oh, I *am* celebrating. And now I'm celebrating this final piece of the puzzle that we've been wanting to add to this project,'" McKenney told me, laughing at the memory.

McKenney was looking for a creative way to obtain this information, because she had discovered that the last time human gut variation was quantified was in 1885. "No way!" I said, interrupting her. Yes, she said, calling this information a "long-lost component" of human anatomy. "We wouldn't know these things if we didn't look. And nobody's been looking, which is why we're reviving this over-one-hundred-years-old line of inquiry," she said.

Because they were both at NC State, Hale, the anthropologist, heard about McKenney's work measuring cadavers, and they teamed up. In a 2023 research paper, they found that among forty-five cadavers, there were significant differences in gut morphology in several areas.[30] One of them was the variability in length of the small intestine. While the large intestine mostly adjusts fluid balance before waste is evacuated from the body, the small intestine is responsible for absorbing nutrients from food.

Among the studied cadavers, the duo found that females' small intestines were "significantly longer" than males'. That matters, because a longer small intestine "can pull more from every digestive cycle, because you have more surface area," said McKenney. Consequently, females' intestines allow them to obtain more nutrition from less food. Hale notes that this makes sense, "because the vast majority of the nutrients you need to replenish your system—especially during reproduction and nursing, like protein and fat—that's what's being absorbed by your small intestine."

Hale said this study might provide an answer to her larger research question of how the female bodies in her study sample were able to withstand life stresses like starvation or disease. The ability to extract more nutrients from food could be one piece of the puzzle as to why women survived more often than men—and it could support the female buffering hypothesis, shedding light on how female bodies endure tough environments.

IS THE MICROBIOME REALLY A MICROGENDEROME?

The bacteria inside those longer small intestines could also offer an advantage. Significant sex-based differences have been found in the gut microbiome, which is composed of all the bacteria, fungi, and viruses that naturally live in our stomach and digestive tracts. (It's important to note that this research is still in its early days, and authors on the topic go out of their way to remind readers that there are still many unknowns in the field.) Other areas of the body, like the skin, mammary glands, and respiratory system, have their own microbiomes. This is an incredibly complex subject because each of our bodies contains nearly as many microbes as human cells, or an even greater number. Understanding our microbiome means learning how each type of microbe works, how they interact with each other, how they react to outside forces, including sex hormones, and how they affect their local environment—skin, lungs, or gut. It's a fast-moving and super exciting area of research because the microbiome is linked to myriad health impacts for all sexes.

What we know so far is that the microbiome is affected by almost every aspect of being alive, both the stuff we can control and the stuff we can't. Those impacts include "diet, ethnicity, antibiotics, stress, psychological factors, maternal health during pregnancy, the method of birth (i.e., vaginal birth versus cesarean section), environmental factors, and exercise," according to Nayoung Kim, a professor at the Seoul National University College of Medicine in South Korea.[31]

It's well established that sex hormones—estrogens and androgens—influence the gut microbiome. But it's a two-way street; the microbiome also modifies how sex hormones are metabolized, in what researchers call a bidirectional impact. These sex-based differences seem to benefit the health of individuals with circulating estrogens. "There are a lot of sex and gender differences in the microbiome," Megan Roche, the Stanford physiologist, confirmed. She said that both fat metabolism and menstrual cycles are affected by and affect the microbiome. This interplay and the associated feedback loops are another aspect of the complex dance that powers female strength.

In just one example of the many research studies in this area, microbiome differences likely play a part in why more men than women get type-2 diabetes, despite the fact that women have more risk factors for the disease. This was confirmed, at least in mice, by a large team of researchers at the Shanghai Institute of Endocrine and Metabolic Diseases in China.[32] They transferred male-mouse microbiota into female mice who'd had their microbiomes removed with antibiotic treatment. With the male gut microorganisms at work, the female mice became more insulin resistant, a precursor for type-2 diabetes. The opposite worked too: When male mice were castrated (depleting their male hormones), their gut microbiome changed to be more like female mice, and their glucose metabolism improved.[33]

The relationship between sex hormones and the microbiome is significant enough, and seen widely enough in various bodily systems, affecting everything from metabolism to immunity and inflammation (and the diseases related to them), that some scientists have given it a name of its own: the microgenderome.[34]

The microbiome is affected by many aspects of life, including exercise. Vigorous activity positively influences it (even in nonathletes), and endurance athletes of all sexes have specifically been shown to have higher gut microbe diversity in ways that benefit their athletic performance, according to a 2021 European study.[35] These specialized microbiome bacteria can specifically help with exercise recovery (due to anti-inflammatory activity) and "provide additional energy substrates for exercise performance."[36] Could this be a reason for female endurance advantage?

It's pretty clear why Roche told me that the microbiome is an area of "exploding research" in sport sciences and especially endurance athletics.

CONSISTENCY IS KEY

In September 2024, Tara Dower became the fastest person ever to complete the Appalachian Trail. Her fastest-known-time (FKT) record of forty days, eighteen hours, and six minutes was four and a half days faster than that of the previous record holder, who was male.[37] "It's not about beating men; it's about finding our true potential," Dower told *Runner's World* magazine. "And, you know. If you beat the men, that's an extra bonus."

Earlier in the book we discussed fatigue resistance or durability. Remember the example of how female students could perform many more reps of medium-hard weightlifting than buff male students could? In endurance sports, the female advantage in muscular durability is about more than just the ability to keep going over time. It also helps with pacing.

For most athletes, but especially those in long or very long races, pacing is incredibly important. Working too hard at the beginning of the race can mean the difference between finishing and dropping out. It's a tricky business for any competitor: You want to go as fast as you can, but without blowing backup energy that will be needed later in a race. "Women have outstanding fatigue resistance," said Roche, which may help them maintain a steady pace over time. Pacing is complex, Roche cautions, and multiple factors—from race mindset and approach, to fueling, personal experience, and fatigue resistance—all factor in. "It's hard to tell which unique factor makes women more consistent (in the timing of their splits) at races," she said, noting that evidence of women's pacing advantage can be found throughout the scientific literature.

This pacing advantage can be connected to the physiological advantages of fat metabolism and fatigue resistance, but it's not all biological. Those physical processes work hand in hand (or foot by foot) with women's psychology, which is formed at least partially by cultural expectations around competition and delayed gratification. The ego-boosting high you get from running past your competitors early in the race is something women are probably better at resisting, and that emotional control is an aspect of women's athletics that so many of the competitors I spoke to talked about. The mental aspect of competition at the highest levels is a subject for another book, but it can't be ignored.

Monahan, the ice-water swimmer, told me, "I think women especially are mentally very strong and can put themselves through a lot to pursue these challenges." So with all these advantages, what is getting in the way of female athletes' achievements?

Chapter Seven

RUNNING ON EMPTY

I was raised with some seriously wonderful messaging around healthful eating—and the food was even better. "Moderation! Everything in moderation! Even moderation!" my grandma would say, smiling and nodding to herself. She loved to eat, and she took a primal joy in feeding animals (for our small pack of pet dogs, she cooked stews made from leftovers) and people alike. There was no "kid food" in our house, but there were a lot of garden-fresh veggies that we grew together. My grandma was a whole-foods advocate before it was cool, and she prepared a wide variety of cuisines. By the time I went to live with her in her sixth and seventh decades, she had quite the repertoire, mixing Lebanese dishes her mother had taught her (like lamb stuffed with whole garlic cloves, m'juderah, a lentil dish I adored as a child, rice pilaf, rolled grape leaves, bursting-with-fresh-parsley tabouli, and lemony hummus) with Armenian foods she had learned from her husband (like lavash, and green beans with almonds). She also loved a few midcentury American favorites like pork chops and applesauce, and meatloaf with mashed potatoes. The last addition to her wide-ranging tastes was what I call "seventies hippie food" from the *Moosewood Cookbook*, like barley bowls, veggie-packed soups, whole-grain breads, and homemade granola bars. It was incredibly delicious food, and it made me feel loved and nourished.

I feel lucky to have never been hungry growing up, a true privilege. I was an active kid. I rode my bike for hours, ran around the woods with my Doberman (literally—I rarely remember walking), rode my neighbor's horses, climbed trees, swam in lakes in the summers, went for multihour hikes with friends to gossip, and played racquetball with my aunt and uncle. And as mentioned in Chapter One, there were many, many chores to keep on top of for our three acres of gardens at home. At school we had gym four days a week, plus a long lunch hour that gave us a solid thirty to forty minutes to play. We were outside for recess in all weather, including on freezing winter days when it was cold enough for us to fashion a slide with our butts. We'd run up a hill and slide down a few times. The ground would freeze, so that by the end of recess, we'd be flying down the hill on an ice luge, snow down the backs of our pants, screaming with terror and glee.[1]

But even with all this positive messaging, food as delicious fuel, and my grandma's strong body from chopping wood and building stone walls as an example, I developed an eating disorder. When I was sixteen, I just wanted to be as skinny as the mid-nineties cool girls in tiny tops and giant pants. "Nothing tastes as good as skinny feels," It-girl Kate Moss said. So I skipped breakfast and lunch, ate a banana, enjoyed my grandma's dinners each night—and then promptly threw most of them up.

Thankfully, that lasted all of about six months before I started getting terrible stomach cramps at odd times and realized I was seriously hurting my body. Around the same time, I started learning more about feminism. I understood right away that a starving woman is not a strong woman, and that a very simple way to remove women's power is to get them to hurt themselves. I saw how women were talked to: Howard Stern and TV hosts subjecting them to public weigh-ins; the shaming of Anna Nicole Smith for daring to be a plus-sized voluptuous woman in tight clothing; the constant conversation about Britney Spears's teenage breasts by forty-something men (to her face!); and Monica Lewinsky, my exact height and wearing a size ten, as I did, being called fat, fat, fat (in addition to her being blamed and shamed for having sex with her boss, the president of the United States).

When I was a teen, my favorite grunge, hip-hop, and punk musicians were anarchic, and finding satisfaction with my body as it was felt that way too. My "fuck that" anger at how women were treated by men stopped my bulimia in its tracks (and led me to take almost enough feminism classes in college for a minor). It wasn't always easy to keep the negative voices out. Fiona Apple's 1998 quote to *Rolling Stone* magazine—"Of course I have an eating disorder. Every girl in fucking America has an eating disorder"—felt frustratingly true.[2] When I started my senior year of high school, I decided, very consciously, that I didn't want to be a part of "all that bullshit," as I wrote in my journal at the time.

I learned then that anger is a useful feeling—it alerts you to something you should pay attention to. I still went through times of undereating and overexercising (see Chapter Five), but eventually I graduated to understanding the "Swole Woman" Casey Johnston wrote about in 2017 on The Hairpin: "Being strong feels better than skinny feels. Eating when you're hungry feels better than not eating. These are truths. Guard them, because you worked so hard to find them. The world is going to lie to you, and people will repeat those lies. You have to hold on tight to your truths, and know that they are true even when you hear the lies, even when they come from your own friends' mouths."[3]

When I was young, I competed in horseback-riding and swimming—and did well, garnering red and blue ribbons—but I didn't find it fun to compete against others. By the time I was fourteen, I had stopped swimming competitively and quit horse shows (though

I continued riding). To this day I like to run by myself, cycle solo, and swim at my own pace. A close friend summed it up when they told me, "You're really ambitious, but not really competitive." Some evidence suggests that how competitive you are is genetic, so maybe it's explained by my genes.[4] I like to think I just enjoy doing what I love without comparing myself to others.

If I had been a more competitive athlete, maybe I wouldn't have stopped making myself throw up when I was a teenager. Maybe my eating disorder would have gotten worse.

HOW UNDERFUELING DISEMPOWERS FEMALE ATHLETES

According to the research, up to 45 percent of female athletes have some kind of eating disorder (ED).[5] (It may be higher or lower depending on the sport. Coach Megan Roche from Stanford's FASTR program thinks that among runners, the stat is over 60 percent.) Most commonly, the ED is what those who study these things call underfueling: They aren't eating enough to provide sufficient energy for their athletic pursuits. The issue is so pervasive in the athletic community that in 1992, a collection of three symptoms defining a medical condition was given a name: the Female Athlete Triad. The criteria were updated in 2007, and the condition remains so pervasive that it's known as the Triad for short.

As the name implies, there are three parts to the Triad: low energy availability, lack of a period, and osteoporosis.[6] One of the most dramatic and long-term effects of the Triad is that even very young, otherwise healthy women get bone stress injuries (BSIs).

The reason why is pretty straightforward: If an athlete isn't eating enough, they will lose weight. Eventually (as we covered in detail in Chapter Three), their period will stop as the body prioritizes organ function over reproductive capacity. Period cessation disrupts estrogen hormones that are important in repairing the microdamages that are a normal part of intensive exercise—especially in trabecular bone (the spongy kind). Trabecular bone is found in the vertebrae and at the ends of long bones like the femurs, ribs, and pelvis. Of the two kinds of bone (cortical is the other type), trabecular BSIs are more common in female college athletes with Triad risk factors.

Endurance athletes tend to have even higher Triad rates. Bailey Kowalczyk is a coach and pro trail runner sponsored by Nike who has run long, tough races all over the planet. She told me she thought that among her cohort of endurance runners, Triad rates could be as high as 75 percent.[7] I was genuinely surprised, thinking that if you were regularly running for three hours as part of training for, say, a 50K (thirty-one miles), that would be encouragement to eat, since the training requires so much energy. Kowalczyk agreed that it made sense to think that way, but for runners, "there's this belief that the smaller you are, the faster you will be"—and that's especially true for female runners. Of course, women also get the message that smaller is better from society at large.

To be clear, this isn't a vague or general message extracted from the cultural ether or traded among athletes. It comes from the top. During college at a Division 1 school, Kowalczyk said her male running coaches told her explicitly that "if I lost a certain amount of weight—was leaner, was smaller—then I could be an NCAA champion. I could be more successful. I could be faster."

Kowalczyk self-identified as a type A personality, a trait she said is common among competitive runners. "I wanted to please the people that I looked up to that I know have my best interests at heart. So of course I'm gonna listen to them. And so I worked really hard to manipulate my body," she said.

The expectation to "look like an athlete"—and to fit into very narrow ideas of what that means—has a profound effect on female athletes in many sports, not just running. Ballet and gymnastics also require significant strength and endurance, and have similarly strict ideas about what their female performers' bodies look like.

In the short term, undereating may enable athletes to fulfill their coaches' notions of how they "should" look and may not directly affect their performance, which is why the custom can be so pernicious. Over the course of a few months, underfueling seems like it's working.

Kowalczyk said that she had two to three weeks of her best training runs ever while she initially was heavily restricting her food intake. But off the track, she had difficulty recovering and felt awful. "I was a shell of myself just living everyday life. I only had energy for those workouts, and then I was just done. For the rest of my runs in the week, it would be like moving my body through cement, and my body didn't want to move. I got muscle spasms and cramps all the time, and little injuries here and there would pop up. Nothing really felt like it was working right." Because her coaches were only seeing the performance aspects of Kowalczyk's new regime, it seemed like their advice was working.

After graduation, she kept running, entering endurance events. She continued to undereat, thinking it would lead her to success. That was what she had been taught. Kowalczyk told me it was grueling to run without enough fuel. "I was falling a lot, and I would fade out really strongly by the end of the run. Maybe the first few hours would feel good. But then, like, my body just hit this crazy wall, and by the end of the run, all of my muscles were aching. I was tripping over things that hardly even existed, like really, really small rocks, and I had no coordination."

And then, in 2019, Kowalczyk broke her sacrum at the end of a run due to repetitive stress. "That should have been an old-person's injury, but it was just from running on really bad bones," she said with the knowledge of hindsight. She did her best to heal but still made sure that she continued to "look like a runner," which meant major food restrictions. She returned to running six months later. "I wasn't fueling enough, my bones weren't strong, I wasn't having periods. And so when I started running again, I broke my sacrum again." This time she recognized that her injury was a warning from her body. She tapped into the knowledge of her running support team and got treatment for what she now recognizes as disordered eating.

"That's kind of when I had this really big shift that I want to treat myself better and see what I could actually do in the endurance world," she said.

When Kowalczyk began training again, she had a radically different approach. Now, powered by nourishing meals pre- and postrun, as well as by fueling during runs (which she hadn't really done at all before), running felt completely different. The extra fuel brought some additional weight, which helped her regain her menstrual cycle—all of which brought her new strength. Instead of feeling like she's going to fall apart as she approaches the finish line, "I'm finishing longer runs than I've ever done in my life, feeling strong and like I could keep going for hours. I'm feeling well throughout the run, and I'm not falling all over the place." She said the world seems brighter when she runs while properly fueled, and she's actually enjoying the experience, noticing the beautiful landscapes she runs through. She said the difference is "pretty wild." And importantly, she hasn't had any more bone injuries. (For more on how estrogen impacts bone health see Chapter Fourteen.)

And yet, as Kowalczyk told me, young female athletes still get the wrong message. "Women in the endurance world are taught that losing your period is actually a good thing, and that having your period makes you slow, which is totally wrong. Losing your period makes you injury prone."

Kowalczyk's journey to health hasn't been a straight shot—she's experienced setbacks along the way. It's taken years to reverse the harmful lessons. "It's been a whole journey of trial and error and making mistakes and trying to work through the mess the young me was kind of put into. I was so impressionable at that age. I wanted to be the best endurance athlete that I could be. So I listened to these coaches. I think it's wild how long it takes to kind of reverse that process, both mentally and physically."

Megan Roche, the Stanford physiologist, said Kowalcyzk's experience is backed up by the research. Eating disorders are a "huge detriment to endurance performance," because underfueling impairs fat metabolism, the very thing that gives female bodies an edge in endurance competition. Not eating enough also "has a big impact on bone stress, injuries, and the body's resilience over time," said Roche—as Kowalczyk's experience shows.

Eating disorders steal athletes' strength, and also their joy. "Underfueling has led to a lot of really talented and even everyday endurance athletes wanting to be in a smaller body than their body is physiologically programmed to be. When you get into that cycle of, 'I need to be smaller, I need to eat less, I need to control this'—nothing is ever good enough," said Kowalczyk.

I couldn't help but wonder whether the rampant nature of eating disorders among female athletes—specifically in the way that they lead to low energy, impaired fat metabolism, and eventually injury and dropping out of competition—could be part of the disparity we see between female and male athletic performance. We know that there is a smaller pool of female athletes for any given sport to begin with, and if half or more of those who make it to that level are underfueling and dropping out earlier than they otherwise would, wouldn't that impact the sport overall?

This is one of the many ways that the science of athletics is affected by cultural gender norms. How can we accurately judge female athletics if over half the competitors are starving themselves?

This smaller-is-better attitude does appear to be changing. "It seems like more and more female athletes are getting bone injuries and speaking out and making this connection between not having their period and getting injured," said Kowalczyk. She observed that younger athletes are looking up to their mentors and seeing that "oh, all the cool people are getting their periods and all the healthy people are getting their periods," and that's what's needed to change this dynamic. "There's definitely a lot that needs to continue to happen," she added.

Thinking about how women's athletics is directly compromised by expectations from both within and outside sports reminded me of a trenchant observation from *Ms.* magazine's column Tools of the Patriarchy: "I think sometimes when we talk about 'the patriarchy' controlling women, then it seems like there is some committee somewhere trying to figure out the best way to do this. But the patriarchy is within us as well.... Social control of women's lives is not just something that someone else does to us, it's something that we do to each other and ourselves as well."[8]

IS IT FAIR TO COMPARE?

Every benchmark we have for athletic achievement—whether it's running or jumping on a track, cycling up or slaloming down a snowy mountain, lapping or diving into a pool—has been based on what men's bodies excel at. Not that women's bodies aren't able to compete in these areas, because they obviously do. But from the get-go, men's bodies have dictated the criteria by which success in physical feats are measured. And they have been the driving force for the development of almost every sport, from training to expectations to what kinds of motions are considered out of bounds and what's fair play.

Because women were purposefully (sometimes forcefully) excluded from athletics until very recently, we are still figuring out what the female body is even capable of. What we learn about female bodies is, even now, based on the types of movement, strengths, events, and sports that have been built around male bodies. What if we were to design new sports around female bodies? What would they look and play like? What would they value and what would they deprioritize?

One definitive answer is that endurance is a female strength both in athletics and outside of it. Even in athletic events designed around male bodies, female athletes win when endurance is a big part of the test—not every race, but often enough that the old assumption of male athletes' superiority in long-distance running is clearly challenged.

Those who pooh-pooh female athletes' endurance advantage point to the many races where male athletes still win. For example, right now there is about a 10 percent difference between men's and women's marathon times, with a male advantage, which is similar to the difference in time for the hundred-meter dash.[9]

Interestingly, marathon-length runs are just about where the significant male advantage seems to end. The longer the event, the better women do. As Christine Yu wrote in her book *Up to Speed: The Groundbreaking Science of Women Athletes*, in 2020, researchers "examined more than 5 million results from nearly 15,500 ultra-running events to determine the average pace and finishing time across all participants. They found that as distance increased, the gap between men and women narrowed. While women were, on average, 11.1% slower than men in the marathon, the percentage dropped to 3.7% for 50-mile races and just .25% for 100-mile races. At distances over 195 miles, women were .6% faster than men."[10]

Female athletes have beaten the guys in quite a few mixed-gender ultra-endurance events, including the 2021 Western States Endurance Run. In that race, three women finished in the top ten, and fifteen of the top thirty finishers were women.[11] As noted in the previous chapter, Jennifer Pharr Davis held the FKT for hiking the Appalachian Trail for all humans for almost five years, and it's currently held by Tara Dower. Camille Herron won the Tunnel Hill 100 in 2017.[12] That same year, Courtney Dauwalter won the Moab 240-Mile Endurance Run; ten hours behind her was the first male finisher. In April 2024, Jasmin Paris became one of only twenty athletes to finish the Barkley Marathons, a 100-mile course that climbs twice the height of Mount Everest and must be completed in sixty hours. Over a thousand ultramarathoners have attempted the feat. Paris also set a faster record than the men's record in the Montane Winter Spine race over 268 miles—while pumping milk for her fourteen-month-old daughter.[13] In other sports, thirty-six-year-old Barbara "Babsi" Zangerl became the first person ever to flash (climb for the first time without falling) El Capitan in November 2024, a three-day climb.[14]

Still, many people argue that pure physiology accounts for the 10 percent difference in marathon speeds and the fact that men still win more endurance races and sports than women. Stephanie Case, a competitive ultra runner, wrote an op-ed for *Outside* magazine that lists quite a few other factors that could account for the difference.[15] For one, starting as kids, girls are socialized away from sports (and even physical activity generally, as our expert Sophia Nimphius pointed out in Chapter One). While gendered sports expectations are a lot better than they used to be, they have still impacted the generation of women who are competing now—and those athletes' coaches are from an even earlier era. Access matters too: Not every kid lives in a nice suburban school district that has the capacity to offer team sports. Outside of school, gender differences that affect sports involvement abound. Among poor girls, the rate of care work and labor outside school is twice that of boys.[16]

It's not shocking, then, that research from the Women's Sports Foundation shows that by age fourteen, girls drop out of sports at twice the rate of boys, which makes sense since there are 1.3 million fewer opportunities for girls to play sports in high school.[17]

As Case pointed out in her op-ed, "This trend continues into adulthood, particularly in strength and endurance sports. Studies have consistently suggested that these sports are likely to be perceived as 'masculine' activities. It's a self-fulfilling prophecy—the more masculine a sport is perceived to be, the lower the percentage of female participants you'll find." She highlights the additional barriers women face in ultra running:

> We hear these things a lot, but they are worth repeating: Women still bear the burden of the majority of household tasks, including childcare, and are too often simply unable to commit the time necessary to train for long races. Those who choose to do so are at risk of being labeled as "irresponsible" or "selfish" for shirking their domestic duties. Women also often contend with different safety and security issues than men, as verbal (and physical) harassment of female runners is all too common. Sometimes, even training in broad daylight can be taxing—every time a man catcalls a woman running by, it shrinks the space for women in this sport.[18]

Put all of this together, and when Case arrives at the starting line for an ultra-running race, she sees a lot fewer female faces than male. When she started her racing season in 2018, at the Transgrancanaria 125K ultramarathon in Spain, just 11 percent of the 886 runners were women. She said that although in shorter ultra runs of fifty miles, about a third of the competitors might be female, in longer ones, women's participation rate drops below 25 percent. "For 100-mile races like the Ultra-Trail du Mont Blanc or Hardrock 100, which impose strict qualification requirements, the female percentage rarely makes it past 10 percent or 15 percent, respectively. Hardrock's lottery system also favors veterans, which perpetuates the historical gender imbalance in the sport."

These factors impact many sports, while others are still catching up to a semblance of equality of opportunity. When the Red Bull Rampage—"considered the biggest and gnarliest mountain-bike competition in the world"—finally opened to female competitors in 2024, Claire Buchar, a mountain bike athlete, told *The New York Times*, "It's bittersweet because we're just a bit behind. The guys have had 20 years to progress to this level. This is Year 1 for us." Although Boucher herself was too old to compete, she found it rewarding to have contributed to making it happen, saying, "It's just the beginning and there's so much potential."[19]

As reported here and by many sports journalists, female athletes receive significantly less financial and training support, adding to the reasons they participate at much lower rates. And recall the maddening statistic that a mere 6 percent of sports research has studied female bodies exclusively. If fewer women are running ultra races (or climbing or mountain biking), that means fewer female bodies are included in the research on those sports. Add to that the high percentage of female athletes who are underfueling, leading to injuries and dropouts.

Considering the myriad factors above, is it any wonder that female athletes are at a consistent disadvantage in multiple areas, and that it shows up in racing times? I find it remarkable that female athletes can and do still win while competing against male athletes, proving it's possible.

Yet scientists still make pronouncements like this one from a 2023 review paper: Though "women may seem to have fewer physiological disadvantages than they typically do in running events (fatigue-resistant muscles, higher reliance on lipid metabolism, and more even pacing strategies than men), these factors do not overcome the physiological advantages of men."[20] I'm not sure it's fair to issue such a statement considering the multiple significant disadvantages that female athletes face. It sure looks like the "advantages of men" are much greater than physiological—not to mention the additional female-specific challenges on performance, like the impact of menstrual cycles, pregnancy, and breastfeeding. I don't think we have come close to seeing what the female body is capable of.

Clearly, some women can best men. What would be possible if all athletes, regardless of gender, genuinely came from a place of equal training, support, research, coaching, funding, and opportunity, starting as kids? The truth is, we have no idea.

So let's imagine a future where those who are born female are equally encouraged to do all kinds of exercise as kids—encouraged to learn what they are good at and what they enjoy. A future where just as many female as male athletes compete in ultramarathons, and when a woman beats a man he doesn't disdainfully claim that he "got chicked."[21]

Let's imagine a world where "run like a girl" isn't an insult. Where it is widely acknowledged that female bodies are very good at lots of things, and that both male and female bodies have real and varying advantages. What if, instead of assuming that a woman could never beat a man in a race due to certain male physiological abilities, we asked what advantages the female body might have?

Chapter Eight

FEMALE PAIN AND DISEMPOWERMENT

My elderly neighbor and I were way up in the mountains, each riding our sturdy mules down a narrow trail, tucked into western saddles. The views of the Cascades were an East Coast kid's dream, dark sapphire and gently spiky, with snowy fingers reaching down into green valleys, a bright blue sky above. It was a perfect day for a trail ride. We were several hours out and had to dismount to descend a super-steep section that would have put our mules at risk of falling if we had stayed on their backs. The footing was tricky, for both mule and human, as we made our way downhill. I slowed, concentrating on not stepping in a hole. When I looked up, I knew immediately that I was exactly where I wasn't supposed to be: just behind my mule, Barney, and a little off to the side.

As all horse people know, this is a danger zone. Ungulates get nervous when something is in their blind spot, and for good reason.

Predators can easily attack their unprotected hind legs. Though they can be acclimated out of it, it's important for animals like the mule I was riding to retain this fear.

I tried to move out of the danger zone, but it was too late. Barney's powerful rear leg came up fast, and his hoof caught me under my chin, launching me into the air. I screamed. I thought that might be it—my last moments. I was happy that at least I was outdoors for my death. The mule was terrified and ran off. My neighbor's mule also got scared and took off, ripping the reins out of his hands. I sat on the ground, blood pouring out of a large cut in my chin, covering my white T-shirt. I couldn't hear a thing, and aside from the initial impact of being hit—equivalent to being struck by a small car moving at twenty miles per hour, or about two thousand pounds per square inch (a human punch to the face is about two hundred PSI)—I didn't feel anything either.

After a couple of scary-difficult minutes, I realized I wasn't dying, which was good. There was no hope of calling for help—we were out of mobile service range—and even if there were, it would have taken anyone hours to get to me. Which was bad. The fastest way to reach medical help was to ride back down the mountain. My freaked-out neighbor, a tough-guy Vietnam vet who was practically in tears, realized the same thing. After quickly assessing and reassuring me, he was off to calm and gather the mules. I sat on the ground, staunching the copious amounts of blood that accompany any face or head injury, a fact I already knew from having been an adventurous kid. The bleeding from my chin didn't worry me, but the words "subdural hematoma" kept running on a loop through my (possibly bleeding?) brain.

I climbed back on Barney—he did seem slightly embarrassed—and rode down the mountain in a daze, my wound eventually clotting. It wasn't until the moment I climbed off the mule at the bottom of the mountain in front of my house that my chin started to ache terribly. I didn't feel the pain until it was "safe" for me to. My partner drove me to the sleepy emergency clinic in a strip mall twenty minutes away.[1]

I'm no superwoman. This phenomenon of pain deferral is common enough to have a name: stress-induced analgesia. Often seen in car accident victims, it's brought on by the body's response to adrenaline, which masks pain while the body is experiencing stress.[2] This important survival mechanism enables an injured person to move away from a dangerous situation. Over the three-hundred-thousand-year history of *Homo sapiens*, this particular adaptation has likely saved many lives.

PAIN IS WEIRD

It's also a great example of how strange pain is. Pain is an inherently subjective experience, reminding us that we each live our life in our own unique body. None of us can ever know what it's like to live in another form or feel its pain. We can only know our own. This inability to really know another's pain makes me feel a bit existentially lonely, but, looking up from my own navel-gazing, it also means that pain is an incredibly difficult thing for researchers to study. Since it's impossible to know exactly how much something hurts, and individuals of any sex vary in their pain reception and tolerance, this subject has baffled and frustrated health practitioners for centuries.

Pain is demoralizing and torturous. It's also an unavoidable aspect of existing in a human body. How an individual is able to deal with pain, whether through prevention, painkillers, or healing, can affect every part of life. Untreated pain can lead to drug addiction, depression, and suicide. Letting pain go ignored or untreated weakens a person, both physically and mentally. The stakes couldn't be much higher.

As the director of the Cognitive and Affective Neuroscience Laboratory at Dartmouth College's Department of Psychological and Brain Sciences, Tor Wager has dedicated his career to understanding how the brain processes and understands pain. Wager told me that there are a variety of receptors in the skin, organs, and peripheral nerves that all carry different messages about pain.[3] Some of these receptors are sensitive to cold, some to heat, and others to chemicals (think of the sting of an acid or the prickly discomfort caused by irritating spices).

We also have "different nerve fibers that respond to deep pressure pain—visceral pain in your organs—and then there are pains that are associated with different body parts, like migraines or other kinds of headaches. Then there's neuropathic pain, which is associated with nerve damage," said Wager.

In addition to pain that has a clear source, such as inflammation or tissue damage (Wager uses the technical term for it: nociceptive pain), and neuropathic pain, researchers now consider chronic pain to be an entirely separate category, and it has a new name: nociplastic pain. This is the toughest pain of all to address, because it may not have an identifiable source. Wager pointed to the frustrating fact that in 90 percent of chronic back pain cases, "there isn't any identifiable pathology in the body." Chronic pain may be heritable. "What we're finding," he said, "is that there's probably something unique about different kinds of chronic pain, but genetic risk for pain is shared across many types of chronic pain. And it's also shared with risk for anxiety, depression, and substance use.[4] If you have the genes that confer risk for one type of chronic pain, they also confer risk for over twenty other types of chronic pain."

WHY FEMALE BODIES EXPERIENCE MORE PAIN

Physical pain disempowers every human body it affects, but a female body, on average, is more likely to experience pain over its lifetime. There are quite a few reasons for this. First, female bodies have to do more—from undergoing a monthly cycle that involves egg battles and a uterine lining's sloughing off so intensely that it causes scarring (see Chapter Three), to the rearrangement, expansion, and contraction of organs involved in growing a fetus. Not to mention pregnancy's effects on bones, joints, and skin, due to the dramatic growth and flexibility demanded of the body before, during, and after childbirth. The female body is capable of significantly altering external and internal form—returning to its former shape and doing so multiple times. All those events can and do cause pain.

Female bodies also experience more pain because, in general, they live longer. They survive illnesses that kill male bodies but leave female bodies disabled and hurting. The female body has a more active immune system too. This can sometimes backfire, causing agonizing autoimmune diseases. In these cases, the physical processes that enable the female body's strength and durability share the same root as the pain.

There's evidence that, in addition to possessing organs and enduring physical processes that can cause pain, premenopausal women experience variations in pain sensitivity over the course of their cycle, suggesting that estrogen affects pain intensity. Vincent Martin, director of the Headache and Facial Pain Center at the University of Cincinnati Gardner Neuroscience Institute, reviewed several dozen clinical studies that tested pain sensitivity in people who were menstruating. His review found that for most pain disorders, from migraines to irritable bowel syndrome to rheumatoid arthritis, pain was greater during the late luteal and early follicular phases of the menstrual cycle (that is, premenstrually and during the period itself).[5]

Martin links this to other studies showing that ovarian hormones affect a number of physical systems, including "inflammation, affective states, stress responses, modulatory pain systems, and afferent sensory systems—which increase or decrease pain reactivity," and these fluctuations can have a "mild to moderate effect on pain response."

Other review studies have found conflicting evidence in humans of these effects but more definite differences in animal experiments.[6] In rats, for example, sex hormones seem to exert a "substantial influence" on how females respond to painful stimuli. Further, luteinizing hormone, which tells the ovaries to release an egg, also seems to make opioid painkillers less effective by desensitizing receptors for opiates in the brain. Understanding the mechanisms behind this interplay between hormones and bodily systems could help researchers understand how ovarian hormones modulate pain and why women experience more pain than men.

At this point in the book, you might be unsurprised to learn that such research is lacking. According to Jeffrey Mogil, a neuroscientist and pain specialist at McGill University, about 80 percent of pain science—including the fundamentals of pain research, medications, and other treatments—has been studied in men or male mice, even though 70 percent of people with chronic pain are women. This isn't just bad science and missed opportunities. "It's actually an ethical lapse," Mogil told the CBC.[7]

Pain's causes and treatments (who gets help for it and who doesn't) are germane to the question of physical strength. Accessing care, rehabilitating injuries, and mitigating pain all affect a body's ability to maintain physical ability and power. As this chapter will show, pain and physical strength are inextricably connected, and if half the population isn't getting the information, resources, or help they need (they're not), then they are weaker for it. That's a systemic problem, not a personal one—and yet pain in female bodies is often treated as an individual issue. That leaves people alone to solve their pain problems, which is a really good way to disempower them—and which has happened over and over again throughout recent (and not so recent) history.

MODERN FEMALE PAIN DISMISSED: PERIODS AND ENDOMETRIOSIS

Ignoring pain in female bodies starts early in humans' lives. A joint French-American-UK study from 2016 tested parents' responses to crying babies.[8] The researchers played various recordings of infants wailing that were manipulated to slightly raise and lower the pitch of the cries. They found that "low-pitched cries are more likely to be attributed to boys and high-pitched cries to girls, despite the absence of sex differences in pitch." The adult men in the study, assuming that lower-pitched cries were male, rated those cries as expressing "more discomfort" than the higher-pitched sounds they (incorrectly) thought were coming from the female babies.

The authors of the compendium "The Girl Who Cried Pain: A Bias Against Women in the Treatment of Pain" wrote about the ways female bodies are regularly treated as less deserving of pain relief, highlighting that it begins in childhood.[9] In a study of postoperative pain in kids, "significantly more codeine was given to boys than girls and . . . girls were more likely to be given acetaminophen."

Until 2016, period pain was not quantified in a way that captured its potential severity. When Professor John Guillebaud of the University College London's Institute for Women's Health said that menstrual pain could be as intense as a heart attack, it made headlines around the world.[10]

Olivia Goldhill, a reporter who interviewed Guillebaud, has written about her severe cramps. She described her pain as similar to the pain caused by a herniated spinal disc, which she has also experienced. She found herself "lying on the floor, unable to move, and literally crying out in agony. My hip and back muscles went into spasm, so that my body was twisted in an S-shape contortion whenever I stood," wrote Goldhill.[11] Others who have experienced both severe cramps and vaginal birth have told me that cramps can be as painful as giving birth.

About 80–90 percent of those who menstruate experience painful cramps (dysmenorrhea) at some point in their lives. Severe dysmenorrhea, the kind that prevents normal daily activities, affects up to 20 percent of women, according to the American Academy of Family Physicians.[12] For some, this level of pain is a regular part of their periods, and for others it's occasional. Enduring extremely painful periods every month is often a sign of an underlying health issue, like uterine fibroids, ovarian cysts, or, especially, endometriosis. Yet despite the now-understood connection between these conditions, menstrual pain is still often dismissed and normalized by health care practitioners. With so much untreated menstrual pain among women, young people often hear from sisters, moms, grandmas, or friends that it is "normal" to have cramps severe enough to keep you out of school or work—so many grow up believing it. When doctors reinforce that assumption, the stories persist across generations, creating a relentless ouroboros of female pain. As my friend Jen said, "I had no idea I had an actual problem. For years, I just thought I had a 'bad period' and overdosed on whatever painkiller I had around for days every month." She was in her late thirties before she was diagnosed with endometriosis.

The agony of endometriosis has been recorded for at least twenty-five hundred years, according to a deep review of historical medical literature from 2012 (and there's some evidence for a reference from four thousand years back).[13] The disease manifests when material resembling the uterine lining grows elsewhere, often around the pelvic organs and bowel—though it's now considered a systemic disease, and the idea that it only affects pelvic organs is outdated. Endometrial lesions can travel to the lungs, eyes, or spine and can trigger far-reaching health effects. "Endometriosis affects metabolism in liver and adipose tissue, leads to systemic inflammation, and alters gene expression in the brain that causes pain sensitisation and mood disorders," according to a 2021 study published in *The Lancet*.[14] When a person's hormones send the monthly signal to shed the uterine lining, that signal also goes out to the endometrial tissue existing in other parts of the body. Pain and inflammation follow, along with cysts and scar tissue (since other organs aren't like the self-healing uterus). This tissue can even bind organs together.

For decades, doctors called endometriosis "elusive" and "mysterious." Moreover, because it's a nonfatal "women's disease," little research was directed toward it. In fact, for most of the twentieth century, the most common prescription doctors gave for reducing the impact of endometriosis was pregnancy. That's why, for decades, it was known as the "career woman's disease." As Maya Dusenbery explained in her book *Doing Harm: The Truth About How Bad Medicine and Lazy Science Leave Women Dismissed, Misdiagnosed, and Sick*, the association of the disease with working women emerged because educated, wealthier professionals "were the ones who had not only the financial ability to get a diagnosis that requires a surgical procedure performed by a specialist, but also enough authority to convince their doctors that their symptoms really were abnormal."[15] Many others had endometriosis; they just lacked the time, money, or influence to obtain a diagnosis.

Dusenbery also pointed out that racist assumptions kept women of color from being diagnosed. Into the 1980s, medical students were taught that nonwhite women didn't get endometriosis. Black women who presented with the same symptoms as white women were told they had pelvic inflammatory disease (PID), not endometriosis. A condition caused by bacteria that spread from the vagina to the uterus, fallopian tubes, or ovaries, PID is usually caused by sexually transmitted infections.[16] Diagnosing PID instead of endometriosis was based on a doctor's assumption of Black women's promiscuity, and it was a racist way to blame and shame the Black women who came to them suffering pain. Of course, those women would rarely return or push for treatment.

Despite decades of doctors' insistence that endometriosis in white women was caused by their deferral of marriage and babies, the data shows that pregnancy doesn't alleviate the suffering for most people with endometriosis.[17] I have two friends (including Jen) who had kids much earlier than they would have preferred to because they were desperate to relieve their endometriosis and figured they wanted kids at some point anyway, so why not sooner. Neither got relief.

According to a 2016 survey, endometriosis is found in about 6 percent of people who menstruate, but the number is probably closer to 10 percent. Because the disease was ignored for so long and diagnosis still takes years of consistent medical care, which not everyone has access to, many women never get diagnosed.[18] "The normalization of severe menstrual pain—in medicine and the larger culture—is in large part a result of the mass under-diagnosis of endometriosis," wrote Dusenbery.[19]

It took a well-known bioengineer who suffered silently with endometriosis for more than thirty years to finally investigate the disease in the serious, multifaceted way that is only possible in a dedicated research lab. Linda Griffith, who earlier had been part of the team that famously grew an ear on a mouse's back, founded the MIT Center for Gynepathology Research in 2009.[20] "I have to say, I'm really shocked that it's the first research center of its kind in America," Padma

Lakshmi, the cofounder of the Endometriosis Foundation of America, said at the center's launch event. "That is stunningly bad news on the one hand, that she's the first one doing it. On the other hand, better late than never."

I could probably use Lakshmi's last line in every chapter of this book.

HOW WE GOT TO THIS PAINFUL PLACE

The neglect of female health, especially reproductive health, clearly isn't anything new. But it's not just a historic lack of interest that has hurt so many; there has also been a denial of the physical impacts, including pain, that result from reproductive health issues (and other kinds of female pain too). Having to endure physical agony when relief is available is a cruelty. Ignoring women's pain goes beyond the physical misery it causes. Dismissing cries for relief sends a clear message: This person doesn't really matter. Telling someone their pain—and, by extension, they themselves—are unimportant is a visceral way to strip away their power. This has long been the case for female bodies, and the message is even more pronounced for queer, nonwhite, and trans bodies. Tell me you don't care about someone without telling me? Ignore, downplay, and dismiss their pain. How did this happen?

Until the early 1900s, the pain of childbirth was minimized and labeled "natural" by Western male medical practitioners. While some of these men were medically cautious, worried that painkillers could slow the progress of childbirth, others adhered to the biblical belief that women *should* endure the agony of childbirth, as willed by God. In Genesis, childbirth pain is presented as a direct consequence from God for Eve's sin of eating the fruit from the tree of knowledge. God says, "I will greatly multiply / Your pain in childbirth / In pain you will bring forth children." (The author of Genesis added, "Yet your desire will be for your husband / And he will rule over you," just in case anyone had any ideas about what you should do after you have survived childbirth.)

Other cultures had different takes. Many types of pain reduction for childbirth were recorded among non-Western communities prior to the influence of Christianity. Although traditional Indigenous knowledge, including midwifery, was decimated by colonialism, some information has survived. In a 2018 review article in the *Journal of Integrated Studies*, author Bob Cole wrote that there is still an "impressive amount of literature" about British Columbian First Nations people's practices of pain relief for childbirth.[21] This despite the fact that "aboriginal midwifery was systematically dismantled, even outlawed in some cases," and "attempts to discredit traditional midwifery practice as archaic and dangerous were highly successful, though disingenuous." Cole details how First Nations people in Canada used black and blue cohosh, herbs that specifically relieved labor pain and helped speed delivery.

Working with Indigenous Knowledge: Strategies for Health Professionals, a book edited by a group of South African nursing professors, details herbs and procedures used for pain relief by people from various parts of the southern African continent.[22] These include Xirhakhari, used by the Tsonga people in southern Mozambique and South Africa, which "helps relieve excessive labour pains." Isihlambezo, a traditional remedy made from various plants by concentrating their salts and minerals into a tea, is consumed by pregnant Zulu-speaking people in the third trimester to provide "anesthesia and analgesia during delivery." In Ethiopia, a butter applied to the belly is used for pain relief. Along with herbal remedies, "females are also encouraged to dance as a way of facilitating the labour process and reducing the associated pain." These are just a few examples of how other cultures have long acknowledged the pain of childbirth and prioritized the well-being of the laboring person alongside that of the baby.

Historically, herbal remedies were used for pain relief in ancient Europe and Britain too—that is, until such practices were considered heretical. The first woman to be burned at the stake in Scotland was midwife Agnes Sampson. Her terrible witchy crime? Reducing pain in laboring women, which—according to the Christian church—went against God's directives.[23] In light of this "moral" imperative, power over both female bodies and their pain was transferred from (mostly) women healers to men and the medical establishment. The influential American obstetrician Charles Meigs, who practiced throughout the middle of the 1800s, specifically referenced his religious beliefs when he spoke about pain relief for childbirth, echoes of which we hear today. Meigs doubted any process that "physicians set up to contravene the operations of those natural and physiological forces that the Divinity has ordained us to enjoy or to suffer."[24]

Those who actually gave birth generally disagreed. Queen Victoria called pregnancy the *schattenseite* (shadow side) of marriage and, with the experience of seven labors and deliveries behind her, requested chloroform for her eighth. After the birth of Prince Leopold in 1853, she described the anesthetic as "blessed chloroform, soothing, quieting and delightful beyond measure" (and she used it again for her ninth childbirth). As the original Victorian-era influencer, she may have encouraged other women to clamor for anesthetic—including prominent Victorians like Emma Darwin, who was chloroformed by her famous husband, Charles. Referred to as "chloroform à la reine," it became commonly used by those who could afford it.[25]

PAIN RELIEF: FIGHT FOR YOUR RIGHTS

Early feminists found male doctors' attitude that women "should" experience the pain of childbirth unacceptable. Elizabeth Cady Stanton—an early leader of the women's rights movement who worked for the abolition of slavery and for suffrage for all women—railed against the "torture of bearing children," which, as a mother of seven children, she knew well. Right alongside the right to vote, pain relief during childbirth became a key aspect of the struggle for women's rights in the UK and the US.

But not for everyone. Pain relief was still in the hands of doctors, not suffragists. Medical practitioners' perspectives about who would receive painkillers were informed by anti-Black views originating in slavery, which rationalized and codified torture of Black bodies. Those biases intersected with Black women's bodies when, after 1808, it was no longer legal to import slaves into the United States, so any new slaves had to be born there. "Black women's wombs are the engines that maintain the institution of slavery," historian Diedre Cooper-Owens told *PBS NewsHour* in 2023.[26] Ensuring that enslaved women had as many babies as possible for their enslavers to use or sell meant that their gynecological care earned new attention from the medical community. It also meant that enslaved women's bodies could be experimented on. The man called the "father" of modern gynecology, James Marion Sims, tried out a variety of gynecological surgical techniques on enslaved Black women without pain medication, based on the false idea that Black bodies feel less pain.

This misinformation still impacts the health of Black and brown women in the United States. Almost two hundred years after Sims's awful experiments, a 2016 study found that half of surveyed white American medical students and residents still held biased views about biological differences between races that affected their use of pain medication.[27] Studies show that Black women receive lower rates of epidurals and other pain relief during childbirth, and are three times more likely to die during pregnancy or childbirth than women of any other race in the United States.[28] This is partly due to poorer prenatal care, but it's also due to medical providers shrugging off Black women's pain and telling them that even extreme pain is normal. (Maternal death rates are significantly lower in other high-income countries: The UK's rate is almost half that of the US, France's is a quarter, and Australia's is one-fifth—but those countries also have higher maternal death rates for Black and/or Indigenous people, so the fundamental racial issues are still present outside the US.[29]) Black babies are also more likely to die or be born prematurely in the US, setting them up for potential lifelong health issues.[30]

How can we change these realities? Educating health care workers about the racial aspects of ob-gyn care is one way, and that's an active area of focus. Implementing changes to on-the-ground care is essential as well. Haben Debessai, a member of the Black ObGyn Project, whose practice is based in Atlanta, suggests that certain simple modifications, even down to renaming medical instruments, can be part of systemic change. For example, instead of referring to the "Sims retractor," the instrument is now simply named after its function—posterior vaginal wall retractor—to decenter Sims.[31] "We're not removing Sims from history, because we know that it's still important for us to know what he did so we don't repeat it in the future," said Debessai.

To that end, medical professionals are also refocusing the history of gynecology away from Sims to the women he experimented on. "Now we're uplifting the three main enslaved women that he did his practices on, Lucy and Anarcha and Betsy," Debessai told me. These three women, along with other, unnamed enslaved women, were forced to undergo multiple gynecological surgeries without anesthesia, naked in front of a crowd of curious doctors. This can only be considered a torturous exploitation of their bodies. But Betsy, Lucy, and Anarcha were more than just victims of the horrors inflicted on them. Sims trained them to be surgical assistants, and they learned to care for each other after enduring some thirty surgeries over a five-year period. "In time, they became skilled medical practitioners in their own right," states the New York Historical Society's website.[32] Debessai notes that, in recognition of these women as the "mothers" of gynecology, there are now scholarships, a powerful trio of memorial statues in Alabama by artist Michelle Browder, and a fellowship honoring them.[33]

For those with access to it, a new form of pain relief during childbirth was developed in the early 1900s when two German obstetricians introduced "twilight sleep." The word-of-mouth phenomenon attracted women from all over the world who were eager to try it. While many were wealthy, some less affluent pregnant women saved up or crowdsourced funds to travel, so great was their desire to avoid the trauma of previous births. Details about this new procedure gained wider attention when journalists Marguerite Tracy and Constance Luepp published the feature article "Painless Childbirth" in *McClure's Magazine* in 1914. It became the most-read article in the publication's history. Mothers-to-be clamored for the anesthetic.

The pushback from physicians varied. As with chloroform, some were legitimately concerned about the short- and long-term health effects of twilight sleep. Others parroted the previous generation's lines about it being "unnatural" or "abnormal" for women to experience seemingly painless births. An unnamed doctor in the *McClure's* article is quoted as saying, "I know of no more pleasing sight than that of a strongly built woman giving birth to a first child, with strong and painful birth-pangs!"

Reading over the lengthy and detailed article, you get the sense that women felt they finally had some control over what was happening to their bodies in a time when they had few other rights. Just months after the article appeared, the National Twilight Sleep Association (NTSA) was founded, with suffragists taking up the cause as something worth fighting for.

Despite the understandable desire to make birth less painful, even Bernhardt Kronig, one of the obstetricians who codeveloped twilight sleep, put the blame for such a need on women. He told *McClure's*, "The modern woman, on whose nervous system nowadays quite other demands are made than was formerly the case, responds to the stimulus of severe pain more rapidly with nervous exhaustion and paralysis of the will to carry the labor to a conclusion. *The sensitiveness of those who carry on hard mental work is much greater than those who earn their living by manual labor*" (italics mine). This isn't a far cry from calling women hysterical, especially those who were educated enough and had enough free time to advocate for themselves.

Like today, in the nineteenth and twentieth centuries, women wanted control over their own bodies. Reading through the editorials and lectures from that era by eminent physicians who opposed painkillers, it's notable—and infuriating—to see how much intellectual reaching they did to justify keeping women in pain. Most of the doctors' objections weren't based on health concerns but rather on various aspects of "morality." When women observed the lengths to which the male-dominated medical establishment went to control them, it radicalized them. Their desire to reduce the pain of childbirth and protect their health and lives united them, and they didn't stop there. After advocating for pain relief, they took up the issues of contraception and sex education, and many who worked on these matters also campaigned for women's right to vote. They viewed suffrage as an extension of the right to experience less agonizing childbirth, and painkillers and birth control were seen as liberators. It is no coincidence that many people who advocated for one of these issues also worked on the others.

DISMISSED AND IGNORED PAIN EQUALS POORER HEALTH CARE

There's a direct connection between the historical and ongoing lack of seriousness regarding the female-specific pains of menstruation and childbirth and the downplaying and ignoring of other types of pain that women experience.

Upon hearing the news that cramps were recognized as being as intense as a heart attack, comedian Sasheer Zamata delivered a hilarious tirade about it during her 2022 comedy tour.[34] During a set about periods and how women are expected to handle them, she bellowed, "If you saw a man bleeding uncontrollably and having a four-day-long heart attack, you wouldn't be like, 'Don't let that stop you! Get on a horse!'" She transitions from an over-the-top cheerleader's voice to a serious one: "And we didn't know that before, because male medical researchers weren't getting periods." Finally, she brings the humorous voice back, but it's edged with anger, the fury born of being ignored, of seeing friends and relatives suffer. "So we've been enduring this pain for years! And just going to work—button it up—going to work. Cause we're trained that way. We're trained to be like 'It's fine, it's fine, it's pain, but it's nothing, I can endure anything.'"

This chapter isn't only focused on whether women are "tougher" than men when it comes to pain tolerance, or how pain in female bodies differs physiologically from that in male bodies for good reason. Understanding these possible sex-based differences requires thoroughly considering the long-standing biases surrounding pain in female bodies. How can we talk about pain without confronting the fact that entire diseases affecting female bodies, like endometriosis, have been ignored, while the existence of others has been outright denied?

Dusenbery presents an example of this denial in her book. In 1983, a medical student named Vicki Ratner suffered debilitating urinary-tract pain, which she described as feeling "like a lit match in her bladder." Seeking relief, she consulted with fourteen different doctors, ten of them urologists. Beyond suggesting she give up medical school and "settle down" into family life, "None of them offered a diagnosis or even relief for her agonizing pain." Ratner spent two days poring over the indexes of medical journals to research her condition. She discovered, via a footnote, that her condition had been described by doctors at Stanford, leading to a diagnosis of interstitial cystitis (IC), which was deemed "very rare." *Campbell's Urology*, a primary textbook in the field, listed it under "psychosomatic conditions," claiming it was caused by "emotional disturbance" and "discharge of unconscious hatreds."[35]

Ratner spent the last two years of medical school in pain. She finally turned to the media as a last resort, sharing her story on *Good Morning America*, which aired a five-minute segment on her plight. She had assumed, as the medical textbooks and doctors had told her, that she was relatively alone in her disease. However, after her *GMA* appearance, she received over ten thousand letters in the first week—and continued receiving correspondence for weeks afterward. People wrote in who had suffered for fifty years with the disease. So did family members of people who had committed suicide due to the unbearable pain of IC. Many shared their experiences of working through intense pain, knowing that IC wasn't a qualifying condition for disability.

Yet even after Ratner's appearance on *GMA*, there was almost a "complete lack of interest" in the illness because, as Dusenbery wrote, urology, "which deals with problems of the male reproductive system and the urinary tracts of both sexes is, to this day, the most male-dominated specialty. Back then, 99 percent of urologists were men; in 2015, according to a WebMD poll, 92 percent were." Today IC is thought to affect eight million to ten million Americans and is recognized as a legitimate medical issue. It has secured some research funding and now has a list of possible treatment options.

There are countless similar stories and examples of women's health issues being minimized or ignored; whole books have been written on the subject (including Dusenbery's). And so we arrive at the ridiculous statistic that while 90 percent of those who have periods experience some degree of premenstrual syndrome (PMS)—which includes physical symptoms like bloating, acne, headaches, stomach pain, and sleep disturbances, as well as nonphysical symptoms like mood swings and anxiety—it is barely studied at all. In contrast, erectile dysfunction, which affects fewer than 20 percent of adult men (mostly those over sixty), receives five times as many studies as PMS.[36]

The notion that women are "the weaker sex" but that it's "normal" for female bodies to endure pain—without the use of painkillers or any effort to address the underlying causes of the pain—is one of the many bizarre cultural ironies that inspired me to write this book. This demented duality has persisted for hundreds of years, permeating both history and our modern lives.

For a "how is this still happening in the 2020s" example, consider that the pain of IUD removal or insertion can be so severe that women have routinely flooded comment sections and Reddit threads about IUDs to share their experiences—because their doctors didn't inform them. "Getting my IUD removed was the worst pain I have ever had—and I gave birth this past year," wrote Rae Nudson in an article for *The Cut* titled "Gynecology Has a Pain Problem."[37] Dozens of comments on her story echo her experience. In a TikTok video, a woman who had recently gotten an IUD suggested that the partner of anyone with an IUD "should get them a fucking dinner, flowers, or ice cream."[38] Over two thousand comments in response included remarks like, "It's by far the most painful thing I've ever experienced," "The pain was so bad I struggled to stay conscious," and "It was as painful as my labor contractions were." One of my trans male friends informed me that top surgery was less painful than getting his IUD removed.

Or consider colposcopy, a common procedure that includes a visual exam of the cervix by an ob-gyn and, if needed, a biopsy, which involves scraping the surface of the cervix to obtain cells to test for cancer. Like IUD insertion, pain relief isn't typically provided for the procedure. "I had a colposcopy with a biopsy six years ago and it was the most painful experience of my life. I was screaming and crying the whole time. It felt like someone was stabbing my insides," shortstakofpancakes wrote on Reddit.[39] There are many, many similar comments.

This kind of medical gaslighting has a real impact on women and those who are assigned female at birth (including trans men), who often feel abused by trusted professionals. "The last [colposcopy] was so violent and horrible I haven't been back to a gyno since," wrote comfortpea on Reddit.[40] There are over twenty forums on Reddit dedicated to colposcopies and pain, some with threads containing a dozen comments and others with hundreds. People share tips on box breathing for managing the pain and discuss how to get insurance to pay for sedation (which is usually not covered in the US, forcing those who want pain relief to pay out of pocket).

Reading these stories made my pulse rise in anger and frustration. I've had a colposcopy, and like many others I didn't experience any pain other than the "slight pinching" I was told would occur. Bodies are different, but gynecological medicine has failed to recognize these variations. When *The New York Times* ran a story in 2022 called "Women Are Calling Out 'Medical Gaslighting,'" it got over three thousand comments, about three times as many as other health-related stories.[41]

The belief that IUD insertions and colposcopies don't require pain relief is based on the assumption that it simply isn't necessary. In a 2022 article published on the University of Nebraska Medicine website, Andreea Newtson, a doctor, stated, "Because the cervix does not have any pain receptors, the procedure does not require anesthesia or numbing of the area."[42] The Cleveland Clinic and even Planned Parenthood echo the advice that the cervix can't communicate pain.[43] That information is "steeped in misogyny and patriarchy," said Nighat

Arif, a doctor with the UK's National Health Service who specializes in women's health and is author of *The Knowledge: Your Guide to Female Health from Menstruation to the Menopause*.[44] She said data shows that the cervix does have a nerve supply. "Three pairs of nerves convey sensation from the cervix to the brain: the pelvic, hypogastric, and vagus nerves. The endocervix has many sensory nerve endings, and the argument that it has few is due to it being underresearched. It's just rubbish to nullify women's pain."

Not taking pain seriously is part of the larger issue of the failure of health practitioners to listen to women. A sample of the evidence: A *New England Journal of Medicine* study found that women were almost seven times more likely to be discharged while having a heart attack because their complaints were dismissed.[45] Two decades later, research shows this is still a persistent issue.[46] When women report pain, they are much more likely than men to receive a sedative instead of pain medication.[47] During ER visits, women with acute abdominal pain wait an average of thirty minutes longer for painkillers.[48] Seniors experience additional wait times, which means elderly women are especially affected.[49] Women are prescribed fewer meds than men for pain—even following a major surgery like coronary bypass, which involves cutting the chest open and cracking the ribs to allow doctors to reach the heart.[50] Women are more likely to be misdiagnosed when they have a stroke, and to wait longer for correct diagnosis of conditions ranging from adrenal insufficiency to various types of cancer and heart disease.[51] It's difficult to find a disease or illness for which the female body is—according to research—treated more seriously or given a higher standard of care than male bodies. I couldn't.

Even for a seemingly nongendered condition like traumatic brain injury (TBI), a 2020 study showed that women were less likely than men to be admitted to the ICU when presenting with such injuries and had shorter hospital stays.[52] Based on this evidence, it's unsurprising that women had worse outcomes six months after bonking their brain hard enough to injure it: They were getting less effective immediate treatment. Even worse, although TBIs are generally considered a "male injury" due to their association with contact sports and risk-taking activities, "a substantial percentage of women experience repetitive TBIs as a result of intimate partner violence," the paper states.

"In women's health medicine, there is a lot of dismissal of pain, and even more so when we break it down by race and ethnicity," admits Debessai. This has been documented in a wide variety of diseases over time and location, but a good overarching example is a 2022 study that found that Black patients' medical records were more than twice as likely to include a negative descriptor like "resistant" or "noncompliant."[53] This tracks with patients' own descriptions of their care: About 12 percent of Black patients and just 2 percent of white patients said in a survey that they thought they had been discriminated against. Patients who experience discrimination often have less trust in providers, or they avoid doctor's visits altogether. This means that women of color are especially affected by problems in health care, including pain relief, encountering a double whammy of poor care.

Transgender women also face disparities in health care, as confirmed by at least a dozen studies since 2015. A 2023 review paper by a large group of Italian and US-based experts specifically examining cancer care found that trans individuals received inferior treatment due to health care providers' "lack of knowledge about gender minorities' health needs," labeling this a "major hurdle."[54] Other barriers in oncology for transgender and gender-diverse people included "discrimination, discomfort caused by gender-labeled oncological services, stigma, and lack of cultural sensitivity of health care practitioners."

UNITING FOR PAIN RELIEF

It's no coincidence that as women have gained cultural power, they've also gained more access to pain relief, but there's still a very real disparity between the care that men and women receive. For example, IUD insertion and vasectomy are both in-office procedures, but Debassai told me that although the two are not perfectly comparable, "there is a lot more pain control and a lot more pain relief offered and routinely used [for vasectomy]." That observation was echoed by Arif. She said in the UK, men undergoing vasectomy are offered local anesthesia, prescriptions for codeine following the procedure, and a sick note for work. She said it's "scandalous" how women are expected to get IUDs implanted without any recognition of the pain. "I even have had women come in their lunch hour and will catch the bus back to work or go and pick up the kids, or even bring their kids to their insertion appointment because childcare costs are so expensive."

In the UK, "strict guidelines for pain relief" were updated in 2021 to acknowledge that IUD insertion and removal procedures can be painful, according to Arif.[55] She uses lidocaine gel, lidocaine injection, and lidocaine spray and prescribes a dose of paracetamol (acetaminophen) or ibuprofen one to two hours before the procedure. Furthermore, she can ask for a referral from the NHS to perform an IUD insertion under general anesthetic.

In August of 2024, the Centers for Disease Control and Prevention (CDC) in the United States updated its guidelines for IUD insertion to suggest more options for pain relief. "Lidocaine (paracervical block or topical) for IUD placement might be useful for reducing patient pain," the update reads.[56] Debessai, citing a study, told me that in practice, getting the shot of lidocaine "can honestly be as painful as the IUD insertion itself."[57] Still, she said it should be an option. She also said taking an antianxiety medication before the procedure can help.

I asked her if simply telling patients about potential pain could be part of bridging the gulf between expectation and experience, so at least people can feel informed and prepared. She replied that until we understand why some people feel more pain from these procedures and how to address it, this kind of open communication around pain is key. Doctors should take time to discuss what will happen both before and during the procedure, and offering a range of pain management options is one way to help patients feel empowered. The CDC guidelines echo this, suggesting that doctors counsel patients about the potential risks (including pain) of getting an IUD and provide alternatives for pain management—as well as a personalized plan for each individual.

Dr. Debessai said she understands women's frustrations with the current state of pain relief, noting that it is an area of active research and change. Credit for this shift is owed in part to the thousands who have shared abut their experiences with painful IUD insertions on social media, prompting journalists (including me) to cover the issue. This is proof that to drive change, yes, we still need to speak up to complain and make our needs known.

Much more work is called for to understand the complexities of the human body and pain, and that's especially true for female bodies. Meanwhile, consider this: According to the CDC, 22 percent of American women experience chronic pain, with rates highest among white women and rural communities. As you go about your life, remember that many people out there are quietly enduring pain.

Chapter Nine

PAIN AND PERSEVERANCE IN ATHLETICS

Bethany Brookshire is a pain expert. She knows both the agony of performing with broken bones and how to inflict pain without causing injury, thanks to her expertise in the Japanese martial art of judo.

Brookshire got into judo after being threatened with a gun on the Washington, DC, Metro. She wasn't hurt, but she was angry at how helpless she felt. She'd been an endurance runner, with reams of medals to show for it, but running had never been for self-defense or strength anyway, so it didn't help her that day on the Metro. "Running was in the service of staying skinny. It was in the service of being hot," she told me.[1] After the gun incident, she found herself looking for a sense of control and capability, which quickly took precedence over willowiness and visible abs. Brookshire considers herself "a peaceable person," and there's no hitting in judo (unlike other martial arts, such as Krav Maga).

"It's basically martial arts wrestling," she said. "No hitting, no strikes, and you can't kick. You can't bring any weapons." She appreciates that the founder of judo, Jigoro Kano, explicitly wanted to train women as well as men, and women trained in the sport starting in the early 1900s. Women weren't allowed to fight in competitions until the 1970s, but Kano "100 percent believed in the power of women fighting," according to Brookshire.

The overwhelming majority of the people practicing the sport alongside Brookshire are men, so that's who she spars with. Whatever physical advantages they have, she has others. Since the sport itself is a "practical application of physics," she can lift men much larger than she is. "If you get the right leverage—you're low enough and he's high enough—he's going over your head." She's gotten a 230-pound guy off the ground and above her.

Though there's no hitting or kicking in judo, pain is a breaking point for competitors. Inflicting it, without causing long-term harm, is part of the game. When she's engaged in competition and she's besting an opponent with an armlock, "I promise, this person is in terrible pain."

Brookshire is now a first-degree brown belt, working toward her black belt. She's teaching others. "It's really fascinating to watch my students, women especially, begin to realize what their bodies can do. With the right techniques—not even hard techniques—I can pin a fully grown man, and he will be unable to get up," she told me.

In judo, size isn't everything, and technique trumps big muscles. It's a sport where a smaller person can fairly fight a larger one, and strength isn't defined by size. Brookshire said female bodies have certain features that provide an advantage in the sport. Wider hips, a trait many people born female share, really helps when pinning her opponent to the mat (one of the main goals of the sport). "Because my legs are pound-for-pound stronger than a man's are on average, I go easily into closed guard," a position where your opponent is "between your legs, and your legs are locked behind them." The point of the position is to make the opponent vulnerable to further painful positions. Then, the final step: "If I get you in my closed guard, my feet and my hands are going to start finding your throat because I've gotten really good at chokes."

Brookshire has learned how to use her body as a weapon in other ways not sanctioned by the sport. "I can actually choke someone with my legs around their diaphragm—that's illegal in judo, but I know how to do it." Placed correctly, Brookshire's "skinny little wrists will find your carotid arteries," located in the narrow space between head and shoulder, and pinch. This move can cause someone to pass out.

What if she's the one hurting during the four-minute match? "You ignore it, right? You grow to recognize that it's a kind of pain that won't hurt you." Being unafraid of pain gives her a competitive advantage in a sport where tapping out equals losing. When she finds herself on the receiving end of an armlock or other hold, she competes through the pain, enabling escapes in situations where "dudes would just tap out."

Judo isn't Brookshire's first dance with pain. She was a ballerina for fourteen years when she was younger. "I have danced on broken toes, numerous times." She thinks this may be why her pain tolerance is particularly high. While overcoming discomfort is part of many, and even most, sports, pain is explicitly part of ballet, one of the most female-associated sports in the world. "If you go to see a ballet, every woman who is on pointe is in pain. It hurts. It never goes away, it never stops hurting."

The knowledge that she can withstand pain and that she has the skills to inflict it has given Brookshire confidence that running never did. When, years after the first time, she was again threatened with a gun, she didn't panic. "I kept my head. I had learned how to deal with male aggression and male anger without panicking. And that's been really, really big for me."

Noticing her extensive tattoos, I point out that she must be great at sitting for them with all her pain experience. "Every time I go to my tattoo guy, he's like, 'Man, men are weak. You're working on a man, he's gritting his teeth, he's pounding the chair, he's having a terrible time. Women fall asleep.'" Though I don't have as many tattoos as Brookshire, I'd heard this when I got mine, too. (Of course, in these conversations and those that follow, they are referring to cisgender men and women. Knowledge-sharing in the transgender community includes many reports of trans men being able to tolerate pain well, like others assigned female at birth.)

HIGHER PAIN TOLERANCE? OR JUST MORE PAIN?

Dozens of male and female tattoo artists have answered the question on TikTok: "Who sits better?"[2] Almost every tattoo artist says women are more pain tolerant.

Tattoo artist Adam Wooddy (@damnitwooddy) made a sketch about this question, doing the classic TikTok thing where he plays both parts: the tattoo artist and the customer.[3] The "customer" asks, "Who sits better for tattoos, men or women?" Wooddy answers, "Women." Customer, looking annoyed, asks, "No, I mean, like, who handles the pain better?" Wooddy again answers, "Women."

Now visibly annoyed, the customer asks, "Who sits longer?"

"Women."

"Who sits more still?"

"Women."

"Who gets better tattoos?"

"Women."

On the clock app, people have theories as to why: Women are expected to deal with pain from a young age; women are biologically "built" to deal with pain; women have more practice with pain; women experience other people's pain as well as their own. Mimsy, a commenter on Wooddy's video, wrote the most upvoted response: "I mean, we're used to going about our day in pain/uncomfortable." Tori Maison replied: "Especially once a month, several days in a row, for literal years."

TikTok (and my friends, random people I've talked to about this book, coaches, my grandma, and my weightlifting crew) have all said that female bodies are tougher when it comes to dealing with physical pain. What does the science say?

Two researchers from medical schools in Poland took direct pain measurements in over one thousand people who were receiving tattoos, both during and after the process.[4] The majority of respondents were women, and researchers found "no difference" in reported pain intensity between men and women during tattooing, though afterward, women reported more pain.

If this data is true over larger populations, it seems that getting a tattoo isn't less painful for women; it's just that they don't complain about it or react to it as much. Why would that be? If you want a one-word answer, "practice" might be it.

THE SUBJECTIVITY OF PAIN: CULTURE AND EXPERIENCE

Pain perception is subjective and complex. Before we can understand if any one group feels pain more or less intensely than other groups, the nonphysiological aspects of pain must be considered. In a well-cited clinical review of sex differences in pain, researchers agreed that a number of influential factors come into play: "Psychosocial processes such as pain coping and early-life exposure to stress may also explain sex differences in pain, in addition to stereotypical gender roles that may contribute to differences in pain expression."[5]

If a given person, whatever their gender, reports on their pain to a doctor, a researcher, or their best friend, how can we objectively understand what they are going through? "We don't know if [pain perception] is about their underlying experience—the pain processing in their brains and their nervous system—or if it's just that they're more willing to *report* pain," said Tor Wager, the Dartmouth pain specialist we met in the last chapter.[6]

Because most research on pain is done using a self-report scale, the psychology behind who reports pain more often is an important part of the equation.

Bonnie Schwartz, a long-distance swimmer who has crossed the English Channel and has swum around the island of Manhattan, is one of those hypercompetent people that you can only watch and admire. I talked to her while she was busy as director of a kids' summer sports camp in upstate New York. She was all smiles as she answered my questions and simultaneously dealt with several camper-related issues, each of which was resolved with a minimum of fuss.

When she's not camp-directing, Schwartz coaches both men and women who are serious about their long-distance swimming. She's also the CFO of a social-justice nonprofit. Did she see sex differences in how her swimmers handled pain? "You think men don't complain, because they're big and strong, but no, it's acceptable for a man to complain when he's doing something hard," she said, adding that her male trainees talk to her about pain issues significantly more often than her female trainees.[7]

For women swimmers, it's a whole different calculus. Schwartz said they complain less, and she thinks that's because they are "afraid of being judged for not being strong enough to do what we're doing. If you're a girl doing athletics and you stop, it's like, 'Oh, it's because she was a girl and she couldn't keep going.'" From behind her desk, she looked thoughtful for a moment. "If you're a woman, you maybe don't want to admit that there's pain."

Schwartz guesses that her female swimmers are doing what she did—"they just power through." Meanwhile there are videos online of men crying out in agony when they are subjected to the equivalent pain of menstrual cramps via electric stimulation.[8] And don't forget "man flu" and all the associated anecdotes shared by exasperated spouses, which is so common that *Time* magazine ran a story with the headline "Why Men Are Much Worse at Being Sick than Women."[9] (The article also reported that, according to a study on a group of mice that had been infected with the same flu, males did actually experience worse symptoms.)

Quite a lot is going on here that's worth acknowledging. There's how much pain any one body experiences from a given input, which can vary with the individual and may also be affected by sex hormones. As we know from Chapter Three, levels of female sex hormones can vary between individuals, over the course of the month and over a lifetime. People often report different levels of pain depending on whom they're speaking to; what you tell your mom, your doctor, and your child may vary. And as Schwartz pointed out, male and female athletes receive varying cultural messages about how to or even if they should talk about their pain.

We each learn personal lessons about pain while growing up, in ways that may or may not be gendered, depending on our parents and community. I not only engaged in stupid tricks on my bike that resulted in ripping my knee open, I fell while climbing big rocks and trees, made it "my thing" to jump off swings at the highest point (spraining my ankle twice), and played imaginary games of "they're out to get you" that sent me careening through the woods at top speed and flinging myself to the ground like I saw in the TV show *The A-Team*. I also played outside all winter and got really, really cold. My grandma never exclaimed over bloody elbows, giant butt-bruises, or frozen fingers, so I didn't worry about them either and learned early how much pain or cold I could deal with. At school, it was an unspoken rule among the kids that you just didn't go to the nurse unless it was very necessary, and most of us had scabs and scrapes that we saw as "cool." Most pain, I found as a kid, dissipates pretty quickly unless there's something seriously wrong. So I sometimes went home bloodied and sore but rarely said anything about it. My grandma only groused about getting the bloodstains out of my clothes. All of this taught me that I was a good judge of my body's limits and needs.

Different childhoods teach different lessons about pain. My friend who cooked from an early age has the professional chef's ability to completely ignore serious burns, whereas I, never interested in cooking beyond baking banana bread, find that burns bring tears and feel impossible to deal with. My partner, once a competitive soccer player, is able to push himself through any muscular pain or strain, but he's much more conscious of anything that breaks the skin. (This led to a fight between us when he insisted I should stop my leg from bleeding following a superficial scrape, and I thought it was fine to let the blood run into my sock and keep hiking.)

If pain is more likely to be dismissed and ignored since infancy for those born in female bodies, it makes sense that they might complain more to seek fairer treatment for their pain. On the other hand, someone whose body has more experience dealing with pain is likely to complain less—and we've already established that female bodies endure more pain over a lifetime. As Schwartz said, some women might prefer to suffer in silence because being seen as weak feels worse than complaining. Perhaps pain perception is mostly about how your parents reacted to your cuts as a kid—and if that's gendered, then so is pain. There are clearly many layers to each person's relationship with pain.

All the above, of course, impacts athletics. Athletes appear on the rugby field or track or in the pool with the weight of their cultures and their experiences on their backs, and they bring those things—for better and for worse—to their sports.

Can anyone of any gender or athletic ability ever separate what they have learned about pain, consciously or unconsciously, from their subjective experience when something just hurts? And how can pain researchers, removed from their research subjects' direct experience of pain, distinguish between what's learned and what's inherent? One way to get at that answer is to bypass what a person says about their pain and look directly at their brain.

THE PAIN-BRAIN CONNECTION

In a 2013 study published in the *New England Journal of Medicine*, Tor Wager and his collaborators conducted functional magnetic resonance imaging (fMRI) scans on 114 participants.[10] Unlike a regular MRI, fMRI can measure oxygen and blood flow in the brain over time—it's like a video instead of a snapshot. The researchers subjected the participants' arms to gentle heat as well as temperatures high enough to produce pain. Each type of sensation generated a distinct brain signature. The researchers could later analyze these signatures to predict which sensation the participants experienced. Specifically, they could expose individuals to the stimuli and, based solely on their brain scan, predict with 93 percent accuracy whether the person felt pain or not.

As the paper states, having this kind of biomarker is incredibly useful for "confirming pain in situations in which patients are unable to communicate pain effectively or when self-reports are otherwise suspect." It could also help identify pain-perception disorders that may underlie chronic pain. Wager said it could definitely help researchers understand the differences in pain perception among individuals; they can observe in the brain that someone is experiencing a specific level of pain, no matter what they report.

When Wager used the brain signature to compare pain in male and female subjects, he found that "women report more pain, but they don't show more activity in that signature," meaning "it's probably not that [women's] sensory systems are hyperactive in response to noxious input, but maybe they're more willing to report it."[11] Of course, this research was conducted in a lab setting, which differs from the athletic field, job site, or home. Pain is complex and has many factors that influence its perception, as mentioned. Still, this kind of empirical data can help cut through some of those influences. While there's much more to learn, the study represents a promising start toward understanding more objective measures of pain.

Pain perception is also impacted by memory—and there's some evidence that female bodies might remember pain differently from male bodies. The reason why might be rooted in childbirth. Wager thinks the pain of labor and delivery and the cocktail of brain chemicals released during those events could have a significant effect on the memory of individuals who have given birth. "You get this big release of oxytocin around childbirth, and it's actually helping you bond, but it's also erasing the memory [of the pain]," Wager said. He has talked to people who work in birthing centers and have assisted in hundreds of births. "There's some proportion of women—larger than you might think—that report no pain after childbirth. They'll say things like, 'Yeah, I remember I was screaming and I know I was in pain—but I can't remember any pain.'" This could be a chemical effect from the oxytocin, or it could be a psychological effect related to the extreme physical trauma of childbirth. Whatever the cause, the lack of (or minimization of) pain memories may extend beyond childbirth and benefit all of us in female bodies.

A 2019 study led by researchers from the Alan Edwards Centre for Research on Pain at McGill University has shown that men and women recall pain differently. Participants were subjected to a painful experience in a lab setting. They returned the next day, knowing it would be repeated. Men were found to be more stressed by, sensitive about, and reactive to the anticipation of future pain. As a result, they rated the repeated pain experience as more intense than women did.[12] In other words, over time, men became more sensitive to pain even when the stimuli remained constant, while women's experience of the same pain held steady.

The experiment highlights the complex processes that follow pain exposure. If, all things being equal, memory of pain varies by physiological sex, this insight could influence pain treatment, noted Jeffrey Mogil, coauthor of the paper, in the accompanying news release. "I think it is appropriate to say that further study of this extremely robust phenomenon might give us insights that may be useful for future treatment of chronic pain, and I don't often say that!" said Mogil. "One thing is for sure, after running this study, I'm not very proud of my gender."[13]

Lead author of the paper Loren Martin said he found it surprising that the men reacted more, because "it is well known that women are both more sensitive to pain than men, and that they are also generally more stressed out."

It's pretty well established that stress and anxiety really do increase the perception of pain.[14] As researchers in a 2023 study from Tel Aviv University found, stress inhibits pain-relieving endorphins.[15] Reducing stress really can reduce pain. Doctors and other health providers must recognize this fact and plan for it—such as in the case of IUD insertion, as discussed in the last chapter. Knowing in advance that you won't be offered pain relief for a procedure that causes significant discomfort can be incredibly stressful for many.

New research that could benefit athletes (and the rest of us)—especially in the realm of chronic pain, which disproportionately affects women—is on the horizon. Wager shared promising findings from a randomized controlled trial where he worked with people who had endured chronic pain for an average of over a decade. In this study, he taught participants to view their pain as "safe and not to fear it," as he summarized it. Remarkably, two-thirds of the participants were either pain-free or nearly pain-free after just one month, and their relief lasted for at least a year, the longest period they were tracked. To Wager, this suggests that the brain plays a substantial role in chronic pain and can be retrained to bring lasting relief.

Underlying neurological differences, including those related to memory, are also an active area of inquiry into sex-based differences in pain experience. Recent findings reveal that the types of cells responsible for sending pain signals to the brain are more diverse than previously understood. In addition to neurons and bacteria—which can attack sensory neurons to prevent pain information from being sent to the brain—immune cells can also generate molecules that trigger pain signals.[16] Microglia, a type of immune cell, appear to be especially important in signaling pain, though possibly only in male bodies.

In a separate study led by McGill University, researchers attempted to reduce a specific type of pain in mice using blockers that targeted microglia. While the treatment was successful in male mice, it had no effect at any dose on the females.[17] The researchers concluded that the female mice must have been using a different pain pathway to send signals from the pain site to the brain. "Males and females are using clearly different kinds of cells and different portions of the immune system to produce what ends up being pain," explained Mogil, a lead neuroscientist on the study, during a keynote speech at the 2017 meeting in Montreal for the Organization for the Study of Sex Differences.[18]

Perhaps there's something "special about pain, which makes it incredibly sex-dependent," Mogil said. "Or maybe everything is sex dependent, and the only reason we don't know it, is that by looking almost exclusively at males, we haven't put ourselves, in any field, in the position to find out."

This research imbalance has led to drugs and treatments that are potentially dangerous for female bodies. You have probably heard about the case of Ambien, which had to be redosed years after its release when it was discovered to affect women much more strongly. Other drugs have been found to be less effective in female bodies. A 2023 Australian review study found that female surgery patients are more likely to wake while under anesthesia, probably because they are less reactive to some anesthetic drugs, so new guidelines are being drawn up.[19] Other drugs cause more side effects in female bodies, at a rate of twice what they cause in men.[20] Estrogen, microglia actions, ratios of fat and water in the body, and "remarkable differences between the sexes regarding the activity of drug-metabolizing enzymes," as a 2021 study put it, all have varying effects on medication efficacy.[21] Now that female bodies are being studied more often, additional sex differences continue to be discovered. In a 2022 study, a new nanotherapeutic drug for neuropathic pain was tested on both male and female rats. The researchers wrote, "Treated males exhibit complete reversal of hypersensitivity, while the same dose of nanotherapeutic in females provides an attenuated relief." They attributed this variability to different actions of macrophages at an injury site for male and female animals. They stated that these observations reinforce "the notion that female neuroinflammation is different than males'."[22]

PAIN SENSITIVITY AS A STRENGTH

Even though most research is conducted on male bodies and male animals, scientists have found that women experience more pain over their lifetimes, in more areas of their bodies, and for longer periods of time.[23] (Wager's fMRI study, discussed earlier, offers one set of data that challenges this pattern.) We've covered many of the reasons why this might be true: additional organs, the tolls of reproduction, hormonal effects, and the fact that women routinely survive illnesses, accidents, infections, and other bodily harms that prove fatal for men, which can cause chronic pain. A 2013 review article in the *British Journal of Anaesthesia* suggests, "Genotype and endogenous opioid functioning play a causal role in . . . influencing pain sensitivity."[24]

Another reason female bodies may experience more pain could be that heightened sensitivity offers some kind of advantage. Why might that be? Could greater pain sensitivity be a long-term adaptation? Could it offer practical benefits?

Pain is usually a blaring alarm that something is physically wrong. During the hundred thousand years that humans were forager-hunters, high pain sensitivity could have nudged people to slow down, rest, and heal, helping them to survive, stay useful to their tribe, and have more children. Slowing down has probably never been easy for busy humans, so having a built-in braking system like pain would be a positive adaptation.

Thinking in these terms, perhaps pain sensitivity is a kind of physical strength of its own. In that way, this chapter shares a theme with the one about menstruation: Both pain and the menstrual cycle have historically been seen as signs of women's physical weakness. But as we did with the female cycle, what if we flipped that idea on its head?

People with a more sensitive pain response simply have a louder alarm system. If feeling more pain forces someone to rest or seek help—which could be as simple as asking someone to bring you food while you recover, or as complex as foraging and preparing plant medicines for healing—that's a net positive.

Feeling and responding to pain signals, along with knowing how to mitigate or heal them—by eating certain plants or using herbs or roots as compresses or disinfectants, and understanding when to rest a swollen ankle versus stretch a tight muscle—formed the foundation of early human medicine. When that knowledge was found to benefit people, protodoctors no doubt gained renown and authority, potentially leading to all kinds of social advantages for them and their families, like better food or shelter. Healers in many cultures often received gifts. Plus, by simply saving the lives of more of their relatives, they could help pass on their genes even beyond their own ability to reproduce.

Communication about pain was likely one of the earliest ways important information was conveyed between humans. Feeling pain is a universal part of being alive, and it can only be understood through "behaviors such as facial grimaces, distinct body movements, and verbal reports," wrote the authors of a 2014 review paper titled "No Pain, No Social Gains: A Social-Signaling Perspective of Human Pain Behaviors."[25] The paper, by a pair of psychologists from the University of New Mexico, details how pain serves the function of self-protection while also signaling important social clues. They argue that communicating about pain isn't just complaining; it's foundational to the evolution of human relationships and connection. When we communicate our pain, we express vulnerability, which helps build trust with others.

Humans today still do the same. When we show others our pain, it can bridge chasms of difference. And so it was in our past: "Pain and pain empathizing behaviors can be understood within the context of a broader framework on the evolution of human expressive behaviors," wrote the authors of the 2014 review. Talking about our pain is a deeply human way we can relate to each other, and it has likely always been so.

Pain is awful, but it's also multidimensional. It can be a way to forge empathy across divides, a reason to develop medical care and expertise, and a way for the individual to understand how physically limited they are or are not.

Reporting pain still benefits women today. Since pain in the female body is more often dismissed or misdiagnosed, the fact that modern women complain about pain—louder and more frequently—probably saves lives in the twenty-first century too. If women need treatment for whatever is causing their pain, feeling it more acutely might be a way to make sure their voices are heard. This could even be one reason women live longer than men. Both in prehistoric times and now, female bodies may experience pain more intensely, which drives them to address it.

Some research has shown that a stronger response to pain also results in a greater natural pain-dampening effect in women.[26] That is, the female body may be more likely to experience pain and is probably more sensitive to it, but also better at modulating it. While experiencing pain more frequently isn't ideal, a heightened sensitivity to it paired with a stronger physical and/or psychological ability to deal with it is a powerful combination.

These results back up what Tor Wager, the Dartmouth pain expert, has found: Female bodies tend to be more tolerant when it comes to pain. "In spite of [feeling more pain more often], which seems like a vulnerability, this might be evidence that women are strong and really resilient to pain," he said. This combination of sensitivity and resiliency is a useful adaptation.

Pain sucks, but in female bodies it could be an overlooked physical power. Maybe greater sensitivity or additional sources of pain aren't the issue; rather, the lack of knowledge, attention, and treatment is the real problem. Understood that way, the core issue lies in the current health care culture and its approach to pain in female bodies—not in the bodies themselves.

THE PAIN OF COMPETITION

Athletes face all kinds of pain. Running a hundred miles in desert heat, cycling five hundred miles across Iowa, or swimming from San Pedro to Catalina Island all bring various forms of suffering. Some pain is shooting, stabbing, or throbbing, while other pain comes with every step as your heel hits the ground, or from being too hot or too cold. Even feeling too tired to go on but pushing through is its own kind of pain.

Most fans, coaches, and athletes regularly echo the idea that how an athlete handles pain is a measure of their strength. This is especially true in endurance sports, where female bodies seem particularly adept, according to the data we have so far. Perhaps it's not so important whether pain tolerance comes from body or mind. The *Oxford English Dictionary* offers several definitions of the word "strength," many that are body-centric, but also one that reads, "the emotional or mental qualities necessary in dealing with situations or events that are distressing or difficult." Mental toughness counts.

Over the decades, in experiment after experiment and in meta-analyses, research has shown that athletes have a higher pain tolerance than nonathletes.[27] Of course, the endorphins activated by regular exercise dampen pain.[28] But having experience with pain, including specific types of pain, helps one deal with it more effectively. For example, athletes who do cold-weather sports, like cross-country skiers and distance runners, tend to have a higher tolerance for cold-specific pain. Similarly, soccer players and endurance runners often show greater heat tolerance.[29]

The available data shows either no difference in pain tolerance between male and female athletes or a female advantage.[30] A highly cited paper from 1981 stated it bluntly: "In general, female athletes had the highest pain tolerance and threshold . . . and were able to use the coping strategy more effectively than the other groups."[31]

Almost all endurance races are held outdoors; swimming, cycling, or running outside means facing the elements, often with extremes of heat and cold. Famously, runners in the 2018 Boston Marathon had to contend with horizontal rain and freezing temperatures—the coldest race in three decades. Welcome to spring in New England! The weather changed the outcome of the race, leading many participants to quit early, especially men. Lindsay Crouse, a journalist for *The New York Times* and a marathon runner herself, noticed this trend and wrote an op-ed about it in which she described conditions as "an example of women's ability to persevere in exceptionally miserable circumstances." The stats were striking: Men's dropout rate was up 80 percent compared with 2017 due to the harsh conditions, while for women it increased by just 12 percent. Overall, 5 percent of men dropped out compared with 3.8 percent of women. The trend held true at the highest level of marathoners. Crouse pointed out that "in good weather, men typically drop out of this race at lower rates than women do, but this year, women fared better. Why, in these terrible conditions, were women so much better at enduring?"[32]

There are tons of theories, similar to those explored in this book, about why women were able to stay the course—body-fat composition, generally higher pain tolerance, and having experience with pain from childbirth. Given the bone-chilling temperatures, the body-fat argument gained traction. Remember, female bodies have a subcutaneous layer of fat that's almost twice as thick on average as men's, and this fat is distributed throughout the body, whereas men's fat tends to be concentrated around their bellies.[33]

But, as Crouse noted, "At the same race in 2012, on an unusually hot 86-degree day, women also finished at higher rates than men, the only other occasion between 2012 and 2018 when they did. So are women somehow better able to withstand extreme conditions?" There's not a lot of science on this topic, and even less that looks at sex differences, so what we are left with is actual race conditions and finishers.

The Western States ultramarathon is a hundred-mile race on the rugged trails of California's Sierra Nevada mountains, held at the end of June every year. It's the oldest race of its length, and Megan Roche, the Stanford researcher, has run it. She described it as one of the "premier endurance events in the ultra-running world—and it's always very hot there."[34] Temperatures can soar to over a hundred degrees during the day, and while that is a dry heat, it's still hotter than human body temperature. The trails require more than eighteen thousand feet of climbing and nearly twenty-three thousand feet of descent over the course of the race. When athletes exert themselves like ultramarathoners do in those sorts of temperatures, they face a whole new set of physiological challenges to keep their bodies cool.

Since 2020, female athletes have regularly ranked among the top ten finishers in the Western States race, with three women making the top ten in 2021. Still, women make up a smaller portion of the overall field; they typically represent about one-third of the racers. In 2024, there were 98 women competing out of 375 entrants.[35]

Roche said female metabolisms are "great for handling heat" from a physiological perspective. A 2021 study led by a team of French scientists bears this out. They looked at cooling and hydration strategies among runners taking part in the Grand Raid de la Réunion (Réunion is an island in the Indian Ocean). It included three races: La Diagonale des Fous (twenty-nine hundred runners over 165 kilometers/103 miles), Le Trail de Bourbon (sixteen hundred runners over 111 km/69 miles), and La Mascareignes (seventeen hundred runners over 65 km/40 miles), with temperatures reaching highs of eight-five degrees and humidity levels ranging from 75 to 90-plus percent. About 17 percent of the runners were women (increasing to 25 percent in one of the races).

The researchers found that, in general, the women had a prior history of fewer heat-related symptoms, such as fatigue and muscle cramping. In the races being studied, women demonstrated significantly lower fatigue due to heat and less muscle cramping, repeating their previous experiences.[36] This study is one of the few larger investigations to examine actual outcomes in real events with enough runners—including 525 female runners across all the races—to provide reliable data.

Physiology isn't everything. Roche thinks women also excel in extreme circumstances due to their mental strength. "I think it's very similar to the pacing conversation, where female athletes often tend to be methodical and are very good about doing cooling protocols," she said.

Pacing is critical in long runs, and women tend to do it better. As discussed earlier, this may be because women's bodies preferentially burn fat over carbs. As the title of one study suggests, "Men Are More Likely than Women to Slow in the Marathon."[37]

But pacing is both "a psychological and a physiological thing," said Roche, and it can make the difference between finishing and not in extreme conditions. Making a plan and sticking to it, keeping a slower pace at the start of a long race, taking time to drink iced water or to cool off in a stream—all these tactics have to do with pacing and mindset and are more common in women racers, she said. This may be why you see higher DNF (do not finish) rates from men in tougher and longer races, like the Leadville Trail 100 Run, where "almost all women make it to the finish whereas more than half the male competitors drop out," as reported in *Men's Running* magazine.[38]

All-or-nothing thinking, which might be more common among male runners, also factors in. Elite distance coach Steve Magness told

Crouse, "Women generally seem better able to adjust their goals in the moment, whereas men will see their race as more black or white, succeed or fail, and if it's fail, why keep going?"[39]

Sleep deprivation is another kind of pain. "Endurance events that involve some element of sleeping, like a four- or six-day race—female athletes are able to handle that sleep deprivation a little bit better," said Roche. Although, once again, there haven't been many studies on this topic in athletics, research in the general population has shown similar results. Studies of all kinds report that women's bodies and minds are less affected when they don't get enough sleep.[40] However, when they do get the chance to rest, women may need more sleep to fully recover, as revealed in a study of agility among university students.[41] "You can think about the evolutionary standpoint of going through labor and not sleeping for sixty hours. It kind of makes sense that female athletes might have evolved to be really good at sleep deprivation," said Roche.

Because testing athletes around competitions is so challenging—especially during endurance events, where participants are already pushing themselves to the limit—lots more research is needed. It's entirely possible that while physiology plays a role in the differences that Roche and other experts have observed in how men and women handle extremes, the psychological aspects could be more influential. Or maybe the opposite is true.

There are also other possibilities as to why the initial evidence seems to point toward female strength in pain tolerance. As Bonnie Schwartz pointed out, female athletes may feel uncomfortable or embarrassed about complaining in male-dominated environments (such as those found in all these races) out of fear of appearing weak. Over time that could lead them to better understand their limits. While women may experience discomfort from heat, cold, and lack of sleep just as intensely as men, they might not remember it as vividly. This could make it easier for them to feel less fear when they tackle a grueling training run that was painful the last time they tried it (the possible neurological mechanisms around this are outlined in the previous chapter). Maybe there's something about the physiology of women who are drawn to endurance running, cycling, or swimming that equips them to handle these challenges better than nonathletes.

"My sense is that women are better in any extreme temperature or any extreme environment," said Roche. "In coaching, I see this anecdotally. Whether it's heat, cold, sleep deprivation, a jungle, desert, or anything extreme, I often see female athletes have the better ability to physiologically and logistically cope."

Pain in any body is complex—difficult to understand, experience, measure, eradicate, medicate, or psychologize. Women's pain in particular has been normalized, ignored, and even used as a badge of honor. We have a long way to go in understanding it, but one thing is certain: Pain is far more multifaceted, and possibly more beneficial, than the simple idea that it makes women weak. In fact, the opposite is probably closer to the truth.

Part Four
OVERLOOKED ABILITIES: SENSITIVITY

Chapter Ten

HELD BACK, HAMSTRUNG, AND HOBBLED, WOMEN STILL NAIL IT

Grey clouds sat low on the landscape, threatening rain, when I bumped off the tarmac and onto the dirt road that leads to Peavy Arboretum. I was just outside Corvallis, Oregon, hometown of Oregon State University (OSU). The college town, surrounded by agriculture and forests, is just ninety minutes from Portland and an hour from the Pacific Ocean, depending on which direction you drive.

Next to me in the passenger seat of my old little red Toyota was Norah Steed, part of the College of Forestry's Class of 2027. OSU's forestry school is ranked first or second in the country, depending on the source. I had met Steed just an hour before at Tried and True, my fave coffee shop in Corvallis, and she defies what many may think of Generation Z; she's determined, deeply thoughtful, and if her wildly energetic but detailed paintings and collages are anything to go by, she has no trouble with focus. Steed was kind enough to take an afternoon out of her schedule on a weekend to teach me about competitive woodchopping. I was super curious about it, because while timber sports seem like one of the most stereotypically male-dominated sports, there's something about chopping wood that has been attracting serious attention from women recently. Competitions are held in all the countries where woodchopping is a sport: Canada, the UK, and the US, of course, but also Spain (specifically the Basque Country), New Zealand, and Australia (where the sport was born, in the state of Tasmania).

Steed and I were both familiar with Peavy Arboretum, a teaching lab for the forestry school. Me because I'd gone on countless hikes there with my partner, his mom, and her pug; Steed because for her it is part classroom, part sports arena. I pulled into a muddy parking spot, and Steed and I headed up a trail. As water started to pelt down, I put my jacket's hood up (revealing myself as a nonnative) but otherwise studiously ignored it. I've lived in the Pacific Northwest long enough to know better than to comment on any kind of moisture falling from the sky. Steed, who grew up about a half hour north of where we were now hiking briskly uphill, proved her local status by being genuinely oblivious to the rain. At the top of the rise, we passed by a football-field-sized pond on our left, and ahead on the right was a person-sized rusted saw-blade with die-cut letters that read, "George W. Brown Arena, OSU Forestry Club Logging Sports Team." In a sawdust- and wood-chip-covered area behind the sign sat piles of giant logs, smaller logs cut into various shapes, long skinny logs balanced on huge ones almost as wide as I am tall, and what looked like a pair of utility poles with spray-painted numbers on them, dug into the ground. It felt a lot like an adult playground, which it kind of was. As the rain pelted down, Steed started describing the purpose of all these giant pick-up sticks and Jenga blocks.

She led me over to a round, two-foot-long log—aka the block—resting on its side on a stand about eight inches off the ground. "Last time I was here doing chopping, I finished my first block, which is cool. And then I loaded this one onto here," Steed said, jumping on top of the wood to show me how to compete in the Underhand Chop event.[1]

Her feet were braced by notches cut into the ends of the log, so she was straddling virgin wood. She mimed slicing down from above with an imaginary axe and said, "You cut between your legs." The pretend axe head landed between her two feet, where chunks had already been taken out of this particular piece. My eyebrows shot up in what my partner calls my "I don't know about that" face, as Steed added, "OK, so it's kind of intimidating for some people, but it *is* fun!"

I'm not totally unfamiliar with chopping wood; my grandma taught me how to swing an axe and, importantly, how to read wood before and while you are chopping. A knot, void space, or rotted part could cause all sorts of trouble if you hit it wrong. I'm not as expert as Steed, but the summer I was seventeen, I knew enough to be hired to do trail maintenance at a local hiking area. I can chop wood, saw small trees and branches, and use a chainsaw to remove a fallen tree from across a trail.

I was also instructed in the safety rules for chopping and chainsawing, which is why I made the face. I was specifically taught never to swing the axe toward your body, and definitely not near your feet. Steed's father taught her many of the same things, but she had to relearn them. I reminded myself that woodchopping is a sport, and one you do around lots of other people, not a chore you do alone in the woods. And safeguards are in place to prevent competitors from chopping into their ankles. Participants wear chain-mail booties and shin guards to protect their legs and feet both while practicing and in competition, which is important since timber sports competitors typically favor flat shoes like Vans or Converse, Steed told me, not chunky boots.

In the Underhand Chop event, the object is to chop one side of the log with an extra-sharp competition axe, then turn around and attack it from the other side to, as Steed put it, "finish your block." Each face of the log has a drive side and a chip side. "The drive side is always fresh cuts. And the chip side is your weaker side, which breaks out when you drive on the other side," she said. Whoever does it fastest wins.

Steed had been practicing this event, and many others, as she got in shape to compete in logging competitions with the OSU team, part of the Association of Western Forestry Clubs conclave. This gathering includes twenty-seven universities and community colleges west of the 100th meridian, which runs through the middle of the Dakotas, down through Kansas and the Oklahoma panhandle, and cuts Texas into thirds.[2]

Next, Steed took me to the Standing Block, an upright version of Underhand Chop with the log raised even higher off the ground, bringing the chopping area to about waist height. As Steed explained, this event is more dangerous because if the axe glances off the wood, "there's a lot more of your body in the way."

These timed woodchopping tests are usually the most challenging events for competitors with female bodies, since on average women weigh less, and the events are pretty much all about hitting the logs as hard as possible—so more weight behind the axe means harder chops and faster times. The average male-bodied competitor has an advantage in these events due to their extra weight and typically larger shoulder muscles. But logging as a career requires much more than short-term brute force to get the job done, and the variety of events in logging-sports competitions reflects that. There are events involving climbing, agility, and chainsaw (and combinations of those), as well as that most recent hipster bar activity, axe throwing, which is mostly about accuracy.

As the rain neared pouring-down status, Steed finally acknowledged its existence when she wiped off her glasses. Then she showed me how the sawing events work. Picture a thin saw you might use to cut down a Christmas tree. Now imagine that it's 5 to 6 feet long (1.5 to 1.8 meters) and has a handle at each end. In both the Double Buck and the Single Buck events, the object is to push and pull the razor-sharp saw through the log, cutting a one- to two-inch-thick round off the end. Double Buck involves two team members working together, one on each end of the saw. They can be same sex (Jack and Jack or Jill and Jill) or opposite sex (Jack and Jill). In Single Buck, a single person saws from just one side. This is Steed's event. "I really like Single Buck because it's kind of all on you. It's very full body," she said.

With my experience using much shorter, thicker handsaws, I know how easy it is for the blade to buckle, even once you have a good rhythm going. Sawing with such long, thin, sharp saws sounded near impossible to me. Steed confirmed that buckling and catching happen easily with the huge saws, and yes, it can be very frustrating. "You have to be strong, but you also have a balance—not pushing too hard and keeping it in a straight line," she said. Sometimes people just use their arms and shoulders to try to force the saw through the wood, and "it doesn't work." This is a good example of how strength isn't the same as short-duration power. To be really good at this event, muscle power must be moderated, focused, and directed in a way that requires several skill sets to translate into useful strength.

"It's kind of been an interesting learning experience to see that brute force isn't everything, in a very concrete way," Steed said, standing barefaced in the downpour that had forced her to put her glasses away. "Sometimes the answer is just kind of like, ease up and focus on smooth movement and full-body participation," she said, bended over the log in sawing position, rain dripping off her nose. She demonstrated the careful movement, slowing down to show how she uses her arms to direct power from her back, abs, hips, and legs into the thin saw edge. I told her it looked almost like a dance movement or ice skating, another sport where power and finesse are directed onto a narrow metal edge. She nodded. "It's very fluid, like tai chi. That fluidity is kind of an important thing. If you're blocky—only using half your body—a lot more force is required than if you're doing it correctly."

Since the technique is so hard to get right, "people kind of hate Single Buck—because you can be strong, but you can still be bad at it," Steed said. This combination of skills made it seem like one of the events in timber sports where women might excel.

THE SOFT SEXISM OF WOMEN'S EVENTS

I looked up the times for Single Buck on the Stihl Timbersports website, which tracks records for timber sports worldwide. It was impossible to compare Single Buck times between men and women at the professional level because there aren't any pro women's times listed on the site. However, at the intermediate level, a bunch of women competitors are listed. In this category, the assisted Single Buck world record for a woman was 14.24 seconds, set by American Martha King in July 2024 while sawing a 15.75-inch (40-centimeter) white pine log. For men, the record is 13.36 seconds, set by Viktor Clarmo of Sweden on the same type of log.[3]

Digging into the data, I found that while the top two times for men are under 14 seconds, there's one at 15.39 seconds (also set by Viktor Clarmo), and all the rest are over 16 seconds. In the women's category, seven different women have set fourteen records, all under 16 seconds. Same wood, same saws, fewer women competitors overall—and yet as a group the best women performed better in this event than the best men (though two men set the best records overall). This raises the question of whether, as the sport includes more women and they gain more experience, it might become an event where women could end up setting the best times overall. Maybe Steed will be one of them. She earned first place in women's Single Buck in her first competition in Idaho a few months after I met her.

Single Buck is unusual in that the challenge for men and women is exactly the same. In timber sports, many of the women's events are modified to make them easier. This troubles Steed. "When you talk about the differences between performance for men and women in this sport, when events have different expectations for women, they're always lower. Like they felt the need to handicap it a little bit," said Steed. From the look on her face, she found this insulting. "Sometimes I think it shouldn't be that way."

When sports have different (that is, easier) standards for women's events, it sends a message to female competitors that they can't possibly beat men on a level playing field. Even when they can. Even when they have.

This idea nagged at me when I followed Steed to a horizontal Douglas fir trunk that was as straight as a telephone pole, but skinnier. One end sat on the damp ground, and the other rested on top of a western red cedar trunk so massive that the end of the Doug fir's trunk was elevated at least five feet off the ground. Like all timber sport events, Obstacle Pole simulates a scenario a woodcutter might come across on the job—in this case, a tree that has fallen and landed on a pile of brush. Steed told me that the challenge starts with the competitor, chainsaw in hand, standing on the ground at the elevated end of the log. When the starting call is made—"Timers ready, contestants ready. One, two, go!"—the aim is to run alongside the log to the end resting on the ground, jump onto the log, and run up it to the elevated end. The competitor fires up their chainsaw and bends over the end of the log they're standing on to cut a disk off the "fallen tree." After cutting the disk, the competitor runs back down the log, jumps off, then races on the ground back to the starting point. They tap the end of the log where they made the cut. If they fall off at any point, they have to start over.

Later, I noticed that the log Steed showed me at Peavy was much, much higher off the ground than the ones being used in the competitions I saw online. Competitors fell regularly. At the time, I thought, "Now this looks like my kind of event!" because I love running along the top of fallen logs I find in the woods during trail runs, and also because I love chainsawing. But running up a log carrying a chainsaw, and starting it while balancing on a log several feet off the ground? Yikes.

Steed said this was another event where the gender rules seemed arbitrarily different, and that surprised me. Of any event I'd seen, this one was about agility, balance, speed, and accuracy—all areas any gender could easily compete in. Steed told me that for a certain competition in Idaho, everything's the same for men and women, but when it comes to the chainsawing part, men perform what's called a match cut. They have to "cut halfway through on one side, take [the saw] out, and finish it on the other side. But for women, they just have to cut straight through," Steed explained. "It's like, why?"

Logging-sports competitions vary in both rules and events. Steed showed me the Pole Climbing pole (the women's maximum climb is shorter than the men's), and then we walked down a muddy trail at the back of the arena, where axe targets were surrounded by tall, young Doug fir trees and birches, which provided some respite from the still-pouring rain. Axe throwing is the same for both sexes—the aim is simply to hit the middle of the target and get the axe to stick there. "You can choose whatever axe you want. And there are some men who choose to use a really light axe, just because they like how they throw better, or women who use heavier ones because they prefer them. So it feels very equalizing," Steed said.

Back up the hill in the arena, I tried the Caber Toss. Several well-saturated logs, each measuring 4 to 5 feet long (1 to 1.5 meters), lay on the forest floor. I hoisted one, barely managing to balance the vertical log upright, using my hands as a brace and leaning it against my shoulder. The move is to "toss" (more like heave) the log from the bottom so it arcs into the air. Winners throw their caber the farthest. This event originated in the Scottish Highlands and mimics tossing tree trunks into water. Heavy weightlifting is great prep, because tossing the caber is similar to a power clean and jerk. I'd just started my weightlifting journey when I visited Steed, so I gave it a shot. Steed is more experienced, but still, the Caber Toss felt awkward for us both. She told me the conference she competes in doesn't have regulations for this event, so every school's logs vary in length, width, and species of tree. There's always a men's caber and a women's, which is supposed to weigh approximately three-fourths of the men's caber. But the lack of regulations means that at OSU, for example, the men's caber is larger but made of a lighter wood, so it's not necessarily heavier. During practice, Steed said men and women often use the caber meant for the opposite sex.

I asked her if feeling that she's competing on equal terms with the guys is important to her. She told me that it helps her "take myself more seriously." She grew up doing track and field, an individual sport except for relays, and she likes the feeling that the competition is "all on you. So there's this large sense of, 'This is what I can do' when I compete." Then she got into what bugs her most: "I don't like when they set these rules that make it feel like, 'Yeah, you got that good time, but if you had been doing what the guys were doing, you would have gotten something worse.' If we were all doing the same cut, then our times would be comparable. I might still get a lower time, but I'd know that's what I can do. Not being allowed to compete equally—it cheapens the accomplishment."

Steed said the culture of the logging-sports team at OSU is "very equal, very, very welcoming, and there's no overt sexism or discrimination or anything like that." She pointed out that the events someone is good at depend on body type, which varies among sexes. A tall, thin guy tends to do well in the agility challenges, while the stereotypically beefy dude is great at the Caber Toss. There are short guys and girls of various weights, and willowy ones. Many people aren't fully grown at age eighteen, so they might even change body shapes over the course of their college years. (I grew another inch during my sophomore year, and that's not uncommon.) Steed said a strong team consists of a balance of people with varying body types and strengths.

At the college level, at least, the competitions are organized in a way that promotes gender diversity. The OSU Logging Sports A Team includes eight members: four men and four women, which allows them to compete in Jack and Jill events. Steed noted that some competitions specify that teams must include at least two people of each sex to qualify, meaning same-sex teams aren't allowed. (So far, there hasn't been a transgender competitor in the sport, and nonbinary people compete in the sex they were assigned at birth.) These competition regulations incentivize female recruitment into the sport.

"There's definitely a strong and growing female presence in the sport," said Steed, whose coach has competed professionally. "My coach said it used to be a lot easier, because there wasn't so much competition—you could do pretty well just by showing up." Now the sport is reaching a point where it's significantly more competitive, essentially on par with men's events, where skill is essential to win.

But how strong can women get if they are constantly expected to do less? To meet a lower standard? Despite proof that women can compete with men, they too often aren't allowed to.

WHAT HAPPENS WHEN WOMEN COMPETE AGAINST MEN?

Shooting was part of the first modern Olympic Games held in Athens in 1896, and today there are events at every summer Olympics involving pistols, rifles, and shotguns. Contrary to the popular assumption that shooting isn't really a physical sport, Suma Shirur, an Indian world-record setter and winner of multiple medals in international shooting events, told Firstpost, an Indian news website, "There is a lot of physical activity that is not seen from outside. Because the most natural aspect of the human body is movement," she said. "If I have to control the movement, hold my body steady, at the same place for a long period of time, I need a lot of muscle endurance. That needs a lot of strength."[4]

In the past, men and women competed against each other in shooting competitions, including in the Mexico City Olympics of 1968 and through the 1976 Montreal Olympics. Over time, however, competitions became separated by sex—often after a woman won. Finding so many examples of this phenomenon in the history of sports led Sheree Bekker, an associate professor of injury prevention in the Department for Health at the University of Bath in England, to detail them in a Tweet thread that discussed figure skating, football (soccer), and shooting.

In the 1992 Barcelona Olympics, a young Chinese woman named Zhang Shan won gold in skeet shooting, setting a world record. An incredibly heartwarming image shows the male silver and bronze winners grinning as they lift Shan off the ground. She's wearing a bright red tracksuit and smiling widely with her arms over her head in triumph.[5]

After Shan's win, the International Shooting Union banned women from competing against men. Since the pool of female competitors in any shooting event has historically been much smaller than the men's, dividing the groups by sex meant that there weren't enough women to form teams in time for the next Olympics. So women couldn't shoot at all in the 1996 Olympics, and Shan wasn't able to defend her title.[6] This series of events set women's skeet shooting competitions back (and discouraged women who might have been inspired by Shan's win). "The men decided to split shooting up into men and women because they didn't like to be overtaken by the girls," Heinz Reinkemeier, who coached 2008 Olympic gold medal winner Abhinav Bindra, told ESPN.[7] Bekker echoes that idea, tweeting bluntly, "Women's sport exists as a category because the dominance of men athletes was threatened by women competing."[8]

When a women-specific skeet shooting event finally debuted in 2000 at the Sydney Olympics, it was made easier, featuring fewer targets to hit and fewer shots. Which raises the same question Norah Steed asked: Why? Even with a separate event, given the history of women's achievements in this area, what's the point of making it easier? There are enough examples of women besting men in shooting competitions to render this decision downright offensive.

In 2002, Anjali Bhagwat won the ISSF Champions Trophy in the women's division of the ten-meter air rifle in Munich. She was invited to compete in the unofficial Champion of Champions, which brought together the five best female and five best male shooters. "It wasn't official or anything but quite competitive. Eventually, one by one, all the guys were eliminated. Then it was just the women, and finally it was just me," she told ESPN.[9] Decades later, Bhagwat has won lots of other competitions, but she's still proud of this singular victory.

There was enough frustration within the shooting community over the disparities between men's and women's events that in 2018, the Olympic regulations were finally equalized, although men and women still don't compete against each other. As part of the Olympics' Gender Equality Review Project, which affected various sports, sex-based differences in shooting were eliminated.[10] Now both men's and women's events require sixty shots, whereas women had previously been expected to complete just forty in the ten-meter air rifle event. The number of shots in other events—including the fifty meter, three hundred meter, and trap and skeet events—were also equalized.[11] This change promoted equality and also provided a great opportunity for researchers to examine how the increase in the number of shots affected performances.

In a joint 2019 study by researchers at the Autonomous University of Madrid, in Spain, and the Universidad de las Fuerzas Armadas, in Ecuador, scientists analyzed shooting performances from the Olympic events prior to and following the 2018 change.[12] Of 292 shooters, split evenly by gender, half shot pistols and half shot rifles. They found that "women's performances did not diminish for the pistol or the rifle category when their number of shots were increased. Men and women shot equally well with rifles, although the men's performance with pistols was higher than that of women." Based on their data, the scientists concluded that standardizing shooting requirements across sexes made sense in the "interest of greater gender equality in sports."

It's interesting to note what the study revealed: negligible differences in men's and women's shooting ability with rifles, but not pistols. Why is that?

Daniel Mon-López, one of the authors of the study, told ESPN it may be related to the highly structured uniform worn during rifle-shooting events. "When you are shooting rifle in these clothes, you are using a very low level of strength relative to your maximum strength. Then your precision becomes more important," he said. Rifle shooters typically use about 30 percent of their maximum strength, an especially interesting point when considering Sandra Hunter's research, which found that women were able to sustain lifting weights for a significantly longer period of time than men—when those weights were below their maximum.

Another explanation for the pistol/rifle differential might be body structure. In rifle events, the tendency for female bodies to carry their weight lower, with wider hips than shoulders, provides a real advantage. A lower center of gravity—particularly for those five feet six and under—gives rifle shooters a stronger standing position. All rifle shooters should aim to brace against their hip, as emphasized by famed coach Heinz Reinkemeier in his YouTube videos. This technique ensures that stability is rooted in skeletal strength, not muscle. Most men's supporting-arm elbows tend to rest on their ribs.[13]

"In the standing position, we [women] . . . use bone structure to hold up the gun. So we literally have a direct line from our hands to our elbow to our hips to the ground," Launi Meili, an Olympic gold medalist in the fifty-meter air-rifle event and current coach of the USA Air Force team, told ESPN. "When you think about that line in women, it's shorter, and generally we have our hip right underneath our elbow."

I was shocked when I read this and saw Reinkemeier's videos—especially since I'm five feet six, and so was my crack-shot grandma, who taught me to shoot. Large gray squirrels were her nemesis because they pilfered expensive seed from her bird feeders. She didn't get mad, she got even. First, she "warned" those that dared set paw on one of her feeders by shooting their tails. Most of a squirrel's tail is fluffy fur, built like a bottle brush, so she aimed for the thin tailbone that runs through the middle. Taking aim with her .22-caliber rifle, she almost always made her shot, resulting in many a cock-tailed squirrel running around our property.

If a squirrel with a crooked tail showed up again at the bird feeder, her next shot was to kill. She thought this was perfectly fair since they had obviously been warned. My friends found this all pretty hilarious and nicknamed her "Grambo," a title my grandma wore as a badge of honor. I was (much) less delighted. By the age of nine, I was in charge of mowing our acre of lawn, which often meant I had to stop the mower, pick up a dead squirrel by its crooked little tail, and chuck it into the woods, yelling "Sorry!" at its lifeless body. I love animals so much that it feels wrong to me to eat them, so needless to say, I didn't approve of my otherwise beloved Grambo's vengeful squirrel-killing. (The poor rodents were just doing what their evolved ecological niche demanded, after all.) But even I had to admit, pacifist feelings aside, I did like shooting—at targets only, please.

I remember impatiently sitting at age thirteen through Grandma's lengthy lessons on gun safety, cleaning, and maintenance before being allowed to take aim for the first time. Finally I got to shoot, but I always kept it to inanimate objects. During several teenage summers, I really enjoyed practicing my shooting from our deck into the woods at a target mounted on a tree about a hundred yards away.

Grandma taught me to brace my elbow on my hip when shooting, a technique I had no idea was also used by female Olympians until I saw Reinkemeier's videos. I had never considered that my height and pear-like body shape, which I shared with my grandma, could be a boon to shooting well.

According to Colonel Kenneth Haynes of the National Rifle Association—a military logician in the US Army who taught both men and women to shoot over a multidecade career—women are better at shooting accurately and faster to pick up the skill. "My units had around 20 percent female personnel in both officer and enlisted ranks. All the women fired Expert their first day, but less than a third of the men did so," Haynes told the NRA Women website. Other instructors backed up Haynes's assertion in the same article.[14]

Some in the shooting community don't think that any physical characteristic truly defines a great shooter, noting the variability in body types among top performers. Others suggest that any advantage (or disadvantage) women have may stem from the gendered relationship to firearms prevalent in most cultures. One argument is that, because guns are highly associated with maleness, men may enter training with an assumption that they'll excel, which can lead to overconfidence, making it tougher to learn. In contrast, women may be more open to instruction. On the flip side, it could also be that cultural exposure has advantages. That earlier statistic about men being better pistol shooters? It might be due to men being more comfortable with the firearms due to exposure to shooting rather than inherent physical advantages. Backing that idea up: A 2008 study of Serbian police officers found that although women needed more training, their proficiency with a pistol following practice was equal to men's.[15]

Firearms aren't the only way to judge strength when it comes to accuracy; darts and archery are slightly less gendered than shooting. As with shooting, there was a lot of overlap between men's and women's scores until professional-level events were separated by sex in 1988. Since body size is so important in archery, some within the community have proposed eliminating sex-based competitions entirely and categorizing participants by ability instead. A commenter called Warbow on the forum Archery Talk, echoing many in the community, argues that the current system "unfairly makes smaller men compete against bigger men and smaller women compete against bigger women."[16] They advocate for categories based on draw weight—the amount of force required to pull the bowstring back—which relates not only to height and weight but also to individual strength distribution. Others suggest that body size is a more appropriate way to divide competitors for a fairer and more interesting competition. This approach, of course, mirrors how wrestling and other combat sports separate competitors by weight class. At the 2020 Tokyo Olympics, a new, mixed archery event debuted, pairing the best male and female archers from each country. While this arrangement allows female and male archers to work together on a team, it still does not involve direct competition between the sexes.

Fallon Sherrock, a champion in women's darts, went all the way to the second round of the British Darts Organisation's World Championship in 2019 after beating top-ranked men.[17] She also reached the final round of the Nordic Darts Championship in 2021 as the only female competitor.[18] This fact—that she was the only female competing—again underlines how little we know about what women are capable of, since there simply aren't enough of them competing at high levels.

Do women truly choose not to compete in as many sports as men? Are they just uninterested in chucking darts and shooting guns? Or are they too busy with responsibilities like caring for siblings and other family members, which often keeps young girls—especially those from lower-income families—from engaging in hobbies and sports as tweens and teens? Is the lack of female competitors by design? Not by a worldwide cabal scheming against women, but rather through societal norms that don't encourage girls to try these activities. By fathers who play with their sons but not their daughters, by brothers who exclude sisters. When girls and women do express interest, are they pushed out and discouraged? Or even sometimes bullied out of competitions?

What else is it, exactly, when women have been disallowed from competing with men as soon as they show they can? In addition to the shooting example from the 1992 Olympics, Sheree Bekker's Twitter thread details how, in 1903, Madge Syers placed second in the Figure Skating World Championships, even though no woman had entered before. By 1905 a segregated women's category was established. "Where women were included (or simply included themselves) it was only when they started threatening men's dominance/entitlement that we were segregated into a separate category," wrote Bekker.[19]

Segregating competition by sex starts looking awfully like an extension of the status quo, in which "women's work"—the mental and physical labor, often unpaid, tied to caretaking and household management—takes up most of their free time. This is compounded by the reality that women's accomplishments, voices, and discoveries in many areas are still often actively quashed. These two facts—that women bear a disproportionate burden of certain types of work, and that their accomplishments are frequently downplayed—help explain why people see men's competition as the default "Sport," while "Women's Sport" is often viewed as a separate (and often lesser) category. It's a power flex. If one group's inclusion is granted on "the terms of those in power," as Bekker notes, it only reinforces who holds the power.

Bekker sends her Tweet thread home: "They didn't want women 'taking opportunities' away from men so they segregated women. It was never about a benevolent (still sexist) aim of supposedly 'giving women a chance to win.' It was about control."

Chapter Eleven

A WELL-BALANCED LIFE AND SPORTING CULTURE

Surfers never fail to mesmerize me. I can almost feel the waves underneath their boards as they glide and bob over the water when they paddle out, and my thigh muscles inadvertently tighten when they pop up to catch a wave. There's something so relaxing about seeing them glide on top of the blue-green water, zipping along (or barely moving fast enough to stay upright), completely governed by the ephemeral roll of water beneath them. Maybe I would have been a surfer too, if I'd grown up full-time in Australia, where I was born.

Now I only watch others surf as I tread water in the deep end of an ocean pool, catching my breath after a freestyle sprint. Sometimes I get drenched as the waves breach the concrete edge of the pool that sticks out into the sea in Austinmer, near where my dad lives in Australia.[1] The surf break that I watch, located on a crescent beach, isn't even special enough to have a name—it's just one of many along the country's eastern coast. After school and after work on a hot summer's day in early February, everyone from kids to teens to adults is in the water, but not as many women and girls as men and boys. For much of the 2000s, the rate of female surfers was just under 20 percent (up from 5 percent in the 1990s), but that has been changing quickly.[2] New data shows that since 2020 the number of female surfers has shot up. A 2024 survey from the Surf Industry Manufacturer's Association shows that 35 percent of surfers in the US are women.[3] In Australia that figure is almost 30 percent—and the majority of new surfers since the COVID-19 pandemic have been women.[4] (Black and Hispanic surfers have also doubled in number in the last decade.)

This growth isn't surprising, considering that the sport is well suited to female bodies' physical advantages, especially the strong balance skills conferred by lower-body strength and a lower center of gravity. As in many other sports, the fact that there have been fewer female surfers over the last fifty years is due to purposeful exclusion. Softball, archery, basketball, and running have all seen women celebrated—only to be later pushed aside or disallowed from competing, so surfing is far from the only sport where women's interests have been quashed. But it's in the running for the most egregious example of women's active ostracism, given the fact that in Polynesia, where it originated, most women surfed. Girls surfed, middle-aged women surfed, and queens surfed, according to Mindy Pennybacker, who grew up in Honolulu. Pennybacker has been surfing for over fifty years, was the longtime surf columnist for the *Honolulu Star-Advertiser*, and wrote the book *Surfing Sisterhood Hawai'i: Wahine Reclaiming the Waves*.

Female Polynesian surfers were the norm precolonization. There's Kelea, a chief of Maui island who lived at Hāna "because of the surf riding there, reveling in the curling breakers of the midmorning," wrote Samuel Mānaiakalani Kamakau, a Hawaiian scholar who recorded this legend in 1865. Kelea was famous among her people for her surfing prowess, and, Pennybacker wrote, she had a "reputation for loving it more than anything."[5]

Pennybacker grew up riding Queen's, a world-famous Waikiki surf break that was named for Queen Lydia Liliʻuokalani—whose own niece Princess Victoria Kaʻiulani famously loved to surf. Peter Puget, a British explorer, called Queen Nāmāhanaʻi Kaleleokalani one of the "most expert" surfers that he saw in Maui when he was there in 1794. He wrote that she rode the huge surf "with wonderful dexterity. On its top she came, floating on a broad board till the break[er] had nearly reached the rocks; then she turned."

Pennybacker recounts a Kamehameha Day surfing contest in 1887 that was recorded in the local paper. Poepoe, a favorite to win, competed against his wife, Nakookoo. She shot like "a flying fish" through the wave, "jostles the champion" and came in first place, winning the competition. The crowd went wild. This is likely the first description of a surfing contest in the English language.

Many sources from throughout the 1800s record surfing as a community sport. Kids, couples, and older people surfed. The "entire female population of Kealakekua" dove naked into the waves atop their surfboards and rode "upon the foaming crest of the surges," according to an 1836 account. Pennybacker's book includes lovely nineteenth-century engravings of large groups of women surfing together, and an account of them riding the barrels of waves, or tube riding, which is often considered a twentieth-century invention.

"Many of the most famous surfers in Hawaiian legend and history were female, starting with the volcano goddess Pele and her sister Hi'iaka, goddess of the hula, the traditional dance closely linked to surfing. Many Hawai'i women today practice both," wrote Pennybacker.

Nothing in the origins of surfing makes it a gendered activity, so the fact that it is today speaks to Western cultural influences alone.

SURFING AND WOMEN'S STRENGTH

Given the history, why *has* surfing been so male dominated in modern history? I asked my father, Gerry Vartan. He started surfing at age sixteen at Rockaway Beach in New York City and didn't stop until his late seventies, when he had long been a resident of Sydney, Australia. One of the most enduring images I have of my father is of him looking out at the ocean and describing exactly at what speed and direction the wind and tides were moving, an ability that stems from his inexplicable (to me) relationship with the waves.

My dad put it bluntly. He said there haven't been as many women surfers in the past fifty years "mainly because the boys always harassed them out of the water. No women allowed!" he said, shaking his head.[6] As someone whose business partners were women, he's long been a feminist, and having been born and raised in NYC, he followed the head shake with his favorite expletive, which starts with an *m* and ends with an *r*.

Pennybacker echoed my dad's experience, sans curse words, writing that when she was younger, she would often be the only girl at a surf break in Diamond Head, and "the boys blocked me and pushed me off waves."[7] She has also been subjected to racism as a hapa woman, a Hawaiian word for a person who is multiracial; Pennybacker is half Asian and half white. This kind of behavior is why male-dominated spaces persist in surfing and other sports. Feeling excluded from any activity or sport is bad enough. Being actively harassed and possibly threatened with violence (which women in even nonphysical arenas, like online gaming, deal with) is something that not everyone can brush off. It's worth stating frankly that racism, transphobia, and sexism continue because violent threats are effective at keeping people out. This is how power structures are maintained, whether in video game forums or local surf breaks few people have heard of.

From his years of surfing around the world, my dad has heard a lot of ugly comments about women. He says they all boil down to the same idea: Surfer chicks are there to look cute and cheer on the men, not to threaten men's skills or egos.

That threat is real. Women have proven they can surf the biggest waves, even though there are many fewer of them competing. In 2020, Brazilian big-wave surfer Maya Gabeira rode a 73.5-foot (22.4-meter) wave in Nazaré, Portugal, the largest wave anyone surfed that year.

She won the women's XXL Biggest Wave Award. The winner of the men's award rode a 70-foot (21.3-meter) wave.[8]

The exclusion of women from a sport they historically were part of isn't just about interpersonal spats and one-on-one threats against women.[9] It's been built into the modern surfing culture and business.

Surfing started to become more popular across genders in the 1950s, with men and women competing against each other in surf contests through the late 1960s.[10] But as surfing grew in popularity, women got pushed out. In the 1966 documentary film *Endless Summer*, women were only seen in bikinis on the beach, not surfing. Excluding female surfers from popular depictions continued. When women were allowed to compete in surf contests, the best and most prestigious waves and tubes were labeled "too hard" for women and deemed off-limits. "For decades, the common male wisdom was that women can't get barreled [ride a barrel wave], despite women's having demonstrated this to the contrary at Pipeline, Teahupo'o, Fiji, and all over the world since the 1980s," Pennybacker told me.[11]

If women weren't allowed to compete in the most popular and lauded surfing arenas, brands weren't interested in sponsoring them. Combine lack of sponsorship with earning much less prize money at the competitions that were open to them (often no money at all), and that means women couldn't afford to travel to the most vaunted waves—nor did they have time to practice. Male surfers' winnings allowed them to dedicate their lives to the sport.

Significant discrimination continued through the 1980s, 1990s, and almost the entire first two decades of the new millennium. "Women were largely barred from 2001 to 2012 from professional competitions on the North Shore of Oahu—Haleiwa, Sunset, Banzai Pipeline, Waimea Bay—whose extra powerful, steep, hollow waves remain the most visible world proving ground for the sport. And there had never been a women's championship contest at Pipeline until 2020," said Pennybacker. In 2016, a research paper that included four months of observations and dozens of in-depth interviews brought together pages of detailed data and quotes about women surfers from both male and female surfers at a beach in California.[12] Its very title, "'We Have to Establish Our Territory': How Women Surfers 'Carve Out' Gendered Spaces Within Surfing," says quite a bit. Is it any wonder that a minority of surfers are women?

As organizations have gotten serious about gender equality and inclusion in surfing, female surfers are carving the waves. In 2017, the World Surf League held its first women's big-wave contest at Jaws break in Maui; in 2019, equal prize money was awarded to male and female surfers; in 2022 equal venues were established. In 2023, for the first time, both women (numbering six) and men (thirty-four) surfed in the "Super Bowl of Surfing," the Eddie Aikau Big-Wave Invitational at Waimea Bay. Surfing was made an Olympic sport in 2021, with equal numbers of male and female competitors. Access and equality have already shown results. Since these changes have been implemented, the number of tubes ridden by women has "increased exponentially," said Pennybacker.

My dad and Pennybacker agree: The skills needed to be a good surfer include, in order of importance, balance, determination, lower-body strength, upper-body strength, flexibility, and patience. Of course, being able to swim is imperative to even trying to learn to surf, and being a strong swimmer will help you catch waves and recover when you get dumped.

Women's greater lower-body strength is an advantage in the sport. "Just watch Carissa Moore power surf her big swoops on a big wave using her strong, muscular legs, or launch into the aerials for which she's been a women's pioneer, and the advantage is so clear," said Pennybacker. A lower center of gravity, strong balance abilities, and fine motor skills can also help female surfers follow the shape of the wave more closely, "because you carve your line with your feet pressing on the board." She said shorter heights or lower weights aren't a disadvantage in surfing since a smaller person can get closer to the face of the wave and deeper into it.

There's also the question of style. Surfing contests, based on decades of male surfers' interests and abilities, have defined the sport, and so that's what surfing ability is judged on. There's an emphasis on attacking or "shredding" the wave and executing high-performance maneuvers. There's also what Pennybacker calls "circusy tricks" taken from skateboarding, like directional reverses and aerial spins. Male surfers are rewarded for these moves, while other ways of surfing have been ignored. "Women for our part tend to surf more organically and rhythmically, taking our cues from the shape and speed of the wave rather than showing how we can dominate and master this natural phenomenon," said Pennybacker. My dad agreed, saying that in his six decades of surfing, the guys tend to "blast around, ripping up the waves. Not that the women cannot do 'airs' and crank cutbacks—but they do it with more fluidity and less ego." Currently, winning style in surfing is defined by what one gender thinks is the most interesting; it excludes other exciting ways of approaching the challenge. What a bummer.

THE SCIENCE OF STRONG SURFERS

Some people still argue that men are "better" at surfing due to greater upper-body strength. The reasoning goes that their big muscles mean they can paddle harder (and therefore faster) into a wave and pop up (move from the prone position to standing) quickly. But every surfer is working with their own body weight, so having giant muscles might look cool, but they also weigh more. When it comes to surfing, it's all about a balanced physique that provides the athlete with maximum power for their particular body size and shape. That's led some sports scientists to test whether the right training could equalize current strength differences across genders.

Joanna Parsonage was interested in exactly this question. As part of her PhD research at Edith Cowan University in Australia, she established that there was only about a 10–12 percent difference in paddling speed between the genders.[13] At the time (about 2017), a majority of women surfers hadn't specifically trained for stronger arms and shoulders, so Parsonage wondered if that gap could be minimized or even eliminated if they built some muscle.

Brawn only goes so far in surfing. "We know there's a plateau where you're not going to get any faster even if your upper body is stronger. That type of strength only helps so much. For men, once they reach 1.3 times their body weight in a pull-up, they don't get any faster in their sprint paddle," Parsonage said.[14] Pennybacker made a similar point. Since women generally weigh less than men, "that's less weight to pull over the lip and into the wave," she said. Put those two facts together, and that means female surfers simply need to strengthen their upper bodies enough to move faster over the wave. It's not necessary to match or exceed anyone else's strength, but rather to have balance in one's own.

Parsonage showed this in another study where she measured the time to pop-up both in real-world conditions and in the lab on force plates (picture giant weighing scales that can sense and calculate force).[15] Instead of dividing the participants by gender, she categorized them into stronger and weaker surfers, with both men and women in each group.

She found that, ultimately, gender differences were less important than how strong the surfers were for their size. Both genders received the same prescription: "We found that the strongest surfers needed to do more explosive work—they already had that baseline strength—and the weaker surfers needed to do more maximal strength work to become explosive and decrease pop-up time."

Parsonage is now the research and innovation manager at Surfing Australia, which represents the country in the International Surfing Association and the Australian Olympic Committee. She helped Australian surfers train for the 2024 Paris Olympics, where they rode the famous Teahupo'o wave in Tahiti, known for requiring superior paddling performance.[16] One of the heaviest waves in the world, it breaks near a shallow coral reef, adding a layer of danger. In addition to strengthening their arms, she and her team of PhD students taught female surfers the correct sprint-paddling technique, because it's "low-hanging fruit to get some real improvement in performance. Surfers are never taught to paddle; they're never—like swimmers—taught technique," she said.

Parsonage grew up surfing in South Wales in the UK. After completing her master's degree there, she moved to Australia for her PhD studies, and she now calls Down Under home. For too long, she said, female sports research only looked at the menstrual cycle and injuries, without asking the more fundamental questions about women's bodies in various sports. After a decade in the field, she's starting to see changes away from focusing on those two narrow (though important) areas and into the larger—and to her, much more interesting and exciting—field of female performance. "If we concentrate on performance—how do we help women perform on the world stage—we're going to help mitigate that other stuff as we go," she told me.

She has seen women surfers' strength and ability improve quickly in the decade since she studied for her PhD. Her data shows that female surfers have definitely gotten faster. "That's probably for a few reasons, but a main one is just strength training. That was new to them ten years ago. Now it's normalized, which is amazing."

The sport of surfing is changing fast, at the professional level and all the way down to amateurs, with women's surf camps proving a reliably popular travel offering that gets women into the waves in places where they don't have to worry about men pushing them around. Ultimately, though, surfing is a solo sport, even if you cruise into the waves with friends. Whether its the little break at Austinmer near my dad's or the giant waves of Waimea Bay, the great joy of riding a wave is ultimately personal. "Surfing, at its best, is a deep, private, gender-neutral communion with a force of nature. In such moments, we are truly free," Pennybacker wrote.

KEEPING AN EVEN KEEL

Balance comes up in plenty of other areas of life besides surfing, or even the world of athletics. On TikTok the chair challenge and the center-of-gravity challenge go viral every so often, with a new crop of guy-girl teams giving them a try. In the center-of-gravity challenge, the participants kneel down and lean forward, putting their elbows on the ground and propping their chins on their palms. Then they each move one arm at a time behind their back. The women balance easily while the men tend to face-plant.

In the chair challenge, a person faces a wall and takes two small steps back. They then bend forward at a ninety-degree angle from the hips. Without moving their feet or bending their legs, they attempt to pick up a lightweight chair that has been placed in front of them. Women can generally return to an upright position while holding the chair, but men usually get stuck in the bent position, unable to rise.

The basic principle behind both these feats is that female bodies generally have a lower center of gravity, making it easier for them to move the upper body from the waist and hips, as explained by experts at the National Academy of Sports Medicine.[17] Of course, this is a generalization, as female bodies come in many shapes and heights, and male bodies, while somewhat less variable in shape, also vary significantly in height. But across a population, the lower center of gravity provides a balance advantage for women in a number of skills.

This isn't terribly newsworthy if you've read other chapters in this book or have a solid understanding of physiology. The fact that these challenges go viral over and over again is that, for most people, the results are genuinely surprising. And they can only be a surprise because of the deep-seated cultural assumption that men are better at everything physical. The "fun" of these videos—and the reason couples are usually giggling as they film themselves—is that the failure of burly men to complete tasks that women easily perform is a shock. We are so used to viewing male muscles as the ultimate measuring stick for strength that a simple balance challenge, effortlessly won by a female body, can still amaze and delight.

The balance beam is known as one of the most difficult apparatuses in the sport of gymnastics, but it's a women's-only event because it requires extreme balance, flexibility, and lower-body strength. There are some seriously hilarious YouTube videos of male gymnasts trying the beam.[18] Although they manage to perform some of the moves, they obviously struggle. These are some very muscular men—actual gymnasts—but still, completing the moves often puts them in visibly painful, borderline dangerous positions.

The men's and women's gymnastic events at the Olympics differ enough that it's difficult to directly compare male and female gymnasts. Both sexes do the floor exercise and vault, which involve flipping and tumbling. However, women perform the floor exercise to music and are judged on the artistic aspects of their routine, while men aren't. (This reminds me of the famous quote used to describe Ginger Rogers's dancing with Fred Astaire: Women dancers have to do everything men do, but "in high heels and backward.") Women gymnasts also compete on the balance beam and uneven bars, while men compete on the pommel horse, rings, high bar, and parallel bars.[19]

It wasn't always divided this way. Women used to compete individually, earning points as part of a mixed-gender team, which included beam and parallel bars. In 1948, the "flying rings," as they were then called, were a compulsory event for both women and men. It wasn't until 1952 that the events were separated by gender into the categories we see today. At the time, a lot of assumptions about women's capabilities informed the gender segregation. Even though some women had already been competing in events that are now exclusive to men, they were deemed too difficult for female gymnasts back then.

Gymnasts in the 1950s were adult women, not the prepubescent and teenage girls who dominated the sport in the 1970s, '80s, and '90s. Adult female bodies tend to carry their weight in their lower halves, giving them a real advantage in an event like the balance beam. This benefit is less pronounced for younger female bodies and the highly athletic gymnasts of the modern era. Many early Olympic gymnasts came from the dance world, which influenced the emphasis on flexibility, artistic flair, and grace in women's events.

This midcentury history has helped to shape the Olympics we see today, influencing everything from competition to judging criteria. Many female gymnasts, especially taller ones, now prefer the "men's" parallel bars to the "women's" uneven bars. Too bad! The segregation in gymnastic events and equipment is based on sex, not height. Frankly, aside from "tradition," what's the point of this? It sure seems that gymnastic standards, events, and judging criteria should be updated from the 1950s.

One thing that has been updated: The age requirement for gymnasts is now 16 and over, so there has been a return to slightly older gymnasts in recent years (the average age in 1992 was 16.45, but it was 22.47 in 2024). Experts say this is healthier for the gymnasts' bodies and minds: They are both physically stronger and psychologically more able to decide to try tricky maneuvers. Janelle McDonald, the head coach of women's gymnastics at the University of California, Los Angeles, told NPR that this new era of gymnasts has shifted the sport toward a high degree of athleticism. "The skills we're seeing right now blow me out of the water," she said.[20]

Clearly gymnastics has changed since former dancers competed in the 1950s, and maybe it will change again. Why not open all events to all gymnasts, allowing those with tall, lithe bodies to compete in the events they might naturally excel at, while shorter powerhouses of any gender can compete against each other? Why not create events focused on grace and artistry and allow those with male bodies to participate in them as well? Human physiques come in a great variety—so why, in athletic competitions, is the dividing line almost always drawn based on biological sex?

TOWARD TRULY BALANCED COMPETITION

There are exceptions to the gender binary in athletics. Equestrian sports, snooker, and sled-dog racing all put men and women in direct competition. Additionally, a small but growing number of teams and sports allow men and women to play or compete together on the same teams, such as relay events in cycling, running, and swimming; racquet sports like badminton, tennis, and racquetball; as well as timber sports, korfball, and ultimate frisbee. But for the most part, there are sports—and women's sports.

With the rise of understanding about gender complexity, some think it's time to challenge the current system, which often excludes trans, nonbinary, and intersex athletes. These individuals often face incredibly stressful abuse due to their identities and the rigid categories in sports.

When *The New York Times* posed the question "Should more sports be coed?" they received some thoughtful responses. Gowri from San Jose, California, suggested that separating the genders based on boys' assumed aggression or physical strength "imposes the mindset that girls SHOULD be afraid of boys . . . setting the wrong mindset for girls in our future." Isabella, from Valley Stream, New York, wrote that coed sports can help kids form a community and noted, "Having kids grow up in coed sports communities can prevent kids from believing in false claims such as 'women are weak,' which is incorrect because women dominate in fields such as control, speed, and endurance."

In reply to the idea that mixed-gender sports would be unfair or less safe, commenter Aparna wrote, "Sports are already inherently unfair, as people are born with different heights and weights, leaving certain competitors with natural advantages. In addition, sports are already incredibly unsafe, with many players in contact sports like football frequently suffering concussions and bone injuries. In order to actually make sports safer and more fair, why not separate by height or weight instead? This change wouldn't just prevent more injuries and make sports fairer, but also help dismantle the idea that one gender is inherently stronger, and solve contentious issues such as the participation of transgender athletes."[21]

So much debate about gender in sports comes down to arguments about fairness, but Aparna's comment really hits the nail on the head. Why is it considered fair to divide sports by gender while disregarding other factors that cause disparities in athletic performance? Anatomical differences encoded in DNA, like height or muscle type, can confer significant advantages or disadvantages. Usain Bolt, who stands six feet five, has a built-in advantage as a sprinter due to his height. (Fewer than 1 percent of people are over six feet four.) Bolt is often head and shoulders taller than those he races against—and he also has uniquely long legs for his height. Is it fair for him to compete?

Michael Phelps has won twenty-eight Olympic medals in swimming, and many commentators attribute his success in part to his unusual physique. Whereas most people's "wingspan" is equal to their height, Phelps's arms are three inches longer than he is tall, giving him a greater reach in the water. Phelps also has an extremely long upper body, relatively short legs, and very large feet, which act as paddles. Additionally, he has a genetic quirk that allows his body to produce half the lactic acid of other swimmers, giving him greater fatigue resistance and faster recovery times. Given all this, is it really fair that he's won medals over other swimmers who don't have these genetic advantages? Why are those advantages considered fair while hormone variations are not?

Bolt and Phelps identify as male. When a female athlete demonstrates strength that approaches what is expected of male athletes, questions arise. Katie Ledecky finished over ten seconds ahead of the runner-up in the fifteen-hundred-meter freestyle swim at the Paris Olympics, and she holds all twenty of the fastest times for that event in the women's category—times that could allow her to qualify for the men's Olympic team. For her talents, Ledecky has been "transvestigated" by internet sleuths who claim she can't possibly be a woman. Same with rugby star Ilona Maher, simply because she's tall and has wide shoulders, traits that benefit a rugby player. Then there's boxer Imane Khelif. When she won a bout in record time at the 2024 Olympics, conservative politicians in the US, author J. K. Rowling, and others accused her of being a man. Their proof? Her jawline and short hair. Another boxer, Lin Yu-ting from Taiwan, faced similar accusations. Both boxers have always identified as women, have a long history of competition, and have won against and lost to other female boxers. The truth is that all these athletes have been accused of being men because they don't look "feminine enough" to some people, leading to scrutiny based on their appearance.

"This whole 'controversy' shows how far the anti-trans movement will push in its pursuit to stamp out gender nonconformity from society. It already pushed trans women out of sports. Now it is going after anyone with a strong jaw or a muscular physique," wrote trans journalist Katelyn Burns.[22]

Very few trans athletes compete in any Olympic sports, but technically they are allowed, if they follow certain rules (which have changed several times in the last ten years).[23] Those who transition from female to male are eligible to compete without restriction. Male-to-female trans athletes have to publicly declare their transition and maintain specific testosterone levels over a period of time, and they are subject to testing.

Here's how that works in practice: Australian surfer Sasha Jane Lowerson is the first and only out trans woman to compete in the World Surfing League, one of the international bodies that oversees Olympic surfing qualifications.[24] The league published its criteria for trans women surfers in late 2022, following guidance from the International Olympic Committee. To qualify, Lowerson must maintain a testosterone level below 5 nanomoles/liter of blood for twelve months (hormone blockers can help trans women achieve these levels). This seems reasonable at first glance: Typical testosterone levels for cisgender women range from .5 to 2.4 nanomoles/liter, so the requirement brings Lowerson close to those levels.

But as we know from earlier chapters, "typical" can be a pretty exclusive concept when it comes to female bodies. It's not uncommon for cisgender women have higher testosterone levels for a number of reasons, including pregnancy. Testosterone levels can reach 3–4 nanomoles/liter in pregnant women, with levels increasing as pregnancy progresses.[25] Polycystic ovary syndrome, which affects up to 13 percent of cisgender women, can push testosterone levels even higher, to over 5–6 nanomoles/liter (and it's not uncommon for levels to reach as high as 8 or 9).[26] In fact, high testosterone is one of the primary criteria for diagnosing the condition. Typical testosterone levels in cisgender males range from 9 nanomoles/liter to 35, but this is primarily in young guys. For the over-fifty crowd, 7 nanomoles/liter is considered low but not uncommon.[27] Additionally, a number of genetic conditions can result in naturally higher or lower testosterone levels.

Based on testosterone levels, wouldn't a pregnant woman or someone with PCOS violate the hormone-level rules for athletes? (It's quite possible to be pregnant without realizing it.) What about people who are intersex or have higher or lower hormone levels but are unaware of it? The truth is there's no way to determine the hormone levels of each athlete unless they are all tested.

Testosterone must have a long and thorough history of providing an athletic advantage for it to be the basis for who gets to compete and who must take hormone reducers, right? It must offer a significant edge to warrant such scrutiny. However, we don't really know that to be true. There are only a few studies connecting testosterone to athletic performance. While high doses can help increase muscle mass, "the link between natural levels of testosterone and muscle mass is not consistent across populations," and "evidence unequivocally linking natural testosterone levels to improved athletic performance remains elusive," wrote Cara Ocobock, director of the Human Energetics Lab at the University of Notre Dame.[28] This uncertainty is part of the reason testosterone is checked in some sports but not others.

Those who argue for excluding trans and intersex people from sports, subjecting them to hormone testing, or otherwise scrutinizing them fundamentally believe that trans women must be better athletes than cisgender women because men are inherently stronger in every way. This book demonstrates that the reality is much more complicated. The idea that trans women athletes are automatically stronger is based on opinion, expectation, and cultural conditioning. "We actually don't know if there's a biological advantage for trans women over cisgender women because the science is not clear," Ada Cheung, who leads the Trans Health Research program, in partnership with the University of Melbourne, told *The Sydney Morning Herald*.[29] "No research has really been done into trans female swimmers or any elite athletes that are transgender. The jury is out."

One proposed solution for maintaining sex categories in sports while being inclusive is the creation of a third category for nonbinary people.

Tokyo is the most recent marathon to implement this change, becoming the sixth event to do so in the Abbot World Marathon Majors (which includes the largest, best-known marathons: Berlin, Boston, Chicago, London, and New York).[30] New York is the only marathon that offers prize money in the new category, a practice it has followed since 2022. The number of participating athletes has grown each year since its introduction. New York had forty-five participants in 2022 and ninety-six in 2023, while London had over a hundred participants in its first year (2023). Separate but equal?

One way to even the playing field (pun intended) is to simply let people of any gender compete, whether on the shooting range, on the ice rink, or in their careers. If sports have contributed to the lowered expectations and underutilization of women, as well as the overexaggeration of men's strengths, perhaps they could also serve to do the opposite. This approach might actually benefit everyone involved.

Gender separation in sports seems natural and normal, but that's only because it's all most of us have ever known. Once you dig into the history and the science, it becomes less clear that this should remain the norm. Bringing everyone together to play sports—while separating individuals by height and weight in contact sports where those factors impact play—might be a way to build bridges across humanity. Because those who play together often develop a different perspective than those who are excluded from one another. (See Chapter Fifteen for an illuminating discussion on this exact topic that I had with korfball players.)

In India, Olympic air-rifle hopefuls train together, and this practice has led to some unexpected advantages. "During the camp, we have controlled matches; sometimes the men shoot better and sometimes the women do. But it helps each of them grow," said Deepali Deshpande, coach of the national air-rifle shooting team, in an interview with ESPN.[31] The equality of skills helps them unite as a team. Deshpande explained, "I like the relationships and camaraderie that form when both genders know they are performing at the same level. Once the boys learn the girls are as capable as them or more, they learn to respect their teammates. That reflects when they go out too. Their general outlook to women changes."

This is where the narrow ideas about gender extend beyond the world of sports and into the rest of our culture. We all lose out when we keep humanity from achieving its best.

Part Five
IMMUNITY: PROTECTION

Chapter Twelve

FEMALE BODIES: GREAT AT DEFENSE

When I was three years old, I lived on the Bowery in New York City with my mother in an apartment that had shiny wood floors, a couple of stools that stood up to a breakfast bar, impossibly tall windows (to a toddler), and not much else. We had come, via London, from her family home on the Sydney coast, where I had gone to sleep and woken up for most of my life listening to the sound of Pacific waves crashing against sandstone cliffs just outside.

To say that I missed home would be an understatement. I had gone from beachy afternoons in the sunshine, playing in tide pools, to the darkness and filth of lower Manhattan during its seediest era in modern history.

Though I later came to love NYC (and especially the Lower East Side and East Village), my most enduring feeling from that time is one of disgust. My mother spent more time at Danceteria catching glimpses of Madonna than with me, and her love of nightlife allowed little room for a toddler's needs. When she didn't leave me completely alone in the apartment, my babysitter was an upstairs neighbor, Dr. John. A tall, slim, and extraordinarily gentle man, John specialized in psychiatry. I remember him as a child would: a towering angel complete with a golden halo of blond hair. More than twenty years later, when I worked in midtown Manhattan at HBO, I learned that my hairdresser, who was based in the Lower East Side, had been part of the crew my mother ran with in the early 1980s. She told me that those dreamy early impressions I had of Dr. John as kindness personified were true. Both of us cried at the memory, her tears falling into my new cut-and-color.

I will always be grateful for the care Dr. John showed me in the worst days of my life, when I was terrified and vulnerable. He is one of many men who have offered me extraordinary kindness.

Thankfully, when I was four, my grandmother took over my parenting. She kept in touch with Dr. John and explained to me that, like another family friend, Dr. John loved other men in the way that my parents loved each other. She always spoke of him fondly. Though she was born in 1918, my grandma always loved people for who they were and found the antigay sentiments commonly held by others in her generation "backward and cruel." I saw her get in fights with friends over the subject; she refused to accept bigotry in her presence.

I never got to see Dr. John when I grew older—never had a chance to thank him for his comfort and care when I was so small and afraid. He died of AIDS in the mid-eighties, like so many gay men (and others) at the time. There's a reason that HIV killed men at much higher rates, and it's not what you may have learned. It wasn't due to rampant promiscuity, or lack of care for themselves or their health. "We used to think that some of the differences in HIV-infection outcomes and treatment were behavioral. Many thought that males weren't responding as well to the HAART regimen [a cocktail of antiviral medications] simply because they were not taking their medication as diligently. But now we know that sex chromosomes play a role in how the body responds to HIV infections," wrote Sharon Moalem, an award-winning neurogeneticist and evolutionary biologist, in his book *The Better Half: On the Genetic Superiority of Women*.[1]

Moalem explained that HIV-positive women have more lymphocytes, which are markers of immune strength, and that they have been found to have lower levels of HIV in their blood compared to men. "This means that women's immune systems may be, at least initially, much stronger at fighting viral infections like HIV," he wrote. A 2015 literature review on AIDS in Africa found that while women are more likely to contract HIV, men are 25 percent more likely to die from it.[2] As the authors wrote, "AIDS prevalence may have the face of a woman, but AIDS mortality has the face of a man." This despite the fact that young female bodies are especially vulnerable to contracting AIDS (young women age fifteen to twenty-four make up a quarter of new infections). Not because of biology, but due to gender inequalities, including "vulnerability to rape, sex with older men, and unequal access to education and economic opportunities."[3] Most women who get AIDS contract it through sex with infected men.[4]

The extent of women's biological advantages in AIDS survival—compared with men's cultural advantages or disadvantages—varies by country and can be hard to tease out of the data. A New York–based study found that women who had AIDS progressed more quickly through the disease. The researchers posited this was probably due to knock-on effects of poverty, like lack of access to health care, poorer care, and less consistent care than men received—all factors, of course, that also affect trans women and women of color.[5] Still, women live significantly longer with AIDS than men do, no matter the age at which they contract the disease, according to data from all fifty states. The data set, which included almost 185,000 people, was published by the HIV Incidence and Case Surveillance Branch of the US Centers for Disease Control and Prevention.[6]

THE FEMALE IMMUNE ADVANTAGE

Female bodies have more powerful immune systems than male bodies do,[7] a fact that's been well understood since at least the 1940s. Long-standing Indigenous medical systems worldwide, including Chinese, African, and Ayurveda, also acknowledge this fact.

The strength of female immunity isn't just a human thing. Among animals, sex-based differences in immune response (advantage: female), have evolved not only in mammals but also in reptiles and birds. Even insects like the humble fruit fly, *Drosophilia melanogaster*—a key player in scientific research—have shown superior female immunity.

The differences in immunity between male and female bodies aren't minor. They are significant and striking, leading the authors of one study on the causes of higher male mortality to conclude with the strongest of statements: "Being male is now the single largest demographic risk factor for early mortality in developed countries."[8] This report takes into account both the male immune disadvantage and the riskier behaviors men engage in (some by choice, like smoking and excessive drinking; others shaped by environment and culture). Still, it's a remarkable conclusion, especially given that female bodies have their own cultural disadvantages when it comes to medical care and lifestyle, like knowledge gaps in health care and more women living in poverty.

Studies from major oncology centers around the world have repeatedly shown that male bodies have almost double the risk of death from cancer.[9] Other research reveals that their antibody response to seasonal flu vaccines is about half that of female bodies, part of the reason that illness and death from flu in older men are more common.[10] According to the Mayo Clinic, female bodies clear pathogens of all types—including fungi, parasites, bacteria, and viruses—more quickly and effectively, and they show stronger responses to almost every common vaccine.[11]

At all ages, the female body has a more robust and mature immune system.[12] In a paper titled "Go Girls! Efficient Female Immunity," the researcher wrote, "It was recently demonstrated in a large cohort of patients suffering severe trauma that both pre- and postmenopausal women have reduced incidence of nosocomial [hospital-acquired] infection. This indicates that ovarian hormones do not account for all the differences between male and female responses to infection."[13]

The immunity difference between male and female bodies starts early. In neonatal intensive care units (NICU) more premature boys die than girls, and even among boys born on time, infant mortality is higher.[14] "This has been explained by sex differences in genetic and biological makeup, with boys being biologically weaker and more susceptible to diseases and premature death," according to the author of a 2012 paper exploring reasons why. At every step along the way—from babyhood to old age—females outsurvive males (exceptions being situations where there is poor maternal health care or female babies are actively abused).[15]

It was a conversation with a NICU nurse that led Moalem to write his book. "Males, I was always taught, are the stronger sex. Yet that's the opposite of what I've seen so far, both clinically and in my genetic research. So why do males seem like the weaker sex in fact?" he wrote. "Maybe you're just not asking the right question," a NICU nurse told him. "Instead of thinking about male weakness, maybe the question is, 'What makes females stronger?'"[16]

To answer that question, let's look at the fundamentals of immunity.

Understandably, we tend to think and talk about the things that make us sick—viruses and bacteria, fungi and parasites—pretty negatively. (Quick reminder that not all bacteria and fungi are pathogens and make us sick; some directly benefit us.) To describe how the body fights them, we commonly use the language of war or law enforcement, as I do below. (Yes, it's a cliché, but it works.)

Just as in human wars it's not always clear who is right and good and who is wrong and bad, so it goes with pathogens. They have helped us evolve while also killing and maiming us along the way. When our bodies fight the organisms that make us sick, our cells are changed. Viruses in particular supply new genetic information via horizontal gene transfer, shaping our very genomes. The contributions of viruses to humanity is so significant that the National Institutes of Health (NIH) called them a "major driver of evolution."[17] So while I'm not suggesting that the invading cells that sicken us are "good guys," it's worth acknowledging that we would not be who we are without them—whatever our sex.

Although we still don't fully understand all the reasons for the stronger immunity of female bodies—partly because they have been understudied for so long—there are some interesting knowns. Understanding these mechanisms could be key to curing a variety of diseases that affect both sexes, from cancer to the flu, and certainly to understanding COVID-19 and its long-term effects.

When it comes to avoiding illness, female sex hormones provide real and significant advantages (see the next chapter for more detail on that subject). There are also underlying cellular differences in how male and female bodies fight disease. But one of the biggest advantages those with XX chromosomes have is an extra set of genes to fight off viruses, bacteria, cancers, parasites, and fungi.

THE POWER OF THE DOUBLE X

Two strong X chromosomes provide a whole host of abilities over the weaker XY combination (possibly including many that are yet to be discovered). As a 2017 joint research paper by an Australian-UK team noted, although the Y chromosome "had long been considered a genetic wasteland," it does contain immune-modulating genes, so it's not entirely accurate to say that all immunity comes from the X chromosome. In those with XY chromosomes, however, the Y seems to serve both advantageous and harmful functions. The researchers wrote that evidence suggests the Y chromosome shows a "genetically programmed susceptibility to diseases with a strong immune component. Phylogenetic studies reveal that carriers of a common European lineage of the Y chromosome (haplogroup I) possess increased risk of coronary artery disease."[18]

The double X gives those who possess this combo myriad advantages in traits ranging from brain disorders to factors that influence longevity (as detailed in Chapter Fifteen)—but especially when it comes to immunity. Scientists are working to learn much more about how these chromosomes work in order to help all human bodies, but the bottom line is that those with XX chromosomes have greater genetic diversity than those with XY. "Females simply have access to realms of biology that males do not have," Huntington F. Willard, the director of Duke University's Institute for Genome Sciences and Policy, told *The New York Times*.[19]

Here's why: Having two versions of a gene carried on the X chromosome—not just one, as a genetic male has—means XX carriers have two sets of every kind of immune cell. In some cases the two genes can compete to produce the strongest version of a given immune cell, and in other cases it means there is a greater variety of cells to fight a particular pathogen.

One way to understand the variability in the types of strength conferred by XX is to do a little imaginary exercise. Working from the outside in, let's pretend an influenza virus is about to invade my genetically female body to wreak its special flu-y havoc. Before that wily virus even makes it into my airspace, my body is on high alert for such interlopers, as part of my innate immune system. Each of my two X chromosomes, one from my father and one from my mother, carries different gene variants not found on the other X. Normally, one X chromosome is turned off, a process called X-chromosome inactivation, but it's not known why my body chooses which X to keep active.[20]

More diversity in the types of cells that are on the lookout for pathogens—such as toll-like receptors, killer-activated receptors, killer-inhibitor receptors, and pattern-recognition receptors—makes for a stronger defense. These receptors reside on the surfaces of neutrophils, the immune system's first responders. Up to 80 percent of all white blood cells are neutrophils, which constantly circulate in the blood looking for invaders.

These beat-cops of the immune system sound the alarm if they come across suspicious interlopers, so having more of them is better. Greater variety in these cells means that as viruses, fungi, and bacteria have evolved to outwit those receptors, so my body's defensive system is better equipped to "spot" them as soon as they hit my nose's mucus membranes or enter through my lips and land in my lungs. Early detection of invaders is a big offensive advantage because timing is crucial when it comes to infections. Our bodies routinely fight off all kinds of bugs, spores, and viruses without our even knowing it.

It has been well-established that female bodies average a higher number of neutrophils compared with male bodies.[21] But the advantage isn't just having more of these cells. Female bodies have smarter, stronger ones. Research published as part of an NIH study in 2020 found that neutrophils in female blood were more mature (giving them full capacity to fight infections) and more responsive than those in male blood.[22] The study attributed this difference to both hormones and the genes within the cells. The XX-advantage effect was especially pronounced in those aged twenty to thirty.

Next in the defensive arsenal are natural killer cells and macrophages—cellular soldiers that eliminate infectious agents. Female bodies also have more macrophages, while male bodies tend to have more natural killer cells.[23]

If the virus that's attacking my cells gets past these defenses, my B cells will quickly make antibodies to fight it. B cells are specialized fighters, part of the adaptive immune system. They are more like the FBI than the patrolling cells of the innate immune system, because they are experts in creating antibodies whose structure specifically matches the invading virus to eliminate it. Creating new antibodies for unknown pathogens is complex work that can take time, but if we can stay alive long enough for our B cells to do their job, we can fight off viruses—even ones we've never encountered before. For example, I contracted COVID-19 at the height of the pandemic and was incredibly sick for two full weeks. It took me two more weeks to recover and even longer to regain my sense of smell. The virus hit me so hard because I was unvaccinated, so my body had no prior experience with the novel coronavirus.[24] Once exposed and infected, I had to make my antibodies from scratch. The next time I got COVID-19, a year (and a vaccination) later, I was only sick for three days and recovered fully within the week.

Once a B cell has met a microbe and created an antibody to fit it, it keeps refining that antibody over time, improving the fit. This process changes the B cells themselves. "When the B cells are being schooled to make better antibodies, mutations begin at a rate that quickly approaches one million times more than is normal, a process called somatic hypermutation. Both male and female cells undergo this process of antibody refinement. Yet it is women who devote more energy to keep educating their B cells through more cycles of mutations, until they are able to produce the best-fitting antibodies, ultimately fighting infections more effectively than men do," wrote Moalem. Theories differ as to why this is, but "one thing is clear: women have immunologically evolved to out-mutate men."[25]

B cells don't just make these snugly fitting antibodies in the moment; they retain a memory of them, which they can use later to re-create the antibodies when needed. This is the basic idea behind vaccination. When you expose a person to small amounts of a pathogen, their B cells learn to fight it, so they are ready the next time they encounter it. In this part of the process, again, the female body outperforms the male, with memory B cells lasting much longer in females.[26] This is one reason that women's bodies often respond more strongly to illnesses, even years after vaccination against them.[27]

Early detection, rapid response to invaders, B cell antibody production, and memory B cells retaining information for future use—all of which female immune systems tend to do better and faster—mean that viruses I encounter are less likely to gain a stranglehold on my body. My immune system is more likely to kill them before they can kill me.

Even older women have stronger immune systems. At the Jackson Laboratory, blood from three similarly sized groups of people—young (under age forty-one), middle-aged (forty-one to sixty-four), and older (sixty-five-plus)—was tested for B cells.[28] There was little change in the number of B cells in women over time, but older men's B cells showed significant loss of function. Interestingly, while both sexes lost immune-cell function in their late thirties/early forties, a second decline happened around ages sixty-two to sixty-four for men, but not until sixty-six to seventy-one for women. That five-year gap is strikingly similar to the difference in lifespan between the sexes.[29]

Having just one X chromosome means the male body is less capable in several ways when it needs to fight viruses, fungi, parasites, and bacteria. As Moalem wrote, "When it comes to males and genes on the X chromosome, their position is one of deficiency when something goes wrong, which it so often does."[30] He suggests that medical research needs to catch up to the genetic superiority of the female antibody response. It's true that information about female immunity could be used to help male immunity through gene therapy, but this isn't about merely understanding the mechanisms of how antibodies work to help all humans fight illness. It's also crucial to ensure that future vaccination protocols take sex into account.

Why is that so important? It might seem like as long as a vaccine trial includes roughly equal numbers of males and females, the bases should be covered. The problem is that when the sexes are combined into one group, any sex-specific reactions can go unnoticed. For example, if most of the males in a group have lower than expected antibody responses, but the women have higher than expected responses, the group as a whole might appear to have sufficient protection. However, because female antibody responses are faster and more intense, clinicians may end up wasting extra vaccines on women who don't need them—or, conversely, not providing enough for men, who might need more booster shots or higher doses to be equally protected.

This is yet another compelling example of why including sex and chromosome makeup as variables in research is incredibly important—to the health of people of all sexes.

Chapter Thirteen

ESTROGEN THE PATHOGEN SLAYER

It was a classic Seattle afternoon as I navigated through the city to meet Caroline Duncombe. By "classic" I mean that it threatened rain but didn't actually rain, and then the sun came out an hour before it set. Duncombe is one of a new generation of scientists studying immunology, and I had about twenty-seven questions that had piled up as I researched how immunity and sex hormones interacted. Another one popped into my head like an annoying mosquito as I walked the last couple of blocks to our meeting spot while just managing to notice that the South Lake Union neighborhood—new to me—was well described by visitseattle.org as a "lively tech zone [that] showcases lake and skyline views."

As I consulted my phone's map, looking for the precise location of the café where I was supposed to meet Duncombe, I wondered if anyone could possibly answer so many questions, and how I could get through them all without seeming obnoxious. I spotted her sitting outside Fresh Flours café despite the cloudy skies. I joined her and barely introduced myself before jumping into my queries. I was suddenly nervous. There were so many things I needed to know.

Over the next hour, Duncombe patiently walked me through her work. The laughter of the women at a nearby table reminded me with every guffaw that some people were using their leisure time to hang out with friends—while Duncombe was taking hers to fill me in on sex-specific immunology and how a young scientist found her way to it.

Duncombe grew up living both in Wisconsin and internationally with her professor mom. She attended college in Minnesota. She's a combo of fast-talking brilliance and "Midwestern nice," as proven by her willingness to take several hours out of her busy research schedule (and also running a very cool improv comedy group for scientists) to meet up. Duncombe got into immunology during the pandemic. It was 2020, and she had just completed a program for young scientists at the National Institute for Allergy and Infectious Diseases (part of the National Institutes of Health), where she had worked in a tuberculosis drug-discovery lab. She wanted to keep focusing on infectious diseases, but inspired by the formative years she'd spent in Sierra Leone (where her mom's job took them), she also wanted to do work that would help people who needed it most. Duncombe's PhD was in pathobiology, "which is the study of host-pathogen interaction—how humans respond to infectious diseases."[1] As she made career plans, she—like the rest of us—had no idea what the pandemic would bring.

Duncombe interviewed for her PhD program just a month before lockdowns for COVID-19 hit. So it was without foreknowledge (but with great timing for her future research) that in the fall of 2020, she walked into Helen Chu's lab to start work. Chu, the first doctor to recognize community transmission of the novel coronavirus in the United States, was named Washingtonian of the Year for her early work in determining how the virus spread. At the time, Duncombe felt lucky, because unlike so many other jobs, which were shut down, hers allowed her to continue going to the lab and working. She found herself in the epicenter of research that had unprecedented global and historical importance.

While Duncombe was working on a project related to predicting long COVID, the differences in transmission of the virus between men and women started making headlines. "Men are more susceptible to severe disease and more likely to die from COVID, and females are more likely to get long COVID," she told me. "Having such a strong observation on the national level brought attention to the issue of sex-specific immunology." Duncombe has brought additional focus to this long-ignored area. She presented on how sex differences affect malaria outcomes at a prestigious conference, which first featured the subject in 2022.

She said that even though these sex-based differences in immunity have been well known for decades, only recently are people "really starting to listen." And do research.

When we had been at the café for nearly two hours, Duncombe asked if I wanted to check out her lab down the street, at the University of Washington School of Medicine research center. I felt like a teenager being invited to the cool girl's house to play records. (Also, I had only gotten through fourteen of my questions, but I didn't tell her that.)

I was expecting the research center to be some monolithic, nondescript edifice, but as we cut between two buildings with undulating glass-curtain exterior walls punctuated by deep orange accents, I was pleasantly surprised by a sense of architectural flourish in a city sorely lacking it. The interiors were even cooler. When we exited the elevator on Duncombe's floor, a gigantic fuchsia slash zagged across the facing wall. (It was actually an extreme close-up of blood cells as seen through a microscope, complete with the Giemsa-staining color scheme.) She swiped her ID badge to let us into her workspace, and the door thunked closed behind us.

Inside, we passed work stations on our left, like those found in any openish-plan office. On our right was a huge, well-lit, enclosed lab space with walls splashed in jewel tones. I had to comment on the carpeting, which featured big blue and turquoise hexagonal shapes at seemingly random intervals. I mentioned how the design resembled cells, and Duncombe pointed out that they looked more like molecules, reminding her of the chemical structure of hormones. But she admitted, laughing, that since that's her field of study, "Everything looks like hormones." At her desk, she pulled up the chemical structures for estrogen, progesterone, and testosterone on her computer.

Comparing the hormone structures on her screen, she said, "See how close they are to each other?" She pointed out how just one bond or atom separated the fundamental structures of the seemingly disparate and powerful hormones. Like human beings, these hormones have far more in common than in difference. It's an interesting piece of information to keep in mind when considering how impactful estrogen is on female immunity.

ESTROGEN, THE WARRIOR HORMONE

That's Duncombe's research focus: She thinks sex hormones in particular have a very significant effect on complex immune responses.

All sex hormones affect us at all stages of life and across sexes—and this effect persists over a lifetime, even when hormones change due to aging or other reasons. "All hormone states can imprint on your immune response," said Duncombe. "When people are kids, they do tend to have very low sex hormone levels, but there are still some, and they are still influencing the immune response." She explained that as hormone levels increase over puberty, and as they cycle in adults (daily for men, monthly for women), they continue to have both real-time effects and long-term effects. "At every phase of your life, not only are your hormones affecting what is happening now, they're setting an imprint for the future as well."

While there are some differences in the immune response between male and female children (for example, little girls have higher numbers of certain T cells), a survey paper on the subject noted the "paucity of studies in this area."[2] It's likely that the immunity differences are explained by the genetic advantages conferred by having XX versus XY chromosomes, as well as by the low hormone levels in kids. But the significant changes that happen during puberty, particularly around female immune advantage (and the fact that many of those advantages wane later in life), point to estrogen and progesterone as major drivers of the larger immunity differences observed in adults.

Sex hormones are complicated. At lower levels, estrogens stimulate the immune system and inflammation (a good thing when that inflammation is working to fight off pathogens). In higher quantities—like during the luteal phase of the menstrual cycle and in pregnancy—estrogen can actually suppress inflammation.[3] It's important for the female body to have a less aggressive immune response during pregnancy in order to accommodate the foreign body of the fetus, which it would otherwise attack. Scientists call this action by estrogens a bipotential effect, since it can go either way. The other primary hormones—progesterone and testosterone—are anti-inflammatory, reducing immune response.[4]

Research has found that, because of the varying amounts of estrogen and progesterone present throughout a menstruating person's cycle, immune strength can vary too. There are "significant changes" in immune cells over the ovarian cycle; they are highest in the first part of the cycle and lower later.[5] (I'm probably not the only one who has noticed that if I'm going to get a cold, it often happens the week before my period, which results in the lovely combination of sniffling, coughing, and bleeding at the same time. Truly a joy.)

Because hormones fluctuate in the bodies of menstruating people (and, as we saw in Chapter Three, not everyone's hormonal changes over the month match expectations), some researchers have suggested that checking blood levels of sex hormones during experiments is the only way to truly understand their impact. In a 2022 research paper from the University of Washington that looked at the subject, the lead author suggested in a release that "researchers should consider the menstrual-cycle stage of patients or study participants when testing a new vaccine or conducting other investigations."[6] In a future of personalized medicine, this could mean that testing hormone levels prior to a vaccination becomes standard practice, or possibly even timing your vaccination schedule to align with your cycle for optimal efficacy.

The estrogen immune advantage is evident in the many types of cells that fight pathogens and across each stage of the immune cell life cycle. When mice or humans are given estradiol, a type of estrogen, it increases neutrophils in the blood and lungs. It bumps up innate immune cells like monocytes, macrophages, and a host of proinflammatory cytokines (messenger cells). It even influences differentiation of the precursor immune cells made in the bone marrow.[7] Across the life cycle of immune cells, estrogen not only increases their numbers but also makes them stronger and smarter.

This has effects on many types of diseases. Estrogen empowers key immune cells to stop HIV, Ebola, and hepatitis viruses from reproducing, which can reduce disease severity and, in turn, lead to reduced transmission to others. The same is true for tuberculosis. And a 2016 study found that estrogen protects female bodies from the flu by preventing the virus's replication.[8] This is such a strong and useful effect that it may even be a check in the "pros" column of taking postmenopausal hormones, said Sabra L. Klein, an associate professor at the Johns Hopkins Bloomberg School of Public Health who conducted the flu study. "If women are taking estrogen-like hormones for other reasons, an added benefit might be less susceptibility to influenza during the flu season," she said in a release. Since flu is a killer for elderly women, "being on hormone replacement therapy could be one way to mitigate the severity of this disease, which is exciting, simple, and cheap."

The same idea is also being considered for older women and COVID-19. During the pandemic, women taking progesterone and estrogen-containing oral contraceptives were found to have fewer cases of COVID and fewer hospitalizations.[9] It's therefore no surprise that a 2021 study hypothesized about whether hormone replacement therapy "may be considered as a viable treatment option for pre/post-menopause women with coronavirus, referring to the fact that sex hormones reduce inflammatory responses and modulate ACE2 expression."[10] At this time, the evidence isn't quite strong enough to suggest hormones to treat COVID, with a 2022 review paper on the subject stating that there are still "unanswered questions" and that "careful patient selection and individualizing the type of drug, route of administration, timing and duration" is necessary, but research is ongoing.[11] Estrogen is powerful.

Despite all this evidence, most studies on immune response factor out biological sex when researchers are trying to understand a chunk of data. And in some cases, they should—it's not always relevant, said Duncombe.

When researchers do this, they say they've "controlled for sex." Plenty of other variables can be "controlled" out. Smoothing out differences in the data and reducing the noise help researchers more clearly determine, for example, that drug A lowers blood pressure better than drug B. But what if drug A lowers blood pressure well in women but not in men? By controlling for sex, a researcher won't see this difference, because every male data point is the same as every female one. So it will no longer be obvious that all the female patients scored higher and all the male lower—these differences will be averaged out. Any obvious sex-based beneficial (or detrimental) effect on women will be lost, mixed into the male data. By controlling for sex, "you're taking away the things that make results different by sex. Sometimes what's different is important information," Duncombe pointed out.[12]

It's a simple but profound question: What differences have we been missing by controlling for sex? This question goes all the way back to the beginning of a researcher's work. "How you design your studies dramatically affects what you can capture," said Duncombe. "If you don't look for the answer, you won't see it. And you have to be asking the right questions in science, but your incentive structures may or may not be completely there to promote this type of research."

THE TESTOSTERONE DISADVANTAGE

Duncombe's current work focuses on sex-specific differences in malaria vaccines. Malaria sickens and kills so many people every year that it's one of the few diseases included in the United Nations' Sustainable

Development Goals in the Good Health and Well-Being category. Many methods of eliminating the disease have been tried and failed. The malaria parasite is incredibly wily, evading the immune system by altering how it appears to the body's killer cells, literally transforming its surface as it travels between blood cells so it won't be noticed by immune cells.[13]

Duncombe's work on a malaria vaccine incorporates sex-based immune differences into her study design. I got a glimpse into this work when I visited the Biosafety Level 1 Lab adjacent to her office (there are four levels of possible danger when it comes to lab-created pathogens, with 1 being the safest). As long as I promised not to touch anything, I was allowed in. To my right, all kinds of lab machinery on long desks (called benches by scientists) jutted out from the main wall. As a bit of a gadget nut, I was dying to ask about the function of each of the bland-looking domed boxes and awkward-looking devices scattered around, but we were already hours into the interview, so I restrained myself. Duncombe led me over to a bench where a microscope with tools and slides was set up.

Since malaria is a complex disease, involving incubation in the liver, most initial testing of potential vaccines is done on mice. Why not just go directly to human testing? "That's a really important question, because you want everything that you do to translate to humans," said Duncombe. "But there are some things that you can't study in a human. You can't take a whole liver out of a human." So, mice it is.

To do her work, Duncombe starts by visiting the insectary, where, weeks before, she's requested a batch of two hundred mosquitos pre-infected with plasmodium, the malaria parasite. She collects the mosquitos with a vacuum gentle enough to avoid killing them, sucking them from their free-flying container-home. "You're trying to get every last one out of there because they're expensive, and then you put them in a tube," she said. I was distracted by the idea that mosquitos were expensive, but of course, scientists can't just use random skeeters. They have to be bred in controlled circumstances, free of other pathogens, and all the same age to ensure that the malaria parasite extracted from them is of similar quality. Even the lowly mosquito is, in this particular way, valuable and integral to saving human lives. We wouldn't have malaria without mosquitoes, but nor would we be able to develop a cure for it.

Now a timed trial begins. With her tube of two hundred mosquitoes, Duncombe sits down at the microscope and—by hand—extracts the malaria parasite from the salivary glands of each insect, one by one. "The fun part is that you have to do everything within four hours because you have to get the dissected parasites into the mice when the parasite is still very fresh and alive," she told me. She sets a timer to ensure she's on track. It takes her about sixty to ninety minutes to extract the parasites from that many mosquitoes. Then she has to purify and inject the parasites into the mice (some of which are vaccinated, some of which are controls) before her alarm sounds. While a lot of science takes weeks, months, or decades, there are also scientific sprints.

In both people and mice, malaria parasites migrate to the liver, where they replicate and then flood the bloodstream with more parasites, which make the host sick with the disease. In humans, that process takes six days, but in mice, it's just two. After a couple of days, Duncombe can take a quick blood sample from the mice to confirm if they have malaria or not. Those who have been successfully vaccinated will test negative.

A vaccine for malaria, which sickens 249 million people every year, killing over half a million of them (mostly babies and children, and mostly in Africa), has been in development for decades.[14] Inspiringly, malaria vaccines are now available, but they are still in the early stages of use. In humans, men and women both get malaria at about the same rates. Women clear it from their bodies faster, so there's some female immunological advantage, but the mechanism is unknown. And nobody has really looked at how biological sex might influence malaria vaccine efficacy, though some previous research indicates a possible hormonal connection.[15] Duncombe thought that if vaccines worked differently in different bodies, that would be important for health workers (and vaccine developers) to consider.

First, Duncombe's research looked at malaria-vaccinated mice, which showed that female mice given the malaria vaccine warded off the disease, but male mice didn't.[16] What was the mechanism behind this difference? To find out, Duncombe removed the ovaries from female mice so they weren't producing sex hormones. These mice stayed healthy when infected with malaria, meaning it wasn't the estrogen or progesterone helping their immune systems respond well to the vaccine, since they weren't able to make any.

Then she turned to the male mice. When their testes were removed, they no longer produced testosterone, and they too were able to benefit from the vaccine's protection. So, estrogen didn't help the malaria vaccine, but removal of the source of testosterone did. This indicated that "testosterone is a dramatic modifier of protection from the malaria parasites," Duncombe said. The male mice's testosterone was likely causing the vaccine to be less effective. She called this a "dramatic effect," adding, "Hormones change so much."

This obviously raises the question of whether suppressing testosterone production might make a vaccine more effective. While I'm not suggesting castrating male humans just to protect them from malaria, what if there were a temporary version of that effect? Duncombe happens to have worked with an endocrinologist who developed a male contraceptive that acutely reduces testosterone in men—but is easily reversed. "This doctor had extra male birth control in his freezer and was kind enough to allow me to use it in my experiments," she told me. She found that when she gave the male contraceptive to the male mice at the same time as the malaria vaccine, they were protected by the vaccine. This is further proof that testosterone "really affects the immune response to the parasite itself," she said.

This might mean that, just as estrogen seems to boost the female immune response for some diseases, testosterone can suppress it, in others, helping to explain the immunity differential between the sexes.

Human males are far from the only animals whose hormones affect immunity. Male lizards' hormones make their immune cells less active than those of female lizards. Female birds have a stronger immune response to pathogens—and the difference is greatest during mating season, when testosterone levels in the male feathered bodies are at their highest.[17] In humans, studies have shown that the higher the testosterone, the more immunity is suppressed.[18] But in many species, including male humans, higher levels of testosterone or other androgens are also linked to higher reproductive success.[19]

The very thing that makes a male animal more successful at breeding may also give him weaker immunity. This is such a common finding across species that it has a name: the immunocompetence handicap hypothesis (ICHH). Why would a species make this tradeoff? According to the ICHH, because it works. Better reproductive success—whether due to a higher sperm count or sexual displays that are more appealing to female mates—shows that the male can "afford" to do it. Only the healthiest males can get away with suppressing their immune systems, and showing off their higher hormone status to females is a way of proving that. Proponents of ICHH call this an "honest" signal for females.[20]

Not everyone agrees with the ICHH. Some studies have tested the opposite idea: that infection or viruses cause the body of an animal to reduce testosterone production.[21] That might happen simply to conserve energy, as an animal can't afford to produce extra testosterone and fight off a pathogen at the same time. The data on this is mixed, with some suggesting that testosterone has a modulating effect but not a suppressive one.[22] Importantly, stress hormones, especially cortisol, likely play a role as well.[23] This is all to say that the work to understand precisely how sex hormones affect the immune system is ongoing and complex. Hormones modulate and are modulated by "these intricate networks of immunity, endocrinology, and physiology," said Duncombe.

What seems clear is that something about male hormones clearly reduces response to the malaria vaccine in mice. How this impacts humans is still unknown, but if Duncombe can better understand the mechanism, it could provide useful leads. As she continues her work, however, she has run up against an issue that has surfaced time and again throughout this book: a significant lack of study.

The basic science Duncombe needs to inform her further studies on a malaria vaccine hasn't yet been done. One example: "No one has studied how malaria parasites respond differently when settling in a male versus a female liver. It's a fundamental biology question," she said. So now she has to backtrack and do it herself.

As covered in earlier chapters, individuals differ in their hormone levels. "We know gender identity exists on a spectrum. But aspects of biological sex also exist on a spectrum, an example being varying hormone levels across individuals," said Duncombe. So, ideally, future research won't just look at biological sex but will also test hormone levels and assess how they are relevant to immune responses—which could lead to more tailored recommendations for vaccinations and treatments.

That's especially important for the many people whose bodies don't produce hormones or who have a modified hormone regime, including those who have had their uterus and/or ovaries removed (about 14 percent of US women have undergone hysterectomies), trans people who are taking hormones or medications that suppress hormones, people with PCOS, and those who have illnesses or take lifesaving medications that affect hormones.

BIOLOGY VERSUS LIFE EXPERIENCE

I thought about immunity quite a bit over the months following my interview with Duncombe, because the specter of a killer disease hung over me and much of the science community even as we all adjusted to life in a post-COVID world. (To be clear, "post" doesn't mean it's over; rather, we now live in a world changed by the virus.) It was reassuring to know that my body might be fundamentally better than my male partner's at fighting off the next viral baddies (and life on earth proves there will always be a next time), but I knew from living through the recent pandemic that in our complex and interconnected human culture, other factors are at play.

It isn't just hormones or genetics that affect humans' immune responses; environment and behavior matter too. Robust fundamental immunity isn't very useful if something in your lifestyle has a greater impact on your health.

Sometimes, factors in a woman's life that are entirely separate from her biology make her more vulnerable to contracting illnesses or dying from them. Tuberculosis (TB), still prevalent in developing countries, is more common in women who cook indoors using biomass fuels like cow dung or wood. Poor ventilation and regular smoke inhalation impair their immune systems.[24] The most common and cheapest screening test for TB involves coughing up sputum to be examined under a microscope. However, studies show that a female body with TB is less likely to have a cough that produces sputum to test. A local health journalist in Pakistan, Shobha Shukla, wrote that women there were often uncomfortable producing the phlegmy substance (honestly, same).[25] Because Pakistan is so highly patriarchal, restrictions on women traveling alone (to a clinic or anywhere else) mean they don't access health care or treatment as early as men do.[26] Add these factors together, and you get more women dying from TB than otherwise might, and none of the causes have anything to do with genes or hormones or immunity at all.

The same is true for cisgender men. Social conditioning and gendered occupations and activities can make men more susceptible to diseases or less likely to seek medical care. Around the world, men engage in more behaviors that expose their bodies to risks, such as alcohol and drug abuse, and violent behavior. Men are also more likely to work in physically dangerous jobs, like mining, roofing, logging, and sanitation work—occupations found on the list of the top ten most dangerous professions.[27] (This gender ratio is skewed because women have been purposefully excluded from those jobs.) In the United States (and most other countries), men are less likely to seek medical treatment than women, regardless of income level or ethnicity. As a result, part of the reason men die from skin cancer at a rate 50 percent higher than women is simply that they tend to go to the doctor too late for lifesaving treatment.[28]

COVID-19 may be another illness where environment plays a significant role in the disparities between who catches it and who it kills.

The evidence that COVID-19 killed fewer women and was especially deadly to older men made worldwide headlines in late 2020—and seems to provide further support for the idea of women's stronger biological immunity, as Duncombe pointed out. But researchers at Harvard University's GenderSci Lab see something else in the data, calling for social factors to be considered.

Sarah Richardson, the director of the lab, started collecting data early in the pandemic and observed sex-based disparities that argued against a purely biological explanation. "In Connecticut and Massachusetts there is no sex difference in confirmed Covid-19 fatalities, while in New York and Florida, men account for about 60 percent of Covid-19 deaths. Globally, the male-to-female death ratio varies from a staggering 2:1 in the Netherlands to 1:1 in Iran and Canada," wrote Richardson and her colleagues Heather Shattuck-Heidorn and Meredith W. Reiches in a *New York Times* opinion piece.[29]

The opinion piece was based on their research paper, which concluded that sex disparities in COVID-19 deaths were "modest" and varied in relation to factors more closely tied to culture than biology, including "health behaviors, preexisting health status, occupation, race/ethnicity, and other markers of social experience."[30]

When I caught up with her more than a year after their research was published, Richardson explained how her lab came to this conclusion. She and her colleagues had quickly and remarkably put together an early COVID-19 tracker and started crunching numbers as the data was collected. "You needed very little training, just the heart and gumption and time to track these sex disparities across geographical space for some period of time to immediately see that the patterns differed wildly," Richardson said.[31]

Looking at the data in more detail, her research team could see that COVID-19 impacted people with "tremendous variation across social groups defined by race and ethnicity, and across geographic location," Richardson told me, noting that "something like 30 percent of the variation was explained by state-specific conditions and policies." Like her previous research into the Spanish flu, MERS, and SARS—which showed that significant social factors influenced who got sick and who didn't, and who died—Richardson's team's work on COVID showed something similar.

Duncombe finds Richardson's work important and carefully considered, agreeing that social impacts can explain some of the effects Richardson details. However, she notes that the clearest sex-based differences in immune response were observed in the initial phases of the pandemic—before vaccines or other precautions were implemented. "Any social disparity in vaccination rates would not have modified these outcomes in the early stages," Duncombe said. That early data offers the best view into the virus's biological effects, since social factors had a bigger impact once the pandemic grew more complex.

Once mask-wearing, social-distancing, stay-at-home orders, and, later, vaccines were rolled out, gendered impacts would have altered the baseline, making it "a lot more tricky to determine the relative contribution of sex-based immune differences," Duncombe said. She pointed to research papers that showed more male deaths in forty-nine countries. While Richardson's work highlights important social factors in disease transmission (a pattern seen in many other viruses as well), the fact remains that biological mechanisms likely also protect women from COVID-19's most lethal effects.

Richardson's work acknowledges this complication. She wrote that a "key factor related to male-female differences in Covid-19 fatalities is that men overall are in a poorer state of health than women." As mentioned above, some of this is linked to men engaging in less-healthy behaviors like smoking, but other analyses have shown that men also have higher rates of diseases that can exacerbate COVID-19's impact, like heart disease, diabetes, and high blood pressure—all of which kill more men on their own. Even before the virus struck, social and biological factors were already affecting men's overall health.

In one paper, two European researchers wrote bluntly, "The degree of cardiac cell death is more pronounced in males than females under various conditions."[32] Dying heart cells are a fundamental weakness, leading to vulnerabilities of various kinds. This is hugely relevant when the inflammatory immune response of COVID-19 hits the body, putting additional stress on an already strained heart.

In a commentary on that paper, three US-based cardiologists pointed out that it's not just underlying cardiac issues in men that contribute to the higher death rate from COVID-19, but also hormones: "Estradiol supports immune system modulation, amplifying innate and humoral immune responses, whereas testosterone is overall an immunosuppressant, in particular inhibiting differentiation of naive CD4+ T cells into T helper type 1 cells, impeding cell-mediated immunity."[33] To translate, female sex hormones strengthen the immune system and testosterone suppresses it, as Duncombe demonstrated with her mice. The cardiologists noted that these "significant" hormonal effects are linked to sex differences in immune response and influence how the heart is affected by the novel coronavirus.

It's not easy for health researchers to tease apart the social and biological impacts of a disease. Magnify those interacting and interlocking effects over a lifetime, and it becomes even more complex. The impact of most human illnesses on the body, and whether they will prove fatal, are shaped by all aspects of our lives. Some are beyond our control, like genes, hormones, or how polluted the air is where we live (there's an association between dirty air and worse COVID-19 outcomes). Others are more within our influence, like vaccinations, the nutritional value of our food, and how much we exercise.[34]

In a chapter about immunity, it must also be acknowledged that the powerful female immune system can save lives in a pandemic, but it can cause harm when that system turns against itself and becomes an autoimmune disease. "Females have higher rates of post-COVID syndrome, which has been linked to higher rates of autoimmune antibodies," Duncombe told me. Long COVID is just one of many examples of the damage a hyperactive immune response can inflict.

Around 80 percent of all autoimmune disease occurs in females. Conditions like lupus, multiple sclerosis, psoriasis, rheumatoid arthritis, and irritable bowel syndrome disproportionately affect women. At the turn of the twentieth century and for many years afterward, doctors believed these diseases occurred equally in men and women, but that women were more vocal about them. It turns out that women actually experienced these issues at higher rates. The assumption by medical practitioners that female patients just complained more obscured the reality. Not recognizing the sex-based aspect of autoimmune diseases meant research on these diseases lagged, delaying progress in understanding and treating them.

The fact that the majority of autoimmune patients are female highlights the cost women bear for a powerful and aggressive immune system. It's a fine balance between protecting our bodies from wily invaders—especially those that mimic our immune systems, as some viruses and bacteria do—and the immune system attacking our own bodies.

Moving forward, a deeper understanding of the female immune system will not only help scientists develop vaccines and treatments for all bodies, but also lead to advances in reducing the effects of immune dysregulation that cause autoimmune diseases.

Researchers like Duncombe and the thousands of others working in immunology are rapidly learning more about how to save lives across all populations. With a growing number of young scientists investigating long-ignored sex differences in the field, immunology is shifting away from its male-centric focus, paving the way to better health outcomes for us all.

Part Six

THE LONG GAME: LONGEVITY

Chapter Fourteen

LIVING LONGER, LIVING BETTER

Kimiko Kamida is behind the bar, mixing me a gin martini, and Mitsu Kakazu, two stools over, smiles as she adds more ice to her already well-diluted whisky. Both chuckle easily at a joke my fixer and translator, Christal Burnette, has made. I've been in the bar less than five minutes, note that this is the fourth time they've all laughed, and decide that these elderly Okinawan women have it all. I'm not in on the joke since I don't speak Japanese, but it doesn't really matter. I smile too.

When I say that these two women "have it all," I don't mean the go-go 1980s version of that idea, coined by *Cosmopolitan* editor Helen Gurley Brown, who proposed that women could and should be able to balance career, kids, Martha Stewart–level housekeeping, and eternal sexiness.[1] No, I mean the "all" that most of us aspire to in the go-low 2020s: that rare combination of health, low stress, friends and the time to enjoy them, plenty of laughs, and a comfy place to hang out.

The Imperial, a bar owned and run by Kamida for fifty years, really is like the *Cheers* bar, where everyone knows your name (and probably has for multiple decades). It's just off the main nightlife street in Naha, Okinawa's capital city. The semitropical island of Okinawa, a prefecture of Japan, was the site of a key battle in World War II, when it was decimated by bombing. The Battle of Okinawa killed fifty thousand Allied soldiers, around a hundred thousand Japanese soldiers, and caused the deaths or suicides of more than a hundred thousand Okinawan soldiers and civilians, or about a third of the population of the island at the time.[2] It also left huge swaths of the southern part of the island flattened and pockmarked with bomb sites. The US military has maintained a presence on the island ever since, even though it's still part of Japan. Most of the island's very elderly people, of which there are many, lived through WWII and remember it.

Today, downtown Naha at night comprises shops with doors flung open into the street, their bright lights shining on young Okinawans sauntering down the sidewalk in their finest going-out gear. If you walk up a small street from the main avenue and ascend a set of stairs to the second floor, the Imperial's door opens onto what feels like a living room that just happens to have a long, curving bar on one side. On the other, three sets of olive velvet loveseats face each other, each one topped by a fresh white lace doily where people's heads might rest after a long draft from their drink. Oil paintings hanging from the brick wall are so distractingly good that I make a mental note to ask about them. The space is cozy and softly lit, and neither too warm nor too cool. A hipster bar in Manhattan would absolutely reproduce this aesthetic, but it would never be as spotlessly clean.

Dressed in a red jacket, Kamida works the bar. Kakazu is Kamida's longtime client and friend. Kamida refers to all the people seated at her bar as friends, and that's not hyperbole; five decades of tending bar will garner you plenty of them. As she places a tiny, midcentury martini glass in front of me (she's never updated to the big, modern version of the glass), I can't help but try to estimate her age. She has a few white streaks at the hairline above her ears, but the rest of her short coif is black, and her cheeks are smooth. If I had to guess based on her face and the physical ease with which she dishes up our snacks into tiny bowls—including sweetened black beans, mildly flavored seitan shreds, and radishes dressed with mayo and spicy horseradish—I would guess she's in her sixties. But she's in her nineties, as is Kakazu.

LIFE THE OKINAWAN WAY

I came to Okinawa to learn about women like these, who are lucky enough to be aging in a "blue zone," one of the regions on earth with exceptional longevity at the population level. In Okinawa, the gap between men's and women's longevity is particularly large—more than seven years. In the 2020 census, Okinawan women lived to an average age of 87.44 and men to 80.27, neither of which is as high as they once were. In 1980, it was 84 for men and 90 for women.[3] Declines aside, Okinawa still has the longest-lived women in the world.[4] Of course, there's more to longevity than just crossing days off the calendar of life. There's also health during old age, and Okinawans are a great example of that as well, as Kamida and Kakazu prove. People, and especially women, don't just live long lives here; they tend to stay healthy and active well into their last decade.

Researchers are investigating why so many people in Okinawa and other blue zones—like Ikaria, Greece; Loma Linda, California; Sardinia, Italy; and Nicoya, Costa Rica—live past age eighty in good health, with many reaching one hundred or more.[5] Since only about 25 percent of longevity is tied to genetics (some experts argue that it's closer to 10 percent), researchers hypothesize that something else is at play in why so many people in these areas live so long.[6] Is there a secret ingredient in what they eat? Some special kind of exercise? Is it something in the air or water?

When I ask Kamida and Kakazu for their thoughts on the matter, they aren't interested in talking about food or exercise prescriptions.[7] Their own healthy longevity is linked to their *ikigai* and their *moai*, they tell me. Both of these ways of living are in decline in Okinawa, asced the influence of Western, and especially American, culture continues to infiltrate the island. I saw that for myself in the street scene outside the Imperial—walking to the bar, I could have been in Oakland, Paris,

Atlanta, Melbourne, or Madrid.[8] But most of Okinawa's longest-lived residents still practice ikigai and moai, which researchers agree are key to longevity.

Ikigai comes from the combination of *iki*, which means "to live," and *gai*, which means "reason." Finding your ikigai, or reason for living, is something most older Okinawans have spent time figuring out for themselves, and it serves as a guide to making life decisions. The knowledgeable and patient Christal Burnette, whose background in longevity science informs her work with the Okinawa Research Center for Longevity Science (she's also a board member of the Okinawa Blue Zone Committee), asked Kamida and Kakazu about their ikigai. She knew she was posing an important and serious question to the two nonagenarians. The merriment in the bar died down for a moment as Kamida and Kakazu grew thoughtful. Ikigai is no joking matter. Kamida, the owner of the Imperial, said, "Coming here and talking to people each night—it makes me feel alive." Kakazu, from her perch at the end of the bar, said, "Talking to people, meeting new people, and singing karaoke!" I knew from prior reporting that singing and dancing are linked to healthy longevity, so the latter part of Kakazu's comment was backed up by a plethora of scientific studies.[9] Burnette told me that many people's ikigai involves socializing and connection with friends, but others' might be the work they do—for a nurse, say, helping and healing others. Ikigai is a combination of passion, vocation, profession, and calling.

Makoto Suzuki, the director of the Okinawa Research Center for Longevity Science and the principal investigator in the Okinawa Centenarian Study, told me that his ikigai was conducting research and caring for his patients. In his sunny living room, seated below the giant, celebratory "Happy Birthday 90" mylar balloons that hung from the second-floor balcony, Suzuki explained the surprising psychological outlook of the centenarians he'd interviewed earlier in his career. When he asked if they were happy, most said they were. He had expected to find psychological problems, including post-traumatic stress from World War II, given the brutal bombings and the many lives lost—but he couldn't find it. Even those living on their own, without family, said they felt good. "I asked them then, *'Why* are you happy?' and most would say something like, 'I have lots of water and good food and friends,'" Suzuki told me.[10]

In Okinawa, friends mean something different than they do in other places, Suzuki said. When Okinawans talk about their friends, they are often referring to their moai—a group that is formed when people are young and that lasts their whole lives. Moais meet officially each month or week; Suzuki, for instance, would be hosting his moai's next gathering a week after we met. Friends in a moai might meet more frequently than that for tea and conversation. While the gatherings include "chatting, dancing, singing," and food, moais aren't just about enjoying the good times. They are a formalized social support system meant to "build trust and social connections," he explained.

Moais began hundreds of years ago as a financial support system; resources were pooled within a village so that when someone needed money to buy land or deal with a crisis, they could tap into the group's funds. Today most moais still have a financial aspect, which serves as a savings bank of sorts. Each member contributes a set amount at each meeting, and the pooled money goes to one member per meeting. At some point in the year, everyone takes a turn receiving the whole pot, getting back exactly what they put in over time. In cases of emergency or hardship, though, a member can access the money when they need it. So a moai offers social support, friendship—and a financial safety net. Traditionally, kids were grouped into their moai as a "second family," with some groups lasting over ninety years.[11] Today, people can join a moai at any age and financial level, and some belong to more than one, depending on their interests and availability. There's lots of evidence that strong social supports are one of the keys to longevity and healthy older age.[12] Moais provide an additional level of assistance so relatives are not the only source of aid for older Okinawans—after all, there are many situations where family is not available.

Suzuki thinks that one of the reasons for Okinawans' longevity is moais, but since both men and women both participate in the social groups, that factor alone doesn't explain why there's a longevity gap between men and women, nor why it's among the largest in the world. Sure, men in Okinawa tend to smoke and drink more than women, so that could be part of the answer, but that's true for mainland Japanese people too. What was Okinawan women's secret?

As Suzuki got up for a third time during our meeting to quickly ascend the stairs and fetch another book for me, I thought about the elderly men who exclusively did the sedentary work of taxi driving in Okinawa, whereas older women often worked in shops and restaurants. I wondered if the longevity difference could be due to activity levels.

When Suzuki returned, he was followed by Miruki, an adorable black poodle puppy who thoroughly distracted us both. After the pup had finished showering me with a face full of delightful lick-kisses, I asked Suzuki if he thought there was a gender difference in daily movement. "Yes, in Okinawa, women are more active than men. You see that in the market—you see the elderly women selling the vegetables and the fish, and they are very active," he said.

This reminded me of something Kakazu had said when we were talking about women's work. When she was a little girl, she noticed that "men would catch the fish, but the women are the ones who would put it in a basket on their head, carry it to market, and sell it, then walk home." Once there, they would perform household chores. And of course there was also childcare, which, especially when kids are young, is physically demanding. Historically, Okinawan women commonly had a half dozen children. "I always thought women did more work," Kakazu said.

Perhaps this Okinawan version of the "second shift"—the Western idea that women who work outside the home come back to a second shift of housework—actually helped them live longer because it kept them more active.

I saw this in action the next day, when Burnette and I drove up the coast to the northern part of the island, which is much more rural than Naha and its suburbs in the south. We passed pristine golden beaches met by aquamarine waves, and the sun blazed overhead as we made our slowish way (speed limits on the island max out at thirty-seven miles/sixty kilometers per hour) to Ogimi. This town is known as the "longevity village" because, even by Okinawa's standards, its residents live exceptionally long lives.

Burnette and I met her friends Quinton Sanicola, his wife, Yohanna, and their one-year-old son, Asa, at Emi no Mise (Emi's Restaurant)—all of us hungry. The front wall of the restaurant was open, and breezes from the forested hills and farm across the tiny street drifted into the high-ceilinged room. We took our seat on benches behind a wooden table and ordered from the menu, which listed just a few options as all the food was local and seasonal. While we waited, we sipped on *karaki*, a tea made from the bark of the local cinnamon tree. Yohanna told me about how her family, like many in Okinawa, celebrates relatives' *kajimaya*—ninety-seventh birthday.[13]

When the baby began fussing in her lap, Yohanna stood, continuing to describe the *kajimaya*. Everyone dresses up "like they would to celebrate a wedding," and they sing and perform traditional dances. It's a huge occasion, something the whole community looks forward to for years—a true celebration of longevity. As she described her grandparents' *kajimaya* celebrations, Yohanna sat down and got up four times to attend to Asa. Throughout our meal, she stood several more times to walk around with him, providing a perfect example of the increased movement women often do, as both Kakazu and Suzuki had pointed out to me.

When the food came, it was colorful and incredibly delicious: mildly sweet purple potato, lightly fatty and salty fried green bitter melon, a purple rice ball, a single tiny fish, and various tangy pickled veggies, all served on a beautiful ceramic plate. Surrounding it were smaller bowls filled with other tasty traditional foods—*mozuku* (thick noodles topped with seaweed); local herbs served as a salad; a parsnip-like root veggie; a tiny piece of what looked like chunky bacon topped by a piece of thick seaweed; and a bowl of miso soup with yet another kind of seaweed. As a lover of veggies and seaweed, I was in heaven!

At the end of the meal, I wasn't stuffed, just very satisfied. Emiko Kinjo, the restaurant's founder and owner, came over to meet us. She told us about her farm, located across the road from the restaurant, where she grew almost all the food we had eaten, since most traditional Okinawan ingredients are no longer available in the island's supermarkets.

Kinjo spoke passionately about her projects, her expressive hands helping me understand what was important even before Burnette interpreted. "I started this restaurant because I wanted to share what the grandmas were making and bring it to the next generation, so I can keep traditional Okinawan food culture alive," she said. Convinced that food is key to Okinawan longevity, she explained that it wasn't just about eating healthy traditional foods; it was the entire process—farming, harvesting, cooking, and serving—that tied food to health. To that end, Kinjo, who is seventy-six and trained as a dietician when she was young, runs regular meals for the very elderly and also offers farming and cooking education for schoolchildren. "I have three faces: I am a chef, a teacher, and an entrepreneur. I love teaching and providing food, and I have to make enough money to pay everyone."

With her aura of good health, her petite frame, and her passion for traditional foods, farming, and cooking, Kinjo reminded me of the pioneering chef and local-foods advocate Alice Waters, famous for the same kind of work on the other side of the Pacific in the Bay Area of the US. Waters once faced an uphill battle connecting local food and farming to the community, but she spawned a movement that still thrives today. Okinawa is lucky to have a sister-in-arms like Kinjo making the case for the island's wonderful local food.

As researchers have found, there doesn't seem to be a single answer to the remarkable longevity on the Japanese island, and different Okinawans I spoke with had various ideas about what contributes to long life. Unfortunately, simply eating the purple sweet potato or having good friends as you age won't guarantee longevity. It's likely a special alchemy of community, good food, useful work, activity, and perspective (and perhaps a lovely climate, which the other blue zones have too) that helps Okinawan women live such long lives.

THE PERSISTENCE OF THE FEMALE LONGEVITY ADVANTAGE

Susan Alberts, a professor of biology at Duke University, and her colleagues proved the female longevity advantage by compiling records for wild populations of nonhuman primates, including gorillas, chimpanzees, baboons, capuchins, and the lesser-known muriquis and sifakas. They also looked at data from six diverse human populations, both modern and historical, in different countries, spanning industrialized societies and hunter-gatherer groups.[14] In all these cases, the oldest individuals were always female, and females as a group outlived males. Even though human lives have changed dramatically since prehistory, the gender longevity gap persists over time and culture. "In spite of the gains in human longevity over the past century, the male-female difference has not shrunk," Alberts told Live Science. This "fundamental regularity," the authors wrote in the study, points to the "deep evolutionary roots to the male disadvantage."[15]

Alberts and her team are not the only scientists to have found these results. Steven Austad, an international expert on aging and the chair of the Department of Biology at the University of Alabama at Birmingham, has kept a huge longevity database throughout his career. He confirms that the female longevity advantage has been consistent for as long as records have been kept and is true among all groups of humans. On average, women live about five to seven years longer than men—and the gap only widens with age. At every stage of life, men are more likely than women to die. There are more women than men in their seventies, and even more women in their eighties. Of the supercentenarians—those verified as over 110 years old by the Gerontology Research Group—all six in that group as of December 2024 were women.[16] Of all the verified supercentenarians, there have been hundreds of women but just thirty-eight men, with only two men among the fifty oldest on record. "Women do not live longer than men because they age more slowly, but because they are more robust at every age," Austad wrote in a 2006 paper.[17]

Why? We don't really know. "No existing explanations account fully for these differences in life expectancy," the authors of a 2021 study on the subject wrote.[18] However, like most experts, they believe the disparity is due to a combination of biological, cultural, and other factors.

Women don't just live longer than men; they also have stronger immune systems into old age, enabling them to fight off diseases better than older men—though that ability does decline over time. This is true "even though many of the recognised social determinants of health are worse for women than men," the authors of the 2021 study wrote. Women often receive poorer health care, or care that's tailored to men, leading to misdiagnoses, incorrect treatment, or no treatment at all. A classic example is the fact that heart attacks present differently in men and women, and doctors have only recently begun taking this into account and incorporating women-specific information into their diagnostic criteria. Many doctors still don't know, for instance, that pregnancy complications like pre-eclampsia and gestational diabetes can predict an increased risk of heart disease, or that women under forty with endometriosis have been found to have a three times higher risk of heart attack later on compared to those without the disease.[19]

Heart attacks are just one mortality factor, and they affect both sexes. It's well known that the US has one of the highest maternal mortality rates among developed countries—so high, in fact, that it exceeds the rate of some countries considered developing. In 2021, the US maternal mortality rate reached a high of 33 per 100,000 births (up from 20 per 100,000 in 2019, where it seems to have returned in 2023).[20] Black and Native American women experience even higher rates of maternal death. This glaring disparity in care is evident when you compare the US to other countries. The UK has a much lower rate of 5.5 deaths per 100,000 births. For comparison, in 2023, Australia's maternal mortality rate was 3.5, Japan's was 3.4, and Sweden's was 2.6.[21] So, in addition to the poorer health care most women receive, they face a unique category of health risks—pregnancy, birth, and the postpartum period—that can, and does, kill them (more or less frequently depending on where they live and their race). And yet, despite all this, women still live longer.

Because my grandma built a stone wall at the new house she moved into at age eighty, I've never had much patience for the idea that older women are, by default, frail. While it's true that aging has very real effects on the human body, the extent to which age weakens an individual depends on a variety of factors, including nutrition, genetics, and especially exercise. Research into how to keep healthy into old age has gotten lots of attention in recent decades, as Boomers enter their sixties and seventies. Weight-bearing activity as well as movement of almost any kind can help maintain muscle mass, strength, and balance. Age doesn't automatically mean infirmity, and plenty of elderly people have enough muscle and stamina to ski, dance, walk, hike, swim, run, and more, for hours at a stretch. (I groove with some of them at community ecstatic [freeform] dances I find wherever I live or sojourn.)

Research has shown that even people who don't start a serious exercise program until their sixties will still enjoy stronger immune systems, fewer falls, and a higher quality of life than those who remain inactive. That means there are lots of opportunities for women to leverage their existing physical advantages to not only live longer but also enjoy better years in old age. For most women in Western societies, it isn't their sex that leads to physical decline in older age; it's their culture.

That culture includes medical care not designed with them in mind, a situation that is slowly improving but is hard for individuals to control. But one thing that is accessible is movement. Across almost every age group, women in developed countries tend to be less physically active than men.[22] Over decades, this inactivity adds up, costing women good, robust years at the end of life—and sometimes shortening their lives altogether.

THE BONE-DEEP COST OF PROMOTING FEMALE WEAKNESS

Growing up in a rural Romanian town, Claudia Tamas regularly carried heavy loads over the mountains. Her farming community had few roads, so people physically hauled everything they needed over rough terrain, as they had for generations. "It wasn't uncommon for me to carry my body weight on my back up and down that mountain," Tamas said of her childhood experience.[23] That was forty years ago, but that kind of work is still a way of life in many parts of the world, where physical labor starts young and continues throughout life, even into old age. Such work would have been the norm for most of our ancestors.

What's most interesting about Tamas's story isn't that she ferried loads up a mountain as a kid, but how a lifetime of similar labor built the strength of the elders in her community. She contrasts this with the people she works with today as a doctor of physical therapy and the director of women's health at the Natural Medicine and Rehabilitation Center in Somerset, New Jersey. The elders in her Romanian community were "pretty much living and dying in their boots, and they never lost their independence, really," she told me.

When she came to the US in 2004 and began her medical and physical therapy education, her first rotation was in a nursing home. What she saw there was "shocking," she said. "I'm seeing so much suffering and loss of independence and quality of life, the vast majority of which can be prevented." That experience motivated her career. She now helps elderly patients heal from acute injuries and works to recondition their bodies. One of the ways she does this is through weightlifting, using what would be considered pretty heavy weights, especially for women.

Tamas follows the protocols established by Belinda Beck, a professor at Griffith University in coastal Queensland, Australia, who has been studying bone health for more than thirty years. Early in her career, Beck hypothesized that the physical frailty often seen in elderly people, particularly women, stemmed more from a sedentary lifestyle than from aging per se, she told me.[24] This frailty frequently involves osteoporosis, which disproportionately affects women.

Although everyone loses about 1 percent of their bone strength each year after early middle age, Beck explained, women lose bone much more rapidly during menopause, when estrogen declines. Motivated by the prevalence and debilitating impact of the condition, she started researching how to help older women (and men) rebuild bone strength.

"It was a really meaningful thing to spend my life examining," she told me.

In addition to the bone loss that occurs at menopause, women often start out with lower bone mass than men because they are typically smaller and have less muscle.[25] Women also tend to do less of the physically demanding work (or workouts) that builds and maintains muscle mass. Muscle loss leads directly to bone loss, Tamas said, and this effect is common in more sedentary cultures, unlike her Romanian community, where physical labor is common at every age and elderly people are stronger.

Over her years working in nursing homes, acute care hospitals, postsurgical units, rehab centers, and now in her own practice, Tamas has seen how low levels of muscle (sarcopenia) impact physical strength and quality of life, leading to fear and loss of confidence. In recent years she's noticed bone weakness appearing earlier than ever—even in women in their forties. Building muscle helps to build bone, so those who do the least activity are at the highest risk for osteoporosis. In fact, traditional women's exercise routines were often specifically designed to avoid gaining muscle, which doesn't support bone health. (Of course, some exercise is better than none, but to build bone at any age, you need to build muscle.) Petite, thin women, who carry less weight over their lifetime, are at greatest risk for osteoporosis. Eating disorders like anorexia are also connected to osteoporosis due to the loss of one's period and low estrogen (see Chapter Seven for more on that topic). Genetics matter too: About 20–30 percent of women over fifty will develop osteoporosis (percentages vary by country and how it's measured), and half of them will break a bone due to low bone mass (osteopenia), with even higher numbers among people with a family history of the disease.[26]

This was my aunt Susan Stuart-Jones's story. She had maintained a very slim physique throughout her life, and both her mother and grandmother suffered from osteoporosis. She watched her vibrant, world-traveling mother (my Aussie grandma) develop the characteristic hunched posture caused by osteoporosis as her vertebrae weakened and compressed over time. When my grandmother broke her hip after a fall, she never made it out of a wheelchair—and eventually became confined to her bed. She wasn't the only one. My aunt's grandmother (my great-grandmother) was called "Little Nana" because she had become so tiny in her later years—but she hadn't started out that way. Her osteoporosis radically shrank her. Both women lived into their nineties but spent those later years in significant discomfort and ill health. My aunt and I have often wondered how much better their last years might have been—and whether they could have lived even longer—if they had known what we now do about osteoporosis and its causes.

Because she'd seen firsthand how osteoporosis had affected the women in her family, my aunt was "scared to death" when she was diagnosed with the condition in her late fifties.[27] Her doctor instructed her to keep doing her light weight-bearing exercises, which she did diligently: walking almost every day (including up and down hills to the beach), swimming in the ocean during summers, and taking regular Pilates classes. She also trained for hiking trips all over the world. But when she had her bones tested again, she discovered that she had continued to lose bone mass. Luckily, my aunt lives in Australia, so when she talked to a physiotherapist about her fears, he was already familiar with Beck's research and recommended she look into it. My aunt started heavy weightlifting immediately, based on the results of Beck's LIFTMOR trial.

While studying for her PhD at the University of Oregon, Beck delved deeply into research on bone physiology. During her postdoc at Stanford University, she focused on how exercise influences bone health.[28] She learned that bone can be built at any age, and the best way to build it is through heavy weight-bearing exercise. That's because, Beck explained, when we lift heavy things, the bone bends a little bit, shifting the fluid within its tiny pores. This movement activates osteocytes (stress-sensing cells in the bones), which then trigger osteoblasts to build more bone. However, the stress on the bone needs to be significant for this process to occur. Lighter workouts might be good for the heart and lungs (and mental health), but they don't stimulate bone growth in the same way.

Beck's big breakthrough came when she applied this knowledge to people who had osteoporosis. She designed the LIFTMOR study, involving 101 subjects over age sixty-five, all with low to very low bone density and almost half meeting the criteria for osteoporosis per their DEXA scans. The randomized, controlled trial divided the participants into two groups. One exercised at low intensity twice a week (walking, stretching, lifting light weights); the other lifted heavy weights twice a week under supervision. (Heavy weights were defined as 85 percent of a person's body weight—for example, 130 pounds for a 155-pound woman.) Beck's trainers ensured that the exercises were performed with minimal risk of injury. The trial ran for eight months, the minimum time needed to observe bone growth, and the results were clear: The weightlifters had significant gains in their bone mineral density, while those in the other group did not, and even continued to lose bone, proving that low-intensity exercise isn't enough to maintain bone strength.[29]

Beck called heavy weightlifting "an incredibly powerful and effective intervention, because it puts new bone exactly where it's needed." She pointed out that she was able to prove this when she showed an increase in bone thickness "right where fractures happen. That's going to have an excellent effect at reducing the risk of fracture at that site." Beck has since expanded her research, demonstrating that her protocol, now dubbed Onero, is effective for men too.

This is the exercise regimen that my aunt started—and still engages in today. She loves lifting weights, and about ten months after she began, I followed suit, heeding Tamas's advice that women should ideally start lifting weights in their mid-thirties to maximize bone strength in older age. Though I'm older than that, I began Olympic weightlifting classes before menopause, which may help me evade my family's osteoporotic curse. Additionally, I've been an avid dancer since my late thirties—an activity particularly beneficial for bone density due to the twisting, jumping, whirling, and stomping movements, which stress the bones in a way that helps build them.

Beyond giving my aunt stronger bones, better posture, and a more streamlined physique, building muscular strength in her late sixties has given her more confidence—something both Tamas and Beck noted in their patients too. Tasks that used to be tough, like moving big plant pots, are now much easier for my aunt. Before lifting something weighty, she sometimes thinks, "Oh no, that could be heavy." But then she realizes, "No, it's not. That really gives you confidence," she told me, smiling. Most importantly, when she recently tripped over a baby carriage while rushing through a crowded Sydney shopping mall and fell on the hard marble floor, she got up, dusted herself off, and walked away—no broken bones to worry about. She's now strong enough to avoid injury when she falls.

For generations, little girls were discouraged from engaging in sports or manual labor, and in many ways they still are. (Though progress has been made in sports, we still have a long way to go.) For hundreds of years, women were told that hard exercise wasn't ladylike and to leave it for the guys—that riding a horse too long might cause their uteruses to fall out, that swimming while menstruating might make them sick, or that running a marathon could damage their ovaries. We now know these claims aren't true. Yet even today, men are often asked to carry (or insist on carrying) heavy items for women, and weighty or bulky products for women are frequently designed to be easy to handle. All of this, research has shown, is bad for women's bodies.

Over a lifetime, this pattern results in older women being significantly weaker than men of the same age. Because muscle strength is directly linked to bone health, this makes women more vulnerable to osteoporosis and fractures in old age. Broken bones can cause other health problems and even lead to death, as was the case with my Australian grandmother. Many elderly people with osteoporosis die from pulmonary issues as their weakened bones collapse around vital organs like the heart and lungs, leaving them more vulnerable to pneumonia.[30] What may seem like a polite or well-meaning removal of physical workload for women ultimately contributes to greater frailty in their elder years.

Anyone with a female body—from girls to young, middle-aged, or older women—should be lifting and carrying heavy loads (or weights) to build muscle and bone and stay healthy for the long haul. Plus, it feels pretty good to accomplish tasks on your own. I can speak from personal experience: Lifting weights makes you feel like a badass.

THE IMPORTANCE OF GRANDMOTHERS

Female longevity is not a modern phenomenon—though if you think it is, you have plenty of company. A persistent bit of misinformation suggests that our ancestors' lives were "nasty, brutish, and short" (to quote Thomas Hobbes), and that everyone in prehistory died by age forty-five. Not true. The "average age of death" is just that—an average. During prehistory and even a hundred years ago, babies and young children died at very high rates, often from accidents, or from illnesses that today are easily treated with antibiotics or prevented by vaccines. It's true that in the past, a baby or child faced a significantly higher risk of death. But throughout human history, as long as you had adequate nutrition, once you made it to adulthood, living into your sixties or beyond was possible, and many did.

When scientists at the University of California, Santa Barbara, and the University of New Mexico examined the lifespans of modern hunter-gatherer societies and excluded infant mortality, they found an average lifespan of sixty-eight to seventy-eight years, similar to that of modern industrialized societies.[31] This extended lifespan is widely accepted among anthropologists. "For people living traditional lives without modern medicine or markets the most common age of death is about 70, and that is remarkably similar across all different cultures," said Christine Cave, a PhD scholar at the Australian National University School of Archaeology and Anthropology, in a 2018 press release about her study of five-hundred-year-old remains from British cemeteries.[32]

While the chances of reaching very old age weren't as high as they are today, human history still included plenty of elderly people. Pythagoras, Socrates, and his teacher, Aspasia, all lived into their seventies. Even the Old Testament of the Bible says that man's allotted time on earth is "threescore years and ten"—seventy years.[33]

Why do humans live so long, and why do women reliably outlive men, especially since their reproductive years end around age fifty? Most animals don't live years (or, like female humans, decades) beyond their reproductive capacity. Only a few other species experience menopause and then continue to live substantially longer, including five species of toothed whales (orcas, narwhals, short-finned pilot whales, false killer whales, and belugas) and maybe chimpanzees—a fact that was just recently documented in wild chimps in Uganda.[34] African elephants and blue whales are very long-lived, but they continue to be fertile into old age (same with some captive chimps). Many don't reproduce even though they can, and they engage in behaviors similar to those of animals with menopause. Many mammals outlive their eggs for a year or two, but menopause is really uncommon.

Since nature provides relatively few examples of this phenomenon, and it occurs in mammals with shared characteristics (elephants and whales are both highly intelligent, social animals), it's reasonable to think there may be a behavioral advantage to living longer than necessary for reproduction. I reached out to menopause expert and biological anthropologist Lynnette Sievert at the University of Massachusetts, Amherst, who has been studying women's midlife health for over three decades.

Sievert explained that part of the answer lies in the fact that eggs have an expiration date. "Almost no mammal has shown viable eggs after about fifty years," she told me from the cozy, book-lined room where she Zoomed with me.[35] "If eggs did go beyond fifty, they'd have all these mistakes. They're so much more likely to have chromosomal abnormality." It's not worth the risk for a female mammal to use resources for a pregnancy that has a high likelihood of failure. Once chromosomal errors become too common—when eggs reach about fifty years—"it's better to invest in the children you already have, not these aging eggs," Sievert said. Since physically healthy postmenopausal women can carry a pregnancy using someone else's eggs—"the uterus is surprisingly robust"—scientists have figured out that the eggs are the limiting factor that stops reproduction, Sievert said.

So why do human females keep living after the fifty-year mark? One of the most popular explanations is the grandmother hypothesis, put forward by Kristen Hawkes at the University of Utah. Broadly, Hawkes suggests that older tribe members, and women specifically, were imperative to our ancestors' flourishing.

To better understand early human communities, today's anthropologists sometimes look closely at their modern-day analogs. The Hadza, in southwest Tanzania, are one of several hunter-gatherer groups that anthropologists believe live much like humans did in the past. Hadza women in their forties, fifties, sixties, and even seventies contribute significant food to younger generations. When anthropologists examined the work these women did, they found that, compared with younger men and women, grandmas gathered more food and shared it more often, especially with their grandkids and great-grandkids, who weren't yet able to forage enough for themselves.[36]

Not only are grandmothers physically able to gather more food, they also serve as invaluable "libraries" of key, lifesaving information. With decades of knowledge about where to find food, plant medicines, water, and other resources across seasons, years, and locations (since hunter-gatherer groups tend to move around seasonally, based on the weather and what's available), elder women are crucial to their communities' survival. This "embodied capital," as the concept is known, is especially important considering that, in traditional societies, women are responsible for hunting and gathering the majority of food supplies (see Chapter One). Research has shown it takes decades to become an efficient forager and hunter, skills that peak in the mid-forties.[37]

When young Hadza women need help, they don't generally turn to men; they ask a more senior female relative. In another of Hawkes's studies, she found that among the Hadza, older women were so important that no nursing mothers were without a grandma or aunt (theirs or someone else's) to help them.[38] These studies showed that grandmothers help feed their grandkids, which allows their daughters or other young women the time, space, and nourishment to have more pregnancies than they might otherwise—enabling populations to grow faster.[39] Ultimately, this means a grandmother's descendants would be more numerous, more likely to survive childhood, and more likely to reproduce, carrying on her genes. Hawkes built an interesting computer model proving just this idea. Starting at a baseline of just 1 percent of grandmas helping female relatives, and creating a conservative positive effect on their offspring, the model showed that over thousands of generations, the percentage of helpful grandmas increased to 43 percent.[40]

This idea has been backed up by research into premodern populations of Finnish and Canadian women, which found that those who lived the longest postreproductive lives had the most grandchildren.[41] (Interestingly, the same is not true for longer-lived men.[42]) We might not be the only species for whom this applies. Research from 2019 suggests that this phenomenon could also explain why orcas live long past menopause.[43] The study shows that grandmother killer whales help their younger relatives find food.

According to Hawkes's hypothesis then, the fact that humans, unlike other primates, can and do live decades beyond their reproductive years isn't a fluke, nor is it a failure of human females to continue reproducing. It's an adaptation selected for over time—one that benefits the whole group by enhancing survival and support for future generations. Hawkes even goes so far as to challenge how anthropologists define reproduction: "You can't really call women who are past their child-bearing years 'post-reproductive' because while they may not be fertile, there is a lot of evidence that they are doing important things for the reproduction of their genes," Hawkes told *Scienceline*.[44]

Sievert thinks that Hawkes's idea might place too much emphasis on the grandmas and not enough on the grandpas. For grandmas to provide extra nutritional help to their grandkids, they first need to live a longer postreproductive life. And that extended lifespan might be due to men. Sievert believes that proponents of the grandmother hypothesis are overlooking an earlier evolutionary driver of longevity. "Maybe longevity is not only about females. Maybe the selection [acted] on males, and females got carried along with it," she said.

How would that work? Unlike women, who are born with a lifetime's worth of eggs, men continually produce new sperm over the course of their lives (though sperm quality does degrade with age). This means that men can become fathers well into their sixties and even seventies. It's not uncommon in many societies for higher-status men to use this ability to reproduce with younger, still-fertile women. This phenomenon might have contributed to longer lifespans for everyone because those healthy older men contributed their genes for longevity to their offspring.

Biologist Shripad D. Tuljapurkar and his colleagues at Stanford University made a strong case for this idea. They used a mathematical model to show that if the mating pattern of older men and younger women has persisted throughout human evolution, it would have preserved longevity genes beneficial to both sexes. They wrote, "Old-age male fertility provides a selective force against autosomal deleterious mutations at ages far past female menopause with no sharp upper age limit." This selection against gene mutations could explain why females don't meet the "wall of death" that previous models predicted should occur after menopause—and which does occur in other animals.[45]

Sievert also questions how many years the grandmother hypothesis could have added to lifespans. After all, she pointed out, in hunter-gather societies, grandmotherhood can start in the thirties and forties and extend into the fifties and sixties, which can obviously benefit younger relatives. But beyond that? "It's great to have a vigorous forty-year-old, fifty-year-old, maybe sixty-year-old grandmother, but that's not selecting for an eighty-year-old lifespan," she said.

Sievert doesn't think grandmothers are unimportant—she's just skeptical that they are a primary driver of the multidecade longevity we see in humans postmenopause. "I don't think that becoming a grandmother selected for menopause and postreproductive life," she said. "But I think that having postreproductive life opened the space for the effectiveness of grandmothers."

I'm the last person who would argue that grandmas aren't important. While nobody can know if I would be alive today if not for my grammy's care, I can say without much doubt that I wouldn't be the relatively happy, functional person I am without her. I'm far from alone. I am one of the 2 percent of American kids (about 2.9 million today) who were or are being raised by a grandparent when their parents were unable to. In the 2020s, that number has increased significantly with the twin scourges of the opioid epidemic, which primarily affects people who are in their childbearing and -raising years (twenty-five to fifty-four), and the COVID pandemic, which has killed a parent or caregiver of over 10 million kids worldwide.[46] If not for human longevity past reproductive age, especially the long lives of women, we would all be worse off, including me.

Beyond my personal experience, though, we still don't fully understand why women live longer than men, even though there are plenty of persuasive hypotheses. We do know that aside from the biological and evolutionary factors outlined above, women tend to take better care of themselves physically and engage in fewer risky activities like smoking cigarettes and drinking heavily. Women are also much less likely to die by suicide and more likely to survive most illnesses at every age, as we've discussed in previous chapters. People from prehistory to today have benefited in many ways from having an involved grandma. Individual situations vary, but older women remain vital to the health and well-being of their younger relatives.

Whether this care might be a driver of human longevity is still unknown. Maybe it simply gives older women a sense of purpose within their families or communities. After all, older women significantly outpace men in terms of time spent working, with or without pay (older women volunteer at twice the rate men do).[47] This sense of purpose could fulfill their ikigai, as Kamida and Kakazu suggest, contributing to their longer lives.

Whatever the reason, the simple fact that women live longer, healthier lives provides some of the most compelling evidence that, despite the health challenges women face, the female body has incredible fundamental strengths that trump those of the male body.

Chapter Fifteen

A MORE EXCITING FUTURE FOR WOMEN'S BODIES

Two of Elaine Harvey's great-grandmothers were known for their horseback-hunting skills, able to take a deer or elk with a single shot. These women lived and rode throughout the drainage of Nch'i-Wàna (Great River), later named the Columbia River by the colonists, which covers large areas of what is today the states of Washington and Oregon. The environments in the territory of Harvey's ancestors vary tremendously, from the snowfields of Pahto (Mount Adams) to sun-drenched, dry grasslands to wet, temperate rainforests.

Harvey is a member of the Kamiltpah and Wilawitis bands, part of the fourteen tribes that make up the Yakama Nation.[1] She still gathers and hunts food on her ancestral lands; her tribe's chiefs secured strong rights to do so in the Treaty of 1855. "We follow the first foods—that is part of our life and religion," Harvey told me.[2]

Her people don't like living in one place year-round. "In the springtime, we're out gathering the roots, and we're also fishing for salmon, and steelhead, and suckers, and Pacific lamprey," she said. "Later, it's small game and more roots for medicines, and in the fall time, we're continuing to salmon fish—the fall runs. And then we go up into the mountains and we gather huckleberries and hunt for elk and deer." Harvey stressed the importance of maintaining her people's traditional ways of roaming with the seasons. "We're always moving, we're traveling, we're working together."

Teamwork means that everyone both hunts and gathers. "Our beliefs around men's and women's roles are not that just women are going to be the root diggers and the berry pickers—the men also do that. And many women fish the Columbia River, including my mom," Harvey said. All work is important for the tribe's health and well-being.

Likewise, both women and men hunt for deer and elk and other large game, as Harvey's great-grandmothers did. She has followed in their footsteps, recently shooting her first buffalo in Yellowstone, as is allowed per her treaty rights. "It's serious, you know. That buffalo had a life. And he gave his life for us. When we take a salmon, when we take a deer or elk, it's a life, and you have to treat it with respect," she said. She went to the buffalo hunt with the right equipment because she knew she'd have to "get in there and clean the guts out and skin the animal. It's a lot of work." She shared the buffalo meat with relatives and filled her freezer, and she stayed up all night with a friend treating the pelt so it would be soft enough to use, a task her grandmother used to do by herself. Harvey learned firsthand that "our elders were really hardcore and strong."

In her community, whether someone hunts is based more on interest than on sex, Harvey said. Wanting to hunt means a person will practice to become proficient. Ability follows desire. "You also have to have that physical strength to walk miles to hunt and pack the meat," she said. In the past it was even simpler: "Everyone had to pitch in and work together because you had to store food for winter." The tribes have survived on this land for thousands of years.

Despite the cultural history of shared labor, Harvey's decision to go into fisheries science—which involved getting a degree in biology, several internships, and plenty of grunt work to learn the ropes—was met with resistance. "People said to me, 'That's a man's job,'" she told me—even though Harvey's own mother was known as a fisherwoman. "I got a lot of pushback," she said, and it was a bit daunting for her to enter the fisheries department, which was very male dominated at that time due to the influence of American culture.

"Everyone was like, 'Yeah, women should just be in the office working at a computer and typing'"—an outlook Harvey took as a challenge. She said her job's labor, especially at the beginning, was both physically and mentally taxing. To track salmon, she jumped from a small boat into a cold, fast-flowing river, holding herself and the boat in place while marking the spot on the GPS. She rafted through remote canyons where she had to grab cables to balance atop the rushing river to open a fish trap. When she told me the stories, she laughed with the bravado of one looking back at fear; the work was definitely "scary" the first few times she did it. She said she often thought of her mom, her grandma, and her great-grandmas when she was faced with fright.

Ultimately, the job gave her confidence. "I was like, well, if I can do that, I can do anything," said Harvey, who also walked miles a day conducting surveys and collecting data she would later analyze, information that went into reports for decision-makers. "I did that for years and years," she said. And so, along with two other women colleagues, "we kind of broke that cycle" of the all-male fisheries department she had walked into in the early 2000s. "After that, more females started coming in, getting hired, and doing the work." Harvey has worked with dozens of young women. She's now the watershed manager for the Columbia River Inter-Tribal Fish Commission.[3]

Harvey believes that women's involvement in fisheries is key to the Yakama tribes' mission to "advocate for the salmon and the sturgeon and all the fish in the rivers, and even the mussels." In her opinion, not only are women good at the work; they also bring a different perspective to the tasks of assessing resources and figuring out how to best protect and restore salmon and improve hatcheries. Like her ancestors, she sees maintaining the health of the fish and the land of the Yakama as the shared responsibility of both men and women. "For the future, I think it's important to have male and female viewpoints working together, collaborating, and it not just be a male thing," she said.

WOMAN THE HUNTER . . .

When thinking about the future of women's bodies and their strengths, at first it might seem strange to look backward in time at Harvey's ancestors—but many of the questions we have today about the capabilities of women's bodies were answered thousands of years ago. We now know that, contrary to what many of us may have learned growing up, the idea of "Man the Hunter"—based on a conference and paper from the late 1960s—is really "people the hunters."[4] Both male and female bodies offer advantages in a hunt. While men might have a slight advantage in explosive short-distance muscular speed and spear-throwing power, women have a small but real advantage in shooting accuracy and, importantly, endurance.

The need to chase prey until it was exhausted is one of the primary hypotheses about why *Homo sapiens* developed the impressive running endurance abilities our species has today. According to a 2024 study published in *Nature Human Behavior* that analyzed 391 ethnographic records from dozens of sources, "persistence hunting" (chasing down prey until it's exhausted or injured) was practiced all over the globe in a wide variety of environments, from forests to savannah.[5] Early humans were able to catch and kill animals that were much larger or faster by tracking and pursuing them over long distances. The researchers even calculated the energy return rate of walking versus running and found that, although running is very calorie intensive, the faster caloric gain once prey was caught and eaten made it worthwhile.

Female bodies are especially well suited to endurance running, as we've covered in previous chapters. They experience less muscle breakdown than males after the same exercise; they preferentially use slow-burning fats over fast-burning carbs; and they have more type-1, slow-twitch muscle fibers, which take longer to fatigue. Together, these three factors are a huge boon in any kind of endurance activity.

The idea that only men hunted is rooted in outdated sexism, as shown by anthropological evidence. "The fossil and archaeological records, as well as ethnographic studies of modern-day hunter-gatherers, indicate that women have a long history of hunting game," wrote Cara Ocobock, of the University of Notre Dame, and Sarah Lacy, a biological anthropologist at the University of Delaware, in a 2023 *Scientific American* article titled "The Theory That Men Evolved to Hunt and Women to Gather Is Wrong." It was the November cover story, so across the front of the print magazine was splashed a gorgeous illustration in reds and oranges of a female body carrying a spear, backgrounded with the simple, powerful words "Woman the Hunter."[6]

Ocobock and Lacy point to the Agta people of the Philippines as a good example. Contrary to the assumptions of the Man the Hunter theory, Agta women aren't sidelined from active work by normal reproductive functions. They wrote, "Agta women hunt while menstruating, pregnant and breastfeeding, and they have the same hunting success as Agta men."

This should come as no surprise after reading the stories in Chapter Eight about modern female athletes who can and do engage in a variety of sports while pregnant and breastfeeding. It has become so normalized that professional soccer (football) player Katrina Gorry timed her breastfeeding between games and training with the Matildas, the Australian national women's team.[7] (Interestingly, Gorry also notes that she feels physically stronger since having had a child, and her training stats back up that claim.) At the 2024 Olympics in Paris, breastfeeding Olympians were, for the first time, provided private spaces to nurse their babies, an accommodation they demanded and received because their numbers warranted the needed space.[8]

This isn't new, even if it might seem that way in mainstream Western societies, which have historically sidelined, diminished, and marginalized female bodies—especially those of mothers, nonwhite people, and older women. Women who are bigger, more muscular, or strong have also been excluded, deemed "lesser" women, or had their gender questioned. But active menstruating, pregnant, and breastfeeding women were more typical throughout human history than cloistered ones. Our female ancestors did what needed doing—including hunting.

A 2023 study of ethnographic data spanning recent history—most of it previously ignored by anthropologists who insisted that only men hunted and women gathered—found that women from many cultures hunt animals of all sizes for food.[9] Abigail Anderson, Cara Wall-Scheffler, and other anthropologists and biologists at Seattle Pacific University analyzed data from sixty-three foraging societies over the past hundred years. Their survey included groups from around the planet, including twelve from Africa, nineteen from North America, six from South America, fifteen from Australia, five from Asia, and six from the Oceanic region. Their findings show that in a vast majority of these societies—79 percent—women intentionally hunted.

The researchers further examined thirty-six of the cultures where women hunted. They found that women hunted game of all sizes, with large animals being the most commonly hunted. Significantly, as they wrote, "In societies where hunting is considered the most important subsistence activity, women actively participated in hunting 100% of the time." They observed that these facts add to the "data that women contribute disproportionately to the total caloric intake of many foraging groups."

. . . AND THE WARRIOR

This modern research supports the fossil evidence. In the past, skeletons found with hunting equipment were typically assumed to be those of male bodies. Even when remains of women were unearthed alongside spears or chipped stone points, they were often considered, as one researcher put it in 1971, "difficult to explain." These artifacts just weren't considered women's tools, even by experienced anthropologists like David Breternitz, at the University of Colorado, who was the one who found himself puzzled.[10] He suggested that the projectile point he'd uncovered must have been "used as a knife or scraper, typically women's tools," to explain its inclusion in the grave. He could not accept the notion that he was looking at the remains of a female hunter, so deep was his gender bias when examining the evidence.

Breternitz was far from alone in his blinkered view of men's and women's roles in prehistory; the Man the Hunter conference had taken place just a few years earlier. Similar misconceptions were pervasive. I grew up learning that the Amazon women of Greek mythology were merely a fun fantasy—representative of an idea, but not real. That was the popular narrative, despite the multiple discoveries of graves containing female hunters and warriors in the very region the Amazons were said to have come from. The anthropological evidence for female hunters was known and building at the same time I was explicitly told they were "just an idea."

In the 1990s, a US–Russian team found over 150 Bronze Age graves containing female bodies that clearly belonged to warriors. According to historian Amanda Foreman, the graves held bronze-tipped arrows and other weapons of war. Some of the bodies bore distinct combat wounds, while others showed evidence of injuries they had survived. "On average, the weapon-bearing females measured 5 feet 6 inches, making them preternaturally tall for their time," wrote Foreman, which aligns with historical descriptions of Amazons as strong and tall. While the find was surprising for some, maybe it shouldn't have been—reports and documentation of female warriors of this kind appear in records from several cultures, including Chinese, Egyptian, and Persian sources.

If Herodotus, the fifth-century-BC Greek historian, got it right, the Amazons lived in northern Turkey and eventually battled the Greeks. Foreman wrote that the Greeks captured the Amazons, intermarried with them, and became the Sauromatians, a nomadic tribe that stretched from the Black Sea to Mongolia.[11] This group is often associated with the Amazons in historical accounts from other cultures.

"The women of the Sauromatae have continued from that day to the present," wrote Herodotus, "to observe their ancient customs, frequently hunting on horseback with their husbands . . . in war taking the field and wearing the very same dress as the men. . . . Their marriage law lays it down, that no girl shall wed until she has killed a man in battle," Foreman recounted. Some linguists suggest that the term "Amazon" may not be Greek but rather may come from "Hamazon," the Persian word for "warrior."

Over the last fifty years, more than three hundred sites containing the remains of female warriors have been discovered, leading to headlines like "Amazons Were Long Considered a Myth. These Discoveries Show Warrior Women Were Real."[12] Like the science behind the "choosy egg" from Chapter Four, this fact continues to be "rediscovered" in the popular press every few years.

Most recently, in 2019, the grave of a woman—an "individual of rank," according to the contents of her burial site—was discovered in what is now Armenia.[13] She had a metal arrowhead embedded in her femur. That same year, in western Russia, the remains of three generations of warrior women were found in a shared grave, surrounded by spears. One was laid to rest in the "position of a horseman."[14] The discovery was hailed as "landmark."

There isn't proof that the Amazon culture was a matriarchy, as they left no written language behind. All that remains are their burials and the stories others have told and recorded about them. It could be that the Sauromatians had both male and female warriors or were female-warrior dominant. Maybe male warriors were once the norm, but after losing many men in battles, women took up arms, changing the culture forever.

A related scenario occurred in Paraguay following the War of the Triple Alliance in the 1860s, during which up to 70 percent of the male population was killed in battle (a figure reaching as high as 90 percent in some areas). As the number of men dwindled, according to local lore, the Paraguayan women became more independent and tough, traits that have persisted into modern times. In 2021, researchers decided to test whether the legends could be true. They analyzed data from regions in the country where the sex ratios were more skewed after the war (with many more women than men) and found "higher [rates of] out-of-wedlock births, more female-headed households, better female educational outcomes, higher female labor force participation, and more gender-equal norms," compared to areas with a higher male population. The researchers concluded that, even more than 150 years after the end of the conflict, "The impacts of the war persist."[15]

Perhaps something similar happened with the Sauromatians, leading women to take on the roles of warrior and protector for generations. Or maybe their culture was like that from the start, and the arrangement persisted as other societies became more male dominated. There's nothing preordained about one gender being in charge. "We can assume that a significant proportion of hunter-gatherer societies weren't patriarchal," said anthropologist Adrienne Zihlman.[16]

In 1878, the ornate Viking tomb of a high-ranking warrior from the tenth century was discovered near Birka, Sweden, prominently located on a raised terrace between a town and a fort. Among the eleven hundred graves unearthed, it was one of only two that held a full set of weapons. It contained a sword, axe, spear, armor-piercing arrows, a knife, the remains of horses, and tokens used in games of strategy. A 2017 paper described these items as "the complete equipment of a professional warrior."[17] By the 1990s, numerous scientific papers had asserted—and generations of students studying the site had been told—that the warrior was, of course, a man.

That 2017 paper wasn't just another rehash of the grave's contents. It was published because new information had come to light. The Viking warrior buried in the fabulous grave was, in fact, female. This revelation unleashed a wave of criticism. Some refused to accept the genomic evidence. Despite the fact that stories, art, and poetry describing "fierce female Vikings fighting alongside men" were common, "women warriors have generally been dismissed as mythological phenomena," wrote Charlotte Hedenstierna-Jonson, an archaeologist at Sweden's Uppsala University.[18] Sounds familiar, right?

Controversy ensued. Some argued the grave couldn't be that of a warrior after all, though that fact had never been questioned before. But for those still clinging to the "Man the Hunter" narrative—a perspective that seems hard for many to let go of—it's brain-bending to accept that the grave could belong to a high-ranking female hunter and warrior.

Some people hypothesized that perhaps the warrior was a transgender man who had socially transitioned. In a follow-up paper to their blockbuster 2017 research, Hedenstierna-Jonson and her colleagues wrote:

> Queer theory also provides a potentially fruitful means of engaging with this individual, and their sense of self may have been—in our terms—non-binary or gender-fluid. . . . It should be remembered that [transgender] is a modern politicised, intellectual and Western term, and, as such, is problematic (some would say impossible) to apply to people of the more remote past. All this is also inevitably speculative, considering the limitations of the archaeological material. There are many other possibilities across a wide gender spectrum, some perhaps unknown to us, but familiar to the people of the time. We do not discount any of them.[19]

WE WERE ALL HUNTERS (AND GATHERERS)

Anthropologists now estimate that about 50 percent of prehistoric hunters were women.[20] This figure comes from a 2020 study that assessed a number of graves. No one is saying that ancient peoples decided to be more "equal." That idea comes directly from the existence of inequality in the first place. We have little understanding of how our ancestors regarded gender, or whether different societies agreed or disagreed with each other on definitions of such. (I'm going with the latter because: Have you met people?) Assuming they regarded gender or sexuality on a binary as we do now is just that—an assumption.

What is clear from anthropological evidence and studies of modern gatherer-hunter societies is that while some of our ancestors may have assigned jobs by gender, not all did. There were practical reasons for this. Excluding anyone who was a strong runner or an accurate spear thrower would have been a waste of valuable talent in the effort to procure food. Including as many hunters as possible just makes sense. "The primary hunting technology of the time—the atlatl or spear thrower—would have encouraged broad participation in big-game hunting," wrote the authors of the 2020 study. "Pooling labor and sharing meat are necessary to mitigate risks associated with the atlatl's low accuracy and long reloading times. Furthermore, peak proficiency in atlatl use can be achieved at a young age, potentially before females reach reproductive age, obviating a sex-biased technological constraint."

It's interesting to think that teenagers—risk-takers who are interested in proving themselves and are full of youthful energy needing to be channeled—might have done much of the big-game hunting in some ancestral communities. This idea popped up again and again throughout the research I reviewed—that the riskiest hunting was accomplished by the youngest people looking to validate themselves. Older adults likely focused on gathering, fishing, and hunting smaller and medium-sized animals, tasks that required less stamina but more patience and expertise.

As anyone who has hunted or tracked animals knows, different species require different skills to find, follow, and kill. Even large animals aren't all the same—the skills needed to hunt the elk Elaine Harvey's great-grandmothers took aim at are different from those required to take down a bear. To hunt efficiently, you need a variety of strengths for different roles. Explosive movements, endurance running, accurate spear-throwing or arrow-shooting, sharp vision in the near-dark, or the ability to smell prey before you see it are all useful hunting skills. Different individuals excel at these tasks depending on sex, age, body type, and genetics.

In modern hunter-gatherer populations where women hunt, they often use a wider variety of tools. As Anderson, Wall-Scheffler, and colleagues observed in their 2023 paper, Agta men tend to stick to hunting with bows and arrows, but Agta women are more likely to have individual preferences that dictate use of a range of hunting implements. "Some women prefer hunting only with knives, a few women use bow and arrows, and others use a combination of the two," they wrote. It wasn't just the Agta—they found the same to be true among women in other groups too. Some women were also more likely to use dogs.[21] It made me wonder if one way dogs became domesticated was through women's work with them for hunting. Even today, about 70–80 percent of professional dog trainers are women.[22]

Of course, male-female couples also hunted together. This has been found not just in one or two populations, but in dozens, from groups in India to the Aka people in the Republic of the Congo, highlighting the teamwork that human beings are known for.[23]

In light of all this data, the recent human history of separating work into "male" and "female" spheres looks like an aberration. Does it happen in hunter-gatherer groups? Of course, and many of those cultures still share power more equally, like the Batek people in Indonesia, who maintain an egalitarian culture.[24] While most of the hunting is done by men using blowpipes to take down monkeys, nothing prohibits women from hunting (and a few do). Similar examples exist in other cultures as well. Dividing work by gender doesn't result in mismatched power dynamics in the tribe though. This is a good example of how there's nothing about doing a certain type of work that necessitates inequality in leadership. That looks like a particular invention of the cultures I grew up within, where certain types of work were deemed more or less valuable along gender lines, thus informing who gets to make the decisions. Looking outside Western culture though, when it comes to humans' survival, there's a wide variety of practices and roles.

For two hundred thousand years of prehistory, through historical times, and into the modern era, men and women have worked together. More- or less-gendered spheres have certainly existed in some cultures, with gender binaries being particularly strict in recent eras in many Western cultures. As we know from the history of science I've touched on in this book, it takes a lot of effort, repression, strict rules, and punishments (even death) to maintain those boundaries. In light of this fact, gender roles among humans hardly seem "natural" or "normal." The reason I've spent so much time looking at the past in a chapter about the future of female bodies is that our future will probably resemble that past, where all genders work and play together. Eventually, I think we'll look back at this time of gender strictures as a weird blip in human development.

That probably goes for the future of sports too.

KORFBALL: THE MOST INCLUSIVE GAME EVER?

It was a moody spring day when I rolled my suitcase over the curb at the Fortuna sports venue after a fifteen-minute ride from the Rotterdam Centraal train station in the Netherlands. I had come straight from a dreamy-zoomy Eurostar ride from London. I'd forgotten how fast trains could go after decades spent living in the US. I was at this sports complex in suburban Delft to watch my first-ever korfball game. It's the only widely played sport where men and women compete together on the same teams.

As I navigated my luggage into the outdoor arena at exactly four p.m., I felt that special travelers' joy when plans made weeks before, from a couch thousands of miles away, come to fruition in an on-time meeting. Jan Jaap van Leusden, the chair of the sports club, beelined over to me. With a warm, wide smile, he showed me where to stash my stuff and treated me to a strong cup of tea. Then we were off for a tour and interview. It wasn't just his well-over-six-feet height that tipped me off to van Leusden's korfball playing experience. He had the bearing of a former athlete. He had begun playing korfball at age four, and so had his son. It's common to start that young, van Leusden told me.[25] The kids in the four- to six-year-old group in the organized leagues are called kangaroos.

Korfball is a popular and important sport in the Netherlands, played by nearly a hundred thousand people in a country of eighteen million. Championship games featuring semiprofessional athletes are broadcast on national television, and both indoor and outdoor courts can be found throughout the country. Kids and adults of all ages play. There are leagues for older people just as there are the kangaroos.

The game was originally developed in 1902 by a teacher from Amsterdam, Nico Breukheysen, who simplified the rules of a Swedish game so that kids could play it easily. Now there are about a million players worldwide in over seventy countries. While many are from Europe (with Belgium ranked second after the Netherlands) and the UK, the sport is increasingly popular in Asia, with Taiwan ranking third and China fifth.[26]

Korfball, like its distant cousin basketball, is fast-moving. Its rules are straightforward enough that I enjoyed watching it after a quick primer from van Leusden. The movement never stops, except when there's a foul, resulting in a free throw from the sidelines. It's an offense-defense game, like soccer or basketball, but players aren't allowed to carry the ball (or dribble or kick it), so an integral part of the game is passing to move the ball up or down the court. There's constant jockeying as players try to open themselves up to receive a pass, block someone from receiving a pass, or set themselves up to shoot.

As in basketball, they're aiming for a basket. It's 11.5 feet (3.5 meters) off the ground—so dunking isn't really a thing, but shooting from behind the basket is (there's no backboard). Korfball is played in two halves of twenty-five minutes each (or four quarters of seven to ten minutes each). Four players from each team play at a time, and they score when they make a basket.

A feature unique to korfball, as mentioned, is men and women competing on the same team. The one gender-based rule is that korfballers are only allowed to directly oppose or block members of their own gender. This mitigates advantages due to height or speed, at least between genders. Women defend against women and men against men, but men and women pass to each other to move the ball around, and both shoot for the basket.

As van Leusden showed me around the older and newer indoor courts at the Fortuna sports venue (the latter even has a cool VIP room), he impressed upon me how important the game is to the community, saying theirs is "blessed" to have the facility. The several-hundred-strong crowd outside—complete with little kids playing on the sidelines and older people sipping beers—clearly demonstrated how korfball brings people together.

Fans and players are focused on making korfball an Olympic sport by the 2032 games in Brisbane, so I asked van Leusden how that was going as we peeked into purpose-built classrooms used by a local school for physical education. He told me that korfball was played once at the Olympics, when the Games were held in Amsterdam at the beginning of the twentieth century (it was also at the Antwerp games in 1920). In recent years, it has gotten lots of attention. "We see more and more countries embracing the sport, because men and women together in one team? It's really interesting." He's right, of course. That's why I was there talking to him.

Korfball is mostly an indoor sport, though there are outdoor matches like the one I planned to watch, as well as a beach-based version. The location of the game changes the rules a bit. Most players prefer to compete indoors because the twenty-five-second shot clock makes for a faster game, and wooden floors versus turf allow players to jump higher, which is more fun. (There's no shot clock for outdoor games.)

The Fortuna team has won national championships as well as the European Cup. The match that had started a few minutes after I arrived was a between-the-seasons practice game, but it looked plenty serious to me. As we made our way to the outside court, van Leusden left me to enjoy the match just as the sun broke through the clouds, especially welcome on a late spring day. The spectators were very involved in the play, their cheers and engagement evident even during a practice game. This fan enthusiasm is part of why the Fortuna team is one of the top teams in the Netherlands, attracting players from The Hague, Rotterdam, Delft, and even other countries.

After the game, I caught up with Michelle van Geffen, who has played korfball since age four and has competed on the national team. She's twenty-eight now and is a member of the highest-level Fortuna team. They lost the game I watched by just one point. She said she was a bit disappointed because she thought her team should have beaten their opponents—but she smiled as she said it. She's known some of her teammates since they were "little guys." I asked her what playing with men is like, but since she's been doing it for most of her life, the question stumped her a bit. She shrugged and replied, "There's no difference for us to play with the boys—we're equal." I asked her if men or women are better in certain positions. "I think there's no difference because we have one of the best players, Fleur Hoek, and she's a goal-scoring person. But we have a guy who's also a goal scorer," she said.[27]

I spoke with Merijn Mensink, a twenty-year-old male team member who has also played korfball since childhood. Our conversation was revelatory. Without prompting, he told me that playing competitively with women provides a great advantage to him on a personal level. "One of the most spectacular things about korfball is that we play together, and we train together, and we see each other every day," he said. "A lot of boys, especially our friends from football, they have difficulties talking to women or approaching women. For us, it's never a problem because our [teammates] are like having sisters. We know how to relate with them, because we do it for ten to fifteen years, right?"

This was such an unexpected answer that I almost got tears in my eyes—but it also felt so real. I remembered how awkward and uncomfortable I felt with boys my own age at ten, sixteen, twenty, even twenty-five, by which time I was living with my boyfriend. Having a dad and going to school with boys wasn't quite the same as playing on a team and working toward a goal together. How might my life have been different if I had played korfball growing up? I felt a stab of jealousy that Mensink and van Geffen got to have such easy rapport with the other gender while also playing a fun sport. As I tried to keep myself together and think of a good follow-up question, Mensink continued, "So communication is much, much better, also in normal life. I think that's beautiful." I thought so too.

Something I've heard about feminism since I was a teenager is that, sure, it's about empowering women. But it's also about giving men a wider variety of options too, so everyone gets to express themselves in the ways that work for them. The patriarchy hurts people, not just because it weirdly attaches certain jobs and hair lengths and clothes to what genitals you have, and discounts the experience of non-cisgender people, but because it limits everyone's life choices and freedoms. Feminism liberates us all to embrace our full humanity. As I talked to Mensink on the korfball field in Delft that afternoon, the sun slowly disappearing behind the clouds, I felt that to be true, yet again.

It's this mindset that led to a *New York Times* headline about korfball: "A Game Men and Women Play as Equals."[28] I wondered: Could there be others?

Before I headed out from the Fortuna sports club, I caught up with Damien Folkerts, who's kind of a big-deal coach, having brought the Fortuna team to national championships in 2019 and 2022, and their European Cup win in 2020. A former player and now a coach for over a decade, he's seen all kinds of players come and go. I asked if he thought men and woman had different strengths in the game. "When it comes to making goals or getting really good at passing, it's just not different. That's the same for all sexes," said Folkerts. When I asked about training, he said weightlifting was now popular with everyone, though some of the particular exercises might vary between men and women depending on body size and center of gravity. "But that's also true between two guys—a tall guy needs different training than one who's a bit more broad and smaller," he said. He kept returning to the point that all training should be individualized to the person based on their strengths and weaknesses, and that gender wasn't the most important factor.

While Folkerts talked about training for the person, not the gender, a sense of déjà vu washed over me, rocking me a bit. I had definitely never been to a korfball game or even been in Delft before, so it felt very strange. I figured I probably needed a good meal, so I headed to the Johannes Vermeer Hotel on Molslaan (which translates to "mole-hitting") street. I was delighted to find that a picturesque canal ran directly in front of my hotel. After checking in, I lugged my bag upstairs then went right back out to find dinner. I walked for a few minutes along canals lined with flowers and took lots of pictures for Instagram before I found a large public square. It was crowded with al fresco tables from various restaurants, so it felt like I was dining with half the town, even though I was flying solo—perfect.

As I waited for my onion soup and sipped a glass of white wine, the happy chatter of dozens of people buzzing around me, I pulled out my notebook. I couldn't shake the feeling of déjà vu. I started taking notes by hand (a reliable way for me to figure things out) and then realized what it was.

TO THE FUTURE AND BEYOND

The conversation with Coach Folkerts reminded me of one I'd had earlier with another (former) coach, with the sun also shining on their face. On the other side of the world, in a busy breakfast café in Sydney, Australia, Sophia Nimphius (now an academic) had shared a similar perspective about how gender might not be the most important factor in athletics. (You may remember Nimphius from earlier in the book. She's the pro-vice-chancellor for sport and a professor of human performance at Edith Cowan University in Perth, Australia.)

When I spoke with Nimphius, she had reached a stage in her research where she was seriously questioning the role of physiological sex differences, especially compared with the vast differences in how genders are treated over a lifetime. All her work pointed to the idea that the huge disparities in training and funding, combined with differing lifetime experiences of doing a sport, impacted male versus female athletes' achievements. To discount all those factors, she argued, was to ignore too many influences on athletic performance.[29]

By way of example, she had cited a study that looked at differences in sensorimotor ability between amateur and elite skiers. In the amateur group, there was a large gap in ability between male and female athletes—but no significant sex difference in the elite group. "If there was a fundamental sex difference in that ability, you should see it in both groups, but you don't," she told me. She observed the same pattern across several different sports and sports injuries, including ACL injuries: large gaps between male and female amateur athletes, but almost no difference at the elite level, where athletes of any sex would have access to the highest-quality coaching and resources.[30] She believed these findings pointed to sex-differences being mostly cultural, not physical.

Nimphius has done her own work on this subject. In a 2019 study, she compared the muscle activity of strength-matched athletes of each gender.[31] By comparing equally strong males and females, she found that their muscle activity was the same, indicating they had the same neuromuscular function at equal strengths. Previous studies had suggested biological neuromuscular differences in men and women, but Nimphius's point was that those differences were actually due to strength, not sex.

Following that work, she published a short but powerful article challenging the idea that differences in athletic ability between the sexes are primarily due to physiological factors, arguing instead that nonphysical aspects of sports play a larger role. In that paper, also from 2019, and titled "Exercise and Sport Science Failing by Design in Understanding Female Athletes," she stated it simply in the first paragraph: "Variables that are modifiable through training (e.g., strength, sporting, and movement skill) likely explain conclusions attributed to gender/sex."[32] She called for a critical reevaluation of the research design of gender-comparison studies to ensure that sports scientists weren't attributing differences to gender when gender had nothing to do with them. This misattribution could have very real and serious consequences for athletes. "Oversimplified gender comparisons may contribute to a persistent belief that female athletes are incapable or simply destined to injury or lower performance. The perpetuation of such a message may lower belief in these athletes," she wrote.

Sports science continues to use language which "assumes that modifiable variables are inherent to female athletes. Instead, the research should be highlighting the modifiable variables of lower strength and motor and sporting skill; not doing so indicates a sporting system (and society) that has failed to provide equal funding, opportunity, and qualified personnel," Nimphius wrote. Along with a lack of belief in female athletic strength, these disparities contribute to a depressingly logical conclusion perpetuated by coaches and sports scientists: that female athletes are inherently weaker.

Devastating.

In the café that day, Nimphius told me that paradigms are hard to shift. "To be blunt, publishing a paper that says men and women are different, it's the easiest thing you can do. And we expect it. We expect it not as scientists, but as people who do science [and] who are people living in this culture."

What if we take Nimphius's advice and move past the long-standing, exhaustingly incorrect narrative that "women are just naturally weaker"? What would that do to the future of women's sports?

She pointed out that a wide variety of sports already test different types of athletic ability. "When I think about creating a new sport, I almost just think about what is it about the sports we currently have that we'd like to see more or less of, and so I remove that gender or sex dynamic and think about what would be great to watch and play." She told me about rugby sevens, which she described as "a shorter format of the game with fewer players—more scoring, higher turnover, more exciting." Like the Netflix feature-film version of a long-form documentary. "There are a lot more tactics to it. It's faster but also more spread out over more games." Rather than creating a new sport, then, existing sports can be updated with fans in mind, an approach that's "human neutral."

She noted that people often assume a "women's" sport needs to be easier, as is the case with many of the Olympic events, wood chopping, and other sports. "People always think women's sports need to scale down, but I bet they haven't thought: Should men scale up?" Basketball is a good example. She suggested raising the basket for all players (hey, korfball does that!), which, she argued, would refocus the game on strategy and tactics, rather than dunking. The absence of dunking is one reason many basketball experts find the women's game a "purer" version—so why not make it the default for everyone?

Why don't men and women play together more? As we collaborate in professional life and share household duties more equally, why are sports still so segregated? That separation is starting to look pretty artificial.

Playing sports together also sidesteps many of the contentious issues around intersex and trans athletes. If sports were oriented around tactics and abilities that any body could possess, and if success were based on skill, training, and experience rather than chromosomes—if we took sex out of athletics entirely—we'd return to a time when everyone worked together toward common goals, whether vanquishing an enemy or taking down a woolly mammoth.

It's time to go back to the future.

EPILOGUE

In late August 2024, Megan Rapinoe retired from professional soccer and her team of a decade, the Seattle Reign. The press materials were effusive, referencing her as a "global legend" and "a queen who always rose to the occasion and rose against the odds." I was finishing up this book, having thought and written extensively about female bodies and women's sports for several years. Now, one of the first mainstream, widely known legends of the game was retiring.

I headed into Seattle via ferry on a perfect summer evening. The skies were clear and still bright, and Tahoma (Mount Rainier) was out. Late-summer locals and tourists made the voyage over to the city with me. After grabbing some vegan junk food and a coffee at the terminal, I boarded the boat and headed to my favorite spot on the sun deck. Lounging in the balmy breeze, I noticed little girls in soccer outfits and tiaras running past, and a blonde mom wearing a T-shirt emblazoned with "Everyone Watches Women's Sports"—a slogan made popular following the female dominance in the Paris Olympic Games and a killer WNBA season.

It seemed like half the ferry was headed in the same direction I was as we all walked from Colman Dock to Lumen Stadium. Spirits were high, and everyone was in a celebratory mood. It wasn't just Rapinoe's retirement; it was a hometown game for the Seattle Reign. I wanted to sit with the crowd, not the press, so I'd bought a ticket and took the random seat they assigned me. I found my way to my spot just minutes after the official retirement event began.

There were fireworks, dramatic music, and lots of crowd participation and cheering. The audience, decked out in black and lavender (Reign team colors) scarves and bucket hats cheered for the little kids walking onto the field, whooped up the band, bellowed at the fireworks, and went absolutely batshit when Rapinoe finally appeared.

Rapinoe's hair was blonde that day, not pink, as I always picture it in my mind's eye, and she wore giant khaki pants and a leather jacket. She looked incredibly cool standing on the stage in the middle of the field. She thanked everyone for making her jersey retirement special—and there it was, with her name and the big number 15. "Obviously, the jersey is going to go up into the rafters somewhere, and when you look up, I guess you'll be thinking of me, cause it's my number," Rapinoe deadpanned, and the crowd laughed. "But just remember," she said, "every time I look at that number, I'm going to see you. I'm going to see every pride flag and every trans flag and every Black Lives Matter flag." The crowd went bananas, and the woman next to me wiped tears away.

Confetti cannons confetti'd. Little girls danced on the Jumbotron, wearing princess ensembles in the "Queen's End" near the Reign's goal. The pomp was almost a bit much. But maybe not for a queen, a legend.

"Thank you to everybody, I love you so much, and thank you for what we've built together," Rapinoe said. Then she was gone, and the game began.

I put my notebook away, bought a Bodhizafa IPA tallboy, and sat back to watch some soccer. It was a great game, with the Reign winning by a goal in the last minutes against the North Carolina Courage.

Here was the future, with Rapinoe's legacy now in the past, part of the history of the sport. She departed as star athletes do, leaving us wanting more. The game went on, with new players ready to make their own history. I'm excited to see how far they—and all of us with female bodies—can go.

ACKNOWLEDGMENTS

A foundational influence on my life and work is the autonomy my grandma provided me with as a child and young adult. That included the freedom to read anything and everything with no censorship or judgment (including Stephen King's *Skeleton Crew* when I was eight) and the physical freedom to explore my world. I was surrounded by neighbors whose fields I hiked, horses I rode, pools and ponds I swam in, and wooded paths I mountain biked. I developed my sense of strength from direct interaction with New York's Hudson Highlands' steep hillsides and deep, rocky gorges, as well as through the example of my grandmother, who died with her muscles on. The community of intelligent, compassionate, and creative people that nurtured me, as well as the (public!) Garrison School—with its school forest, intensive science education, extensive school library, and brilliant educators (shout out to Mrs. Impelliterre aka Miss Carlson, my first- and second-grade teacher)—provided me with the kind of education and community that as an adult I can only refer to as a blessing.

Also libraries. As a kid, stepping through the Alice and Hamilton Fish Library's big white doors was always exciting. My grandma truly believed in the idea of open inquiry for kids, and since no topic was off-limits for our conversations, I brought many ideas to her from books that some might deem inappropriate for children. We talked about them. Her openness and those conversations are some of the richest and most valuable memories of my childhood. At age fourteen,

I was employed at that library as a page, a job that involved reshelving returned books (a great lesson in observing what people actually read) and spending improperly long times in corners voraciously reading whatever looked interesting.

Other libraries that have figured into my life and the writing of this book include a duo in Connecticut, where I lived for over a decade: the gorgeous Westport library, which allows for views and walk breaks along the Saugatuck River, and the cozy, Edwardian-era South Norwalk branch of the Norwalk Public Library, which still has wonderfully high ceilings, great light, and a disused fireplace in one of the main rooms. The Thirroul Public Library, near my father's home in Australia; the Philomath Public Library, near Corvallis, Oregon; the gorgeous, community-rich, and inspiring Berkeley Public Library, in California—all have served as homes away from home when I have lived in those places. In Washington State, the architectural marvels that are the Seattle Public Library and the Bainbridge Public Library have proven to be wonderful places for work and research.

Librarians, I so admire and appreciate you! If I weren't a writer, I think I would be one of your tribe.

Syracuse University was my undergraduate alma mater, and whenever anyone calls into question the usefulness of a liberal arts–focused program, I will go on at annoying length about how important it was to me. Here, I'll attempt brevity. After a thoroughly disappointing high school experience, I was a kid in a nerd's candy store when I got to SU, which offered a huge variety of courses. I devoured most of the ones I took. Many were (way) outside my majors, but all those "extra" classes? I ended up using them—from a suite of anthropology classes that fulfilled no requirements but fascinated me, to a semester of mostly Spanish art and culture classes at SU's Madrid study-abroad program (still a life highlight!). I am so happy to have overstuffed my course load every semester. Professor Catherine Newton kindly assisted me in navigating my Bachelor of Science degree in geology after years in the pre-med program. My BA in English (which I got "for fun" and so I had an excuse to "read lots of novels") was undergirded by concentrations in feminism, Shakespeare, and British literature, because those were what interested me. My roommates were all art majors, which was an additional education for me. (I'll admit right here that I was part of many art and film projects, thankfully pre-internet.) I can only say that I loved it all (including lots of amazing off-campus experiences in the city of Syracuse). Is it strange to write such a long passage about one's university education? Probably, but it was that important to me, and as such, I will be a proud Orangewoman forever.

Enrolling in the MFA program at Columbia University might be one the worst financial decisions I've ever made. It was also the best in all the other ways, from my understanding of myself in my late twenties to how I define myself as a writer. I had worked for seven years between college and grad school, and so going back for my master's felt like a supreme indulgence. To be able to think and practice writing for hours a day? With other people who cared about it as much as I did? Taught by professors even more obsessed? It was nothing short of dreamy. Professor Lis Harris was my champion before and during my time there, and I'll never, ever forget the phone call I received from her offering me a spot in the program. So many of my classmates inspired me (and continue to do so). That includes the nonfiction writers Brook Wilensky-Lanford, Rachel Aviv, and Cat Bohannon, and fiction writers Alexandra Kleeman, Lincoln Michel, and Catherine Lacey. Thanks for the kick-in-the-ass inspiration when I needed it.

The Northwest Science Writers (and especially fellow author Michael Bradbury) have lent support and encouragement during my most frustrated days—thank you. Jim Motavalli has acted as both writing mentor and good friend through my, let's say, "adventurous" style of running my writing career and our shared enjoyment of cars. Jim, I'm indebted to you for all the van-buying help over the years! Science-writing colleague Jenny Morber was a generous and thoughtful thinker as I wrapped my head around this book, especially in its earliest stages. I'm indebted to Lost River Writer's Workshop colleagues Betsy Mason, Kat McGowan, and Emily Sohn for their patience and support in the painful last weeks of the writing of this book. It's not hyperbole to say that our teacher (and scribe/goddess) Jacqui Banaszynski swooped in at the last minute with some editing instruction that I desperately needed.

I'm not sure where this book would be if not for Christie Aschwanden's online Make a Book class during the depths of the pandemic (and her indefatigable encouragements), not to mention the fact that she connected me with my beloved agent, Alice Martell. Alice is simply the best agent I can imagine, and her energy and appreciation for this project felt special—but I have a feeling that's something all her authors feel!

Thank you to editor Alexandra Sifferlin for green-lighting (and editing) the feature article that first allowed me to explore this topic—your edits and the positive response I got were key to my pursuing the topic as a book. Thanks to editors Laura Helmuth, Jeffery DelViscio, and Rachael Bale, who have supported me and taught me so much during my science-writing career.

Without the guidance of my editor T.J. Kelleher at Basic Books—and his faith in this project—I don't know where I'd be. Thanks to Kristen Kim for her early read, and to Gillian Sutliff and Shena Redmond for doing all the nitty-gritty work to get this book on its way to readers. So much appreciation to Jessica Breen, Katherine Robertson, and Angie Messina for sharing this book with the world! I especially appreciate the incredibly detailed and thoughtful read copyeditor Kelley Blewster gave this book. I'm also grateful to my team at Atlantic Books in London, including editor Erika Koljonen, and Laura O'Donnell and Anna Carmichael. I hope we get to have another fun pub lunch in the near future.

The Sloan Foundation Book Grant greatly increased the depth and accuracy of this book as it allowed me to pay for rigorous fact-checking, expert readers, and travel. *The Stronger Sex* is richer and more thorough than it would have been without the significant financial support granted to me by Doron Weber and his team. Thank you.

Speaking of teamwork, my fact-checkers, Spoorthy Raman and Annika Robbins, ensured accuracy and were both such pros, as well as being kind and flexible when I threw new material at them—thank you. Expert readers Kaedyn Nedopak and Katelyn Burns, thank you for your much-needed insights and gender critiques. To my science readers, Professor Kate Clancy at the University of Illinois, Merri Wilson, and Emma Verstraete, I so appreciate your feedback. A special extra thanks goes out to Professor Lynnette Sievert at the University of Massachusetts, who jumped in as a science reader and who has now blessed my life twice. (Proving that life is stranger than fiction, the first time Sievert crossed my path was when I read through her syllabus for a class attended by a close friend at UMass Amherst in the late nineties. The syllabus was so interesting it inspired my own dive into anthropology as an undergrad at Syracuse. Decades later, I was contacting interviewees when I thought I recognized a name, and I figured out why.)

Bethanee, Lance, Andrea, and Abe at Elevate, thanks for teaching me how to lift weights up and put them down, and for creating a community where I am celebrated for my strength—and also feel totally comfortable falling on my ass.

To the faithful and energetic participants of the worldwide ecstatic dance community, I wouldn't have kept half as mentally (or physically, of course!) healthy while writing this book without you. My musical muses and influences get a shout-out too. A few powerful women's voices in particular have encouraged me time and again as I wrote. Thank you first and always to Tori Amos, and also Agnes Obel, Queen Latifah, Florence Welch, Sade, Barbra Streisand, Ani DiFranco, Björk, Julie Andrews, and PJ Harvey.

My best friend since first grade, Liz Wagner, has always offered a willing ear and incredibly insightful, thoughtful perspectives on my life and work. I'm so lucky to have not only a deeply loving friend in her, but also a truly brilliant one. I also must thank Liz's late father, Eric Wagner, a celebrated mathematician, artist, and IBM lifer who was a direct and inspiring example of how science can be a part of one's work, life, and perspective on the world. Mr. Wagner not only helped develop the personal computer; he also always took the time to discuss interesting new ideas in science with teenaged me (which I would segue into by ostentatiously perusing the copies of *Science News* always piled up in the corner of the Wagners' house). Mr. Wagner took my interest and my intellect seriously, a true gift. His kindness, peace activism, and ethos of care for people and the planet is carried on by his daughter. My thanks to Liz will always go beyond this book or any particular project, as she has been there for me over the decades in the way that we mean when we say "family."

On that note, thank you to my aunt Susan Stuart-Jones for holding all the pieces together over the years. There are some women who do the hardest work in families and earn the moniker of matriarch, and she is one. I loved hearing her insights into aspects of this book—and she gets full credit for inspiring my Olympic weightlifting journey! I hope we have many hikes to come in the mountains of the world!

My father, Gerry Vartan, has always been incredibly supportive of my career ambitions, even when I was doing weird, experimental, early-on-the-internet things. Nothing has ever been too out-there for him, which has provided both my life and work with tremendous freedom (if you are sensing a strong family value here—yes!). My dad has consistently advised me to pursue my ideas, offered critiques, and been a real cheerleader when things got rough. He also washed countless dishes during my long sojourns in the Sydney area while reporting and writing this book, and he has been a fun and dynamic conversation partner, even when we disagree—sometimes very loudly. (I won't ever agree with you about capitalism or dishwashers, Dad!) An artistic soul through and through, with a long creative career, my father has always been genuinely interested in science and connected to the natural world as a surfer. Figuring out how to communicate complex scientific ideas with him—someone curious and knowledgeable but not trained in science—has been helpful as I think of the Ideal Reader while composing my articles and this book.

My good friend Amanda Freeman—in her wonderfully practical, straightforward New England way—has always been the best sounding board in my writing life. I have admired her since freshman year at Syracuse University, and still do. Her husband, Trey Ellis, also offered some key assists in making this book the best it could be. Thank you

Joanne for putting me up and feeding me so well during my travels for this book. Appreciations must go out to Georgia and Bing for providing such a peaceful and animal-filled place to write and Zen-kitty energy.

My grandmother Doris Ross taught me what strength was and never doubted mine. She stood up to her physically abusive father at sixteen, won ballroom-dancing competitions in Manhattan, flew her Cessna up and down the East Coast, raised her sons and me, chopped her own firewood, sewed my clothes, grew and cooked our food, saved lives as an EMT, taught a generation of EMTs how to do emergency medicine, raised countless dogs and cats, built perfectly laid stone walls, traveled the world on Pan Am's around-the-world ticket thrice, delivered babies in Papua New Guinea—and looked chic as hell doing it all. It's a blessing to have been raised by her, and I know, deep in my bones, that she is incredibly proud of me.

To my Aussie nana, Adeline Stuart-Jones, a beauty queen and fashion plate with an intense wanderlust (I think I inherited that gene), I wish I'd had a chance to dance with you more. I treasure the memories of picking out mangoes at the farmer's market and keep you in my heart. Your legacy is the command to follow your dreams, and I have.

To my beloved black cat Josephine, who listened patiently and wisely to my articles and book chapters read out loud to her for a decade, I will always miss you. To my kittens, Coppélia and Wollstone, you make me proud to be a child-free cat lady!

To my partner, Simon, I'm lucky to have had your attentive and compassionate eye on this book—and on my life—for the past fifteen years. You never doubt my ability to do what I set my mind to, and I'm so excited for all our adventures to come.

Lastly, I'm incredibly grateful to all the scientists and experts that I spoke with for this project—busy people all, who took time out of their schedules for me and this book. There were the interviews (often several), plus more time spent providing me with papers and evidence, edits and ideas. I consider it a great privilege to be able to reach out into the world and speak to brilliant people, ask questions, and be The Curious Human I am.

NOTES

Introduction

1. N. Anderson, D. G. Robinson, E. Verhagen, et al., "Under-Representation of Women Is Alive and Well in Sport and Exercise Medicine: What It Looks Like and What We Can Do About It," *BMJ Open Sport and Exercise Medicine* 9, no. 2 (January 2023), https://doi.org/10.1136/bmjsem-2023-001606.

2. Amethysta Herrick, Zoom interview with author, July 1, 2024.

3. Mihaela Pavlicev, Anna N. Herdina, and Günter Wagner, "Female Genital Variation Far Exceeds That of Male Genitalia: A Review of Comparative Anatomy of Clitoris and the Female Lower Reproductive Tract in Theria," *Integrative and Comparative Biology* 62, no. 3 (2022), https://doi.org/10.1093/icb/icac026.

4. Hida Viloria, "How Common Is Intersex? An Explanation of the Stats," Intersex Campaign for Equality, April 1, 2015, www.intersexequality.com/how-common-is-intersex-in-humans/.

5. Sharon DeWitte, interview with author, October 2023.

Chapter One: The Culture of Women's Muscles

1. E. Ramos, W. Frontera, A. Llopart, and D. Feliciano, "Muscle Strength and Hormonal Levels in Adolescents: Gender Related Differences," *International Journal of Sports Medicine* 19, no. 8 (1998), https://doi.org/10.1055/s-2007-971955.

2. "About Our Age Divisions," Pop Warner, revised January 2024, www.popwarner.com/Default.aspx?tabid=2676344.

3. "High School Athletics Participation Survey: Based on Competition at the High School Level in the 2021–22 School Year," National Federation of State High School Associations, accessed October 15, 2024, www.nfhs.org/media/5989280/2021-22_participation_survey.pdf.

4. "Beyond Biology: A Gendered Approach to Injury with Joanne Parsons and Stephanie Coen. Ep 478," *BMJ Talk Medicine*, SoundCloud, June 4, 2021, soundcloud.com/bmjpodcasts/beyond-biology-a-gendered-approach-to-injury-with-joanne-parsons-and-stephanie-coen-ep-478.

5. Sophia Nimphius, interview with author, March 25, 2024.

6. Tom Wigmore, "Why Are Great Athletes More Likely to Be Younger Siblings?," FiveThirtyEight, December 1, 2020, https://fivethirtyeight.com/features/why-are-great-athletes-more-likely-to-be-the-younger-siblings.

7. April Heinrichs and Matt Robinson, "The U.S. Women's Youth National Teams Program Finding the Next Mia Hamm and Alex Morgan," *Soccer Journal*, November 2014.

8. Cassie Tomlin, "ACL Tears in Girls and Women," Cedars-Sinai, June 23, 2023, www.cedars-sinai.org/blog/acl-tears-in-girls-and-women.html.

9. Australian Rules Football, which has roots in both rugby and Gaelic football, was formalized and codified in 1858. Some say the sport predates other forms of football, like American, Canadian, various types of rugby, and soccer. It's a fast-paced, full-contact sport, but players don't wear padding or helmets (soft headgear is allowed), and it involves more running than other types of football, except soccer. Aaron Fox, Jason Bonacci, Samantha Hoffmann, Sophia Nimphius, and Natalie Saunders, "Anterior Cruciate Ligament Injuries in Australian Football: Should Women and Girls Be Playing? You're Asking the Wrong Question," *BMJ Open Sport and Exercise Medicine* 6, no. 1 (April 2020), https://doi.org/10.1136/bmjsem-2020-000778.

10. Fox et al., "Anterior Cruciate Ligament Injuries in Australian Football."

11. David Goldstein, Diane Haldane, and Carolyn Mitchell, "Sex Differences in Visual-Spatial Ability: The Role of Performance Factors," *Memory and Cognition* 18, no. 5 (September 1990), https://doi.org/10.3758/bf03198487.

12. One set of researchers even posited that men's superior skills in spatial tests are tied to navigational ability, and then directly linked that with male reproductive success: Men who can travel farther can father more children. This is a real reach considering that there doesn't seem to be evidence that only men travel long distances; generally, hunter-gatherer societies move seasonally (or permanently into a new area) as a group. And it also disregards how important navigational skills are to women in hunter-gatherer societies, who travel significant distances to forage. There's also evidence from some communities, tracked using DNA, that it was women who were more likely to switch into new communities, and travel distances to do so. Isabelle Ecuyer-Dab and Michèle Robert, "Have Sex Differences in Spatial Ability Evolved from Male Competition for Mating and Female Concern for Survival?," *Cognition* 91, no. 3 (April 2004), https://doi.org/10.1016/j.cognition.2003.09.007.

13. Melanie S. Rohde, Alexandra L. Georgescu, Kai Vogeley, Rolf Fimmers, and Christine M. Falter-Wagner, "Absence of Sex Differences in Mental Rotation Performance in Autism Spectrum Disorder," *Autism* 22, no. 7 (August 2017), https://doi.org/10.1177/1362361317714991.

14. Carole K. Hooven, Christopher F. Chabris, Peter T. Ellison, and Rogier A. Kievit, "The Sex Difference on Mental Rotation Tests Is Not Necessarily a Difference in Mental Rotation Ability," ResearchGate, 2008, www.researchgate.net/publication/228978237_The_Sex_Difference_on_Mental_Rotation_Tests_Is_Not_Necessarily_a_Difference_in_Mental_Rotation_Ability.

15. Angelica Moè, "Are Males Always Better than Females in Mental Rotation? Exploring a Gender Belief Explanation," *Learning and Individual Differences* 19, no. 1 (2009), https://doi.org/10.1016/j.lindif.2008.02.002.

16. Dena Crozier, Zhaoran Zhang, Se-Woong Park, and Dagmar Sternad, "Gender Differences in Throwing Revisited: Sensorimotor Coordination in a Virtual Ball Aiming Task," *Frontiers in Human Neuroscience* 13 (2019), https://doi.org/10.3389/fnhum.2019.00231.

17. Maggie Mertens, town hall event, Seattle, WA, June 24, 2024.

18. Excerpted in Robin Vealey and Melissa Chase, "Reasons for Gender Differences in Youth Sport," Human Kinetics, 2022, us.humankinetics.com/blogs/excerpt/reasons-for-gender-differences-in-youth-sport.

19. Kate Clancy, Zoom interview with author, August 2022.

20. D. Peterschmidt, "Why Are Female Athletes at a Higher Risk of ACL Injuries?," Science Friday, July 2022, https://www.sciencefriday.com/segments/female-athletes-acl-injuries/.

21. Anne M. Peterson, "U.S. Men's and Women's Soccer Teams Formally Sign Equal Pay Agreements," *PBS NewsHour*, September 6, 2022, www.pbs.org/newshour/economy/u-s-mens-and-womens-soccer-teams-formally-sign-equal-pay-agreements.

22. Jennifer Pharr Davis, interview with author, April 2019.

23. Mara Jameson (alias), Zoom interview with author, September 1, 2023.

Chapter Two: Centering Women's Muscles

1. "More Girls Entering the Ring," ABC News, March 13, 2024, https://abcnews.go.com/WNN/video/girls-entering-ring-108074713.

2. AP, "Girls' wrestling has become the fastest-growing high school sport in the country," Facebook reel, 2022, www.facebook.com/reel/780686160172814.

3. Audrey Jimenez, Zoom interview with author, June 11, 2024.

4. "Tucson teen makes history as first girl to earn Arizona wrestling title," ABC15 Arizona, video posted February 20, 2024, YouTube, www.youtube.com/watch?v=X7L8FSha9XY.

5. "Women's Wrestling Team Achieves Success Despite Limited Funding," ABC News, May 28, 2024, https://abcnews.go.com/US/video/womens-wrestling-team-achieves-success-despite-limited-funding-110624839.

6. "Muscular Strength and Endurance," Health Link BC, last reviewed November 2016, www.healthlinkbc.ca/healthy-eating-physical-activity/being-active/health-benefits-physical-activity/muscular-strength.

7. Definition of "strength," Merriam-Webster dictionary, accessed December 11, 2024, www.merriam-webster.com/dictionary/strength.

8. Specifically, argue some researchers, male humans' areas of stronger muscles are for combat, not hunting. Like other mammals, male humans have upper-body muscular strength that is "greatest in the structures that are used as weapons." In a study from the University of Utah, larger strength differences were found in the muscles used for punching compared to those needed for throwing weapons, leading one researcher to tell ScienceDaily, "This is a

dramatic example of sexual dimorphism that's consistent with males becoming more specialized for fighting, and males fighting in a particular way, which is throwing punches." University of Utah, "Why Males Pack a Powerful Punch," ScienceDaily, accessed November 25, 2024, www.sciencedaily.com/releases/2020/02/200205132404.htm.

9. Cara Wall-Scheffler, "Women Have Stronger Legs and Other Side-Effects of Human Body Proportions," *American Journal of Human Biology* 36, no. 5 (2023), https://doi.org/10.1002/ajhb.24034.

10. Birgitta Glenmark, Maria Nilsson, Hui Gao, Jan-Åke Gustafsson, Karin Dahlman-Wright, and Håkan Westerblad, "Difference in Skeletal Muscle Function in Males vs. Females: Role of Estrogen Receptor-β," *American Journal of Physiology-Endocrinology and Metabolism* 287, no. 6 (2004), https://doi.org/10.1152/ajpendo.00098.2004; Paul Ansdell, Kevin Thomas, Glyn Howatson, Sandra Hunter, and Stuart Goodall, "Contraction Intensity and Sex Differences in Knee-Extensor Fatigability," *Journal of Electromyography and Kinesiology* 37 (December 2017), https://doi.org/10.1016/j.jelekin.2017.09.003.

11. Sandra Hunter, interview with author, August 18, 2022.

12. Sandra K. Hunter, Daphne L. Ryan, Justus D. Ortega, and Roger M. Enoka, "Task Differences with the Same Load Torque Alter the Endurance Time of Submaximal Fatiguing Contractions in Humans," *Journal of Neurophysiology* 88, no. 6 (2002), https://doi.org/10.1152/jn.00232.2002.

13. Sandra K. Hunter et al., "Men Are More Fatigable than Strength-Matched Women When Performing Intermittent Submaximal Contractions," *Journal of Applied Physiology* 96, no. 6 (June 2004), https://doi.org/10.1152/japplphysiol.01342.2003.

14. Ashley Critchlow et al., "Fatigability During Submaximal Intermittent Contractions with Arm Muscles Differed for Strength-Matched Men and Women," *Medicine and Science in Sports and Exercise* 36, supplement (May 2004), https://doi.org/10.1249/00005768-200405001-00566.

15. James L. Nuzzo, "Narrative Review of Sex Differences in Muscle Strength, Endurance, Activation, Size, Fiber Type, and Strength Training Participation Rates, Preferences, Motivations, Injuries, and Neuromuscular Adaptations," *Journal of Strength and Conditioning Research* 37, no. 2 (2023), https://doi.org/10.1519/jsc.0000000000004329.

16. Thomas Beltrame, Rodrigo Villar, and Richard L. Hughson, "Sex Differences in the Oxygen Delivery, Extraction, and Uptake During Moderate-Walking Exercise Transition," *Applied Physiology, Nutrition, and Metabolism* 42, no. 9 (2017), https://doi.org/10.1139/apnm-2017-0097.

17. Ana Sandoiu, "Battle of the Sexes: Are Women Fitter than Men?," *Medical News Today*, December 5, 2017, www.medicalnewstoday.com/articles/320263.

18. Sandra K. Hunter, "Sex Differences in Human Fatigability: Mechanisms and Insight to Physiological Responses," *Acta Physiologica* 210, no. 4 (2014), https://doi:10.1111/apha.12234.

19. B. Knechtle, G. Müller, F. Willmann, K. Kotteck, P. Eser, and H. Knecht, "Fat Oxidation in Men and Women Endurance Athletes in Running and

Cycling," *International Journal of Sports Medicine* 25, no. 1 (2004), https://doi.org/10.1055/s-2003-45232.

20. Gregory C. Henderson and George A. Brooks, "Women Utilize Lipid as Fuel More than Men During Exercise—Is There a Paradox?," Physiological Society, July 2008, www.physoc.org/magazine-articles/women-utilize-lipid-as-fuel-more-than-men-during-exercise-is-there-a-paradox/; Jennifer Wismann and Darryn Willoughby, "Gender Differences in Carbohydrate Metabolism and Carbohydrate Loading," *Journal of the International Society of Sports Nutrition* 3, no. 1 (June 2006), https://doi.org/10.1186/1550-2783-3-1-28.

21. Anne-Marie Lundsgaard and Bente Kiens, "Gender Differences in Skeletal Muscle Substrate Metabolism—Molecular Mechanisms and Insulin Sensitivity," *Frontiers in Endocrinology* 5 (November 2014), https://doi.org/10.3389/fendo.2014.00195.

22. Nicole Golden, "Fast-Twitch vs. Slow-Twitch Muscle Fiber Types + Training Tips," National Association of Sports Medicine, 2022, blog.nasm.org/fitness/understanding-fast-twitch-vs-slow-twitch-mucle-fibers.

23. Robert O. Deaner, Rickey E. Carter, Michael J. Joyner, and Sandra K. Hunter, "Men Are More Likely than Women to Slow in the Marathon," *Medicine and Science in Sports and Exercise* 47, no. 3 (2015), https://doi.org/10.1249/mss.0000000000000432.

24. Chandra Shikhi Kodete et al., "Hormonal Influences on Skeletal Muscle Function in Women Across Life Stages: A Systematic Review," *Muscles* 3, no. 3 (2024), https://doi.org/10.3390/muscles3030024.

25. Sophia Nimphius, Jeffrey M. McBride, Paige E. Rice, Courtney L. Goodman-Capps, and Christopher R. Capps, "Comparison of Quadriceps and Hamstring Muscle Activity During an Isometric Squat Between Strength-Matched Men and Women," *Journal of Sports Science and Medicine* 18, no. 1 (2019), www.ncbi.nlm.nih.gov/pmc/articles/PMC6370970/.

26. Sophia Nimphius, "Exercise and Sport Science Failing by Design in Understanding Female Athletes," *International Journal of Sports Physiology and Performance* 14, no. 9 (2019), https://doi.org/10.1123/ijspp.2019-0703.

27. Jamie Wareham, "New Report Shows Where It's Illegal to Be Transgender in 2020," *Forbes*, September 30, 2020, www.forbes.com/sites/jamiewareham/2020/09/30/this-is-where-its-illegal-to-be-transgender-in-2020/.

28. Clare McFadden et al., "Determinants of Infant Mortality and Representation in Bioarchaeological Samples: A Review," *American Journal of Biological Anthropology* 177, no. 2 (2021), https://doi.org/10.1002/ajpa.24406.

29. David A. Raichlen and Daniel E. Lieberman, "The Evolution of Human Step Counts and Its Association with the Risk of Chronic Disease," *Current Biology* 32, no. 21 (2022), https://doi.org/10.1016/j.cub.2022.09.030.

30. Tom Metcalfe, "This New Tool Will Help Upend Everything We Assumed About Ancient Gender Roles," *National Geographic*, July 6, 2023, www.nationalgeographic.com/premium/article/proteomics-archaeology-sex-dna-protein.

31. Alison A. Macintosh, Ron Pinhasi, and Jay T. Stock, "Prehistoric Women's Manual Labor Exceeded That of Athletes Through the First 5500 Years

of Farming in Central Europe," *Science Advances* 3, no. 11 (2017), https://doi.org/10.1126/sciadv.aao3893.

32. Sabrina Strings, *Fearing the Black Body: The Racial Origins of Fat Phobia* (New York University Press, 2019), 17.

33. Ben van Beneden and Amy Orrock, *Rubens & Women* (Dulwich Picture Gallery, 2023), 159.

34. Van Beneden and Orrock, *Rubens & Women*, 160.

Chapter Three: The Power of Periods

1. User name and original post deleted, "AITA for telling my brother-in-law to get over...," r/AmItheAsshole, Reddit, 2023, www.reddit.com/r/AmItheAsshole/comments/14oxpef/aita_for_telling_my_brotherinlaw_to_get_over_my/.

2. Elinor Cleghorn, *Unwell Women: Misdiagnosis and Myth in a Man-Made World* (Dutton, 2021), 20–24.

3. Ariadne Schmidt, "Labour Ideologies and Women in the Northern Netherlands, c.1500–1800," *International Review of Social History* 56, no. S19 (2011), https://doi.org/10.1017/s0020859011000538.

4. Rachel E. Gross, *Vagina Obscura: An Anatomical Voyage* (W. W. Norton, 2022), 130.

5. Emily Martin, "The Egg and the Sperm: How Science Has Constructed a Romance Based on Stereotypical Male-Female Roles," *Signs* 16, no. 3 (1991), www.jstor.org/stable/3174586.

6. Martin, "The Egg and the Sperm."

7. This cultural impact can go even further than laughable texts anthropomorphizing bodily cells. As Martin wrote, "Endowing egg and sperm with intentional action, a key aspect of personhood in our culture, lays the foundation for the point of viability being pushed back to the moment of fertilization." Yes, this is how our cultural beliefs intertwine with the science and then bounce back into the culture. Characterizing bodily reproductive functions as "intentional action" in gendered stories will "likely lead . . . to new forms of scrutiny and manipulation, for the benefits of these inner 'persons': court-ordered restrictions on a pregnant woman's activities in order to protect her fetus, fetal surgery, amniocentesis, and rescinding of abortion rights, to name but a few examples."

8. National Institutes of Health, "RePORT: Research Portfolio Online Tools," May 14, 2024, https://report.nih.gov/funding/categorical-spending#/; National Institute of Child Health and Human Development, "How Common Is Infertility?" February 2, 2018, www.nichd.nih.gov/health/topics/infertility/conditioninfo/common.

9. Katie Zhang, Staci Pollack, Ali Ghods, et al., "Onset of Ovulation After Menarche in Girls: A Longitudinal Study," *Journal of Clinical Endocrinology and Metabolism* 93, no. 4 (2008), https://doi.org/10.1210/jc.2007-1846.

10. Marla Broadfoot, "NIEHS Launches 'A Girl's First Period' Study," *Environmental Factor*, September 2019, factor.niehs.nih.gov/2019/9/feature/2-feature-study.

11. A refresher on the simplified cycle: Day one is the first day of bleeding, or the menses phase, which can last from three to seven days. The follicular phase lasts until ovulation, around day fourteen (yes, the menses and follicular phases overlap). During the follicular phase, estrogen rises, which causes the uterine lining to thicken. At the same time, another hormone, FSH (follicle-stimulating hormone), causes the eggs in the ovaries to grow, with one forming a mature egg from day ten to fourteen. Around day fourteen, yet another hormone, LH (luteinizing hormone), causes the ovary to release the egg, a process known as ovulation. This is when pregnancy can happen. After ovulation comes the luteal phase, during which the egg travels through the fallopian tube and into the uterus. Unless the egg meets a sperm and becomes fertilized (pregnancy), estrogen and progesterone levels drop, and the uterine lining is shed (a period). The whole cycle is considered within the norm if it takes between twenty-one and thirty days. "Menstrual Cycle," Cleveland Clinic, last reviewed December 9, 2022, my.clevelandclinic.org/health/articles/10132-menstrual-cycle.

12. Committee Opinion, "Menstruation in Girls and Adolescents: Using the Menstrual Cycle as a Vital Sign," American College of Obstetricians and Gynecologists, December 2015, www.acog.org/clinical/clinical-guidance/committee-opinion/articles/2015/12/menstruation-in-girls-and-adolescents-using-the-menstrual-cycle-as-a-vital-sign.

13. Anne Trafton, "Study Reveals How Egg Cells Get So Big," *MIT News*, March 4, 2021, news.mit.edu/2021/study-reveals-how-egg-cells-get-so-big-0304.

14. Gross, *Vagina Obscura*, 132–133.

15. This really helped me understand how the stresses one's grandparents went through can have epigenetic and health implications for both their children and their grandchildren. The egg that made me—the one that, once fertilized, became my mother—was first created inside my Australian grandma's body when she was a newly married twenty-four-year-old in Sydney.

16. The other day I experienced my monthly bout of mittelschmerz, the German term for when you can feel yourself ovulating. About 30–40 percent of women can feel the unique sharp, pinching pain for a few hours during the middle of their cycle. Knowing what I now do about the intense competition that egg had just gone through to be released, my first reaction was to jump up, throw my arms overhead, and yell to my stomach, "You got the GOOOOOOLD!" Then I started laughing at myself. Who says it's not fun knowing what's going on inside your own flesh machine? (Also, yes, I don't work in an office!)

17. Rhona Lewis, "How Many Eggs Does a Woman Have? At Birth, Age 30, 40, More," Healthline, June 26, 2020, www.healthline.com/health/womens-health/how-many-eggs-does-a-woman-have.

18. Dong Zi Yang, Wan Yang, Yu Li, and Zuanyu He, "Progress in Understanding Human Ovarian Folliculogenesis and Its Implications in Assisted Reproduction," *Journal of Assisted Reproduction and Genetics* 30, no. 2 (2013), https://doi.org/10.1007/s10815-013-9944-x.

19. O. J. Ginther, M. A. Beg, F. X. Donadeu, D. R. Bergfelt, "Mechanism of Follicle Deviation in Monovular Farm Species," *Animal Reproduction Science* 78, no. 3–4 (2003), https://doi.org/10.1016/s0378-4320(03)00093-9.

20. Kate Clancy, *Period* (Princeton University Press, 2023), 63–68.

21. Angela R. Baerwald, Gregg P. Adams, and Roger A. Pierson, "Characterization of Ovarian Follicular Wave Dynamics in Women," *Biology of Reproduction* 69, no. 3 (2003), https://doi.org/10.1095/biolreprod.103.017772.

22. Angela Baerwald and Roger Pierson, "Ovarian Follicular Waves During the Menstrual Cycle: Physiologic Insights into Novel Approaches for Ovarian Stimulation," *Fertility and Sterility* 114, no. 3 (2020), https://doi.org/10.1016/j.fertnstert.2020.07.008.

23. Yang et al., "Progress in Understanding Human Ovarian Folliculogenesis."

24. The Final Girl is a horror movie cliché with academic roots that's been turned on its head in recent years, especially by Jamie Lee Curtis in her roles in the *Halloween* film franchise. (There's even a slasher film called *Final Girl*.) A Final Girl is defined as the last survivor of a group of people who are stalked and killed by the movie's monster/murderer, a common trope beginning in the mid-1970s and continuing through the heyday of horror in the 1980s. Identified as a cultural phenomenon by UC Berkeley film and history professor Carol Clover in her fun-to-read book *Men, Women, and Chainsaws*, the Final Girl "alone looks death in the face, but she alone also finds the strength to either stay the killer long enough to be rescued (ending A), or to kill him herself (ending B)."

25. Lynnette Sievert, interview with author, March 11, 2024.

26. Jay M. Baltz, David F. Katz, and Richard A. Cone, "The Mechanics of the Sperm-Egg Interaction at the Zona Pellucida," *Biophysical Journal* 54, no. 4 (1988), www.academia.edu/58588401/The_egg_and_the_sperm_How_science_has_constructed_a_romance_based_on_.

27. Martin, "The Egg and the Sperm."

28. Carrie Arnold, "Choosy Eggs May Pick Sperm for Their Genes, Defying Mendel's Law," *Quanta Magazine*, November 15, 2017, www.quantamagazine.org/choosy-eggs-may-pick-sperm-for-their-genes-defying-mendels-law-20171115/.

29. Josh Barney, "Fertilization Discovery Reveals New Role for the Egg," *UVA Today*, November 13, 2019, news.virginia.edu/content/fertilization-discovery-reveals-new-role-egg.

30. Kalindi Vora, "Evelyn Fox Keller's Research Reminded Us Women Make Science Better," *Ms.*, March 25, 2024, msmagazine.com/2024/03/25/evelyn-fox-keller-women-minorities-science/.

31. John L. Fitzpatrick, Charlotte Willis, Alessandro Devigili, et al., "Chemical Signals from Eggs Facilitate Cryptic Female Choice in Humans," *Proceedings of the Royal Society B: Biological Sciences* 287, no. 1928 (2020), https://doi.org/10.1098/rspb.2020.0805.

32. Carmen Leitch, "Human Eggs Can Choose the Sperm They Prefer," LabRoots, June 20, 2020, www.labroots.com/trending/genetics-and-genomics/17865/human-eggs-choose-sperm-prefer.

33. Gross, *Vagina Obscura*.

34. Hiroshi Tsuda, Yoichi M. Ito, Yukiharu Todo, et al., "Measurement of Endometrial Thickness in Premenopausal Women in Office Gynecology,"

Reproductive Medicine and Biology 17, no. 1 (2017), https://doi.org/10.1002/rmb2.12062.

35. Jemma Evans, Giuseppe Infusini, Jacqui McGovern, et al., "Menstrual Fluid Factors Facilitate Tissue Repair: Identification and Functional Action in Endometrial and Skin Repair," *FASEB Journal* 33, no. 1 (2019), https://doi.org/10.1096/fj.201800086r.

36. Caroline Gargett, Zoom interview with author, August 2023.

37. Lijun Chen, Jingjing Qu, and Charlie Xiang, "The Multi-Functional Roles of Menstrual Blood–Derived Stem Cells in Regenerative Medicine," *Stem Cell Research and Therapy* 10, no. 1 (2019), https://doi.org/10.1186/s13287-018-1105-9.

38. Yu-Liang Sun, Ling-Rui Shang, Rui-Hong Liu, et al., "Therapeutic Effects of Menstrual Blood–Derived Endometrial Stem Cells on Mouse Models of Streptozotocin-Induced Type 1 Diabetes," *World Journal of Stem Cells* 14, no. 1 (2022), https://doi.org/10.4252/wjsc.v14.i1.104.

39. Natasha Strydhorst, "New Stem-Cell Cultivation Procedure Boosts Hope for Cures," Yale School of Medicine, May 13, 2019, medicine.yale.edu/news-article/new-stem-cell-cultivation-procedure-boosts-hope-for-cures/.

40. Heather Boonstra, "Fetal Tissue Research: A Weapon and a Casualty in the War Against Abortion," Guttmacher Institute, February 9, 2016, www.guttmacher.org/gpr/2016/fetal-tissue-research-weapon-and-casualty-war-against-abortion.

41. Meredith Wadman, "Trump Administration Restricts Fetal Tissue Research," *Science*, June 5, 2019, www.science.org/content/article/trump-administration-restricts-fetal-tissue-research.

42. Ann Devroy, "Bush Called 'Adamant' on Fetal Tissue Research Ban," *Washington Post*, May 19, 1992, www.washingtonpost.com/archive/politics/1992/05/19/bush-called-adamant-on-fetal-tissue-research-ban/a987c0f2-772f-462e-ac2d-ebd7657050b4/.

43. Chen et al., "The Multi-Functional Roles of Menstrual Blood–Derived Stem Cells in Regenerative Medicine."

44. M. A. Perlstein, "Allergy Due to Menotoxin of Pregnancy," *Archives of Pediatrics and Adolescent Medicine* 52, no. 2 (1936), https://doi.org/10.1001/archpedi.1936.04140020046005.

Chapter Four: Hacking the Female Body to Win at Sports

1. Katie Kindelan, "USWNT Used Innovative Period Tracking to Help Player Performance at World Cup," *Good Morning America*, July 15, 2019, www.goodmorningamerica.com/wellness/story/uswnt-innovative-period-tracking-player-performance-world-cup-64339368.

2. Stacy T. Sims, *Roar*, rev. ed. (Penguin Random House, 2024).

3. Stacy Sims, Zoom interview with author, February 2023.

4. Kieran Pender, "Ending Period 'Taboo' Gave USA Marginal Gain at World Cup," *The Telegraph*, July 13, 2019, www.telegraph.co.uk/world-cup/2019/07/13/revealed-next-frontier-sports-science-usas-secret-weapon-womens/.

5. Pierre-Hugues Igonin, Isabelle Rogowski, Nathalie Boisseau, and Cyril Martin, "Impact of the Menstrual Cycle Phases on the Movement Patterns of

Sub-Elite Women Soccer Players During Competitive Matches," *International Journal of Environmental Research and Public Health* 19, no. 8 (2022), https://doi.org/10.3390/ijerph19084465.

6. Jessica Bahr, "Female Athletes Are Calling for More Research into Periods. Here's Why," SBS News, September 4, 2022, www.sbs.com.au/news/article/female-athletes-are-calling-for-more-research-into-periods-heres-why/qb39a4sje.

7. Emma S. Cowley et al., "'Invisible Sportswomen': The Sex Data Gap in Sport and Exercise Science Research," *Women in Sport and Physical Activity Journal* 29, no. 2 (2021): 146–151.

8. Iñigo Mujika and Ritva S. Taipale, "Sport Science on Women, Women in Sport Science," *International Journal of Sports Physiology and Performance* 14, no. 8 (2019), https://doi.org/10.1123/ijspp.2019-0514.

9. Mikaeli Anne Carmichael, Rebecca Louise Thomson, Lisa Jane Moran, and Thomas Philip Wycherley, "The Impact of Menstrual Cycle Phase on Athletes' Performance: A Narrative Review," *International Journal of Environmental Research and Public Health* 18, no. 4 (2021), https://doi.org/10.3390/ijerph18041667.

10. Maggie Lange. "Why Don't We Know How Periods Affect Exercise?" *The Cut*, October 8, 2021, www.thecut.com/2021/10/how-periods-affectexercise.html.

11. Home page, FitrWoman website, accessed October 22, 2024, www.fitrwoman.com.

12. Jess Freemas, interview with author, July 2023.

13. "Premenstrual Dysphoric Disorder (PMDD)," Johns Hopkins Medicine, 2019, www.hopkinsmedicine.org/health/conditions-and-diseases/premenstrual-dysphoric-disorder-pmdd.

14. "Harness the Power of Your Menstrual Cycle," Nike, last updated August 22, 2022, www.nike.com/gb/a/syncing-training-and-cycle.

15. Amanda Brooks, "Racing and Running on Your Period: What You Need to Know," Run to the Finish, last updated June 3, 2024, www.runtothefinish.com/women-runners-natural-pms-remedies/.

16. Carmichael et al., "The Impact of Menstrual Cycle Phase on Athletes' Performance."

17. Brianna Larsen, Zoom interview with author, March 2023.

18. Carmichael et al., "The Impact of Menstrual Cycle Phase on Athletes' Performance."

19. "The Science of Athletic Performance Across the Menstrual Cycle," GenderSci Lab, September 10, 2019, www.genderscilab.org/blog/period-power-or-wrong-period; Jacquelyn N. Zita, "The Premenstrual Syndrome 'Dis-Easing' the Female Cycle," *Hypatia* 3, no. 1 (1988), https://doi.org/10.1111/j.1527-2001.1988.tb00057.x.

20. Meredith Reiches, "Period Power or Wrong, Period?," GenderSci Lab, Harvard University, September 10, 2019, www.genderscilab.org/blog/period-power-or-wrong-period.

21. "About Us," Oova, accessed October 4, 2024, www.oova.life/about-us.

22. "Polycystic Ovary Syndrome," World Health Organization, June 28, 2023, www.who.int/news-room/fact-sheets/detail/polycystic-ovary-syndrome.

23. Amy Divaraniya, interview with author, August 3, 2024.

24. Carmichael et al., "The Impact of Menstrual Cycle Phase on Athletes' Performance."

25. Jia Tolentino, "The Hidden-Pregnancy Experiment," *The New Yorker*, May 4, 2024, www.newyorker.com/culture/the-weekend-essay/the-hidden-pregnancy-experiment.

26. "ACLU v. Department of Homeland Security," ACLU, last updated July 18, 2022, www.aclu.org/cases/aclu-v-department-homeland-security-commercial-location-data-foia.

27. Shiona McCallum, "Period Tracking Apps Warning over Roe v. Wade Case in US," BBC, May 7, 2022, www.bbc.com/news/technology-61347934.

28. Mitchell McCluskey, "A Nebraska Mother Who Provided an Illegal Abortion for Her Daughter and Helped Dispose of the Fetus Gets 2 Years in Prison, Report Says," CNN, September 23, 2023, www.cnn.com/2023/09/23/us/nebraska-abortion-pill-jessica-burgess/index.html.

29. Lisa M. Malki, Ina Kaleva, Dilisha Patel, Mark Warner, and Ruba Abu-Salma, "Exploring Privacy Practices of Female mHealth Apps in a Post-Roe World," *Proceedings of the 2024 CHI Conference on Human Factors in Computing Systems*, May 11, 2024, https://doi.org/10.1145/3613904.3642521.

30. Shiona McCallum and Tom Singleton, "Period Trackers 'Coercing' Women into Sharing Risky Information," BBC, May 15, 2024, www.bbc.com/news/articles/cmj6j3d8xjjo.

31. User name deleted, "If this is related to recent USA events," comment on post by gullibletrout, r/selfhosted, Reddit.com, 2022, accessed December 2, 2024, www.reddit.com/r/selfhosted/comments/vk17oe/comment/idnrwcf/.

Chapter Five: Female Fat Is Fundamental

1. L. Shauntay Snell, "I'm a Plus-Size Runner and I Got Heckled at the NYC Marathon," *The Root*, November 30, 2017, www.theroot.com/im-a-plus-size-runner-and-i-got-heckled-at-the-nyc-mara-1820797012.

2. Home page of Latoya Shauntay Snell's website, accessed August 12, 2024, www.iamlshauntay.com/.

3. Latoya Shauntay Snell (@iamlshauntay), "My body size and disabilities don't make me inexperienced," Instagram, August 12, 2024, www.instagram.com/p/C-kvVn8OkxO/?hl=en.

4. Latoya Shauntay Snell, interview with author, August 16, 2024.

5. C. S. Larsen, "Equality for the Sexes in Human Evolution? Early Hominid Sexual Dimorphism and Implications for Mating Systems and Social Behavior," *Proceedings of the National Academy of Sciences* 100, no. 16 (2003), https://doi.org/10.1073/pnas.1633678100.

6. Scott Flynn et al., "How Much Fat Is Needed?," *Concepts of Fitness and Wellness*, 3rd ed., LibreTexts Medicine, accessed December 14, 2024, med.libretexts.org/Bookshelves/Health_and_Fitness/Concepts_of_Fitness_and_Wellness_(Flynn_et_al.)/06%3A_Body_Composition/6.03%3A_How_Much_Fat_is_Needed.

7. K. Karastergiou, S. R. Smith, A. S. Greenberg, and S. K. Fried, "Sex Differences in Human Adipose Tissues: The Biology of Pear Shape," *Biology of Sex Differences* 3, no. 1 (2012), https://doi.org/10.1186/2042-6410-3-13.

8. Herman Pontzer, Yosuke Yamada, Hiroyuki Sagayama, et al., "Daily Energy Expenditure Through the Human Life Course," *Science* 373, no. 6556 (2021), https://doi.org/10.1126/science.abe5017. "Metabolism" refers to the sum total of reactions that occur throughout the body within each cell, providing the body with energy. This energy gets used for vital processes and the synthesis of new cellular material. Every living organism uses its environment to survive by taking in nutrients that act as building blocks for movement, growth, development, and reproduction. All these processes are mediated by enzymes, which are proteins with specialized functions in anabolism and catabolism. The rate of energy production, called the basal metabolic rate, is affected by factors such as sex, race, exercise, diet, age, body mass, and body composition, and by diseases like sepsis or cancer. Arturo Sánchez López de Nava and Avais Raja, "Physiology, Metabolism," StatPearls Publishing, 2022, www.ncbi.nlm.nih.gov/books/NBK546690/.

9. Gina Kolata, "What We Think We Know About Metabolism May Be Wrong," *New York Times*, August 12, 2021, www.nytimes.com/2021/08/12/health/metabolism-weight-aging.html.

10. Stacy Sims, Zoom interview with author, February 2023.

11. Alexia Severson, "Testosterone Levels by Age," Healthline, November 7, 2018, www.healthline.com/health/low-testosterone/testosterone-levels-by-age#adolescence.

12. Lainbert-Adolf-Jacques Quetelet, "A Treatise on Man and the Development of His Faculties," *Obesity Research* 2, no. 1 (1994), https://doi.org/10.1002/j.1550-8528.1994.tb00047.x. This is a facsimile reprint; Quetelet published the original in 1842.

13. Iliya Gutin, "In BMI We Trust: Reframing the Body Mass Index as a Measure of Health," *Social Theory and Health* 16, no. 3 (2017), https://doi.org/10.1057/s41285-017-0055-0.

14. There is also evidence that carrying too much body fat can create excess estrogen, which might be at the root of the higher infertility rates among women categorized as "obese" by the BMI scale. Alternatively, the infertility could be due to another imbalance that results in higher body weight.

15. Gemma Breen and Allison Branley, "Belinda Could Barely Walk, but Was Told to Lose Weight for Life-Changing Knee Replacement Surgery," ABC News, September 25, 2023, www.abc.net.au/news/2023-09-26/bmi-impact-on-patient-outcomes/102898372.

16. "Mayo Clinic Offers Kidney Transplant to Patients with High BMI Who Were Previously Considered Ineligible," Mayo Clinic, February 5, 2021, www.mayoclinic.org/medical-professionals/transplant-medicine/news/mayo-clinic-offers-kidney-transplant-to-patients-with-high-bmi-who-were-previously-considered-ineligible/mac-20507217.

17. Jonathan Thomas Evans, Sofia Mouchti, Ashley W. Blom, et al., "Obesity and Revision Surgery, Mortality, and Patient-Reported Outcomes After Primary Knee Replacement Surgery in the National Joint Registry: A UK Cohort Study," *PLOS Medicine* 18, no. 7 (2021), https://doi.org/10.1371/journal.pmed.1003704.

18. Lisa M. Brownstone, Jaclyn DeRieux, Devin A. Kelly, Lanie J. Sumlin, and Jennifer L. Gaudiani, "Body Mass Index Requirements for Gender-Affirming Surgeries Are Not Empirically Based," *Transgender Health* 6, no. 3 (2021), https://doi.org/10.1089/trgh.2020.0068.

19. B. Hassan, C. R. Schuster, M. Ascha, G. Del Corral, B. Fischer, and F. Liang, "Association of High Body Mass Index with Postoperative Complications After Chest Masculinization Surgery," *Annals of Plastic Surgery* 92, no. 2 (2024), https://doi.org/10.1097/sap.0000000000003737.

20. Emma Betuel, "Scientists Discover Why Males and Females Store Fat in Different Body Parts," *Inverse*, January 22, 2019, www.inverse.com/article/52635-differences-in-body-fat-in-men-and-women-genetics.

21. Maartje Klaver, Daan van Velzen, Christel de Blok, et al., "Change in Visceral Fat and Total Body Fat and the Effect on Cardiometabolic Risk Factors During Transgender Hormone Therapy," *Journal of Clinical Endocrinology and Metabolism* 107, no. 1 (2021), https://doi.org/10.1210/clinem/dgab616.

22. Sharon Kirkey, "'BMI Is Trash': Why So Many Doctors Say It's Time to Ditch Body Mass Index," *Montreal Gazette*, October 24, 2022, montrealgazette.com/health/why-so-many-think-the-bmi-is-trash.

23. Sonia Anand, interview with author, July 30, 2024.

24. X. Zhang, N. Ma, Q. Lin, et al., "Body Roundness Index and All-Cause Mortality Among US Adults," *JAMA Network Open* 7, no. 6 (2024), https://doi.org/10.1001/jamanetworkopen.2024.15051.

25. Roni Caryn Rabin, "Time to Say Goodbye to the B.M.I.?," *New York Times*, September 6, 2024, www.nytimes.com/2024/09/06/health/body-roundness-index-bmi.html?searchResultPosition=1.

26. Ben Watts, "Ilona Maher: 2024 Bellport, New York," *Sports Illustrated*, August 28, 2024, https://swimsuit.si.com/swimsuit/model/ilona-maher-2024-si-swimsuit-photos.

27. A. J. Tomiyama, J. M. Hunger, J. Nguyen-Cuu, and C. Wells, "Misclassification of Cardiometabolic Health when Using Body Mass Index Categories in NHANES 2005–2012," *International Journal of Obesity* 40, no. 5 (2016), https://doi.org/10.1038/ijo.2016.17.

28. S. Bryn Austin and Tracy Richmond, "Opinion: It's Time to Retire BMI as a Clinical Metric," MedPage Today, October 19, 2022, www.medpagetoday.com/opinion/second-opinions/101296.

29. Sonia Anand, "Ethnicity and the Determinants of Cardiovascular Disease Among South Asians, Chinese and European Canadians" (PhD thesis, McMaster University, 2002), macsphere.mcmaster.ca/handle/11375/8884.

30. Anoop Misra, "Ethnic-Specific Criteria for Classification of Body Mass Index," *Diabetes Technology and Therapeutics* 17, no. 9 (2015), https://pmc.ncbi.nlm.nih.gov/articles/PMC4555479/; and Ji-Hee Haam, Bom Taeck Kim, Eun

Mi Kim, et al., "Diagnosis of Obesity," *Journal of Obesity and Metabolic Syndrome* 32, no. 2 (2023), https://doi.org/10.7570/jomes23031.

31. S. B. Heymsfield, C. M. Peterson, D. M. Thomas, M. Heo, and J. M. Schuna Jr., "Why Are There Race/Ethnic Differences in Adult Body Mass Index-Adiposity Relationships? A Quantitative Critical Review," *Obesity Reviews* 17, no. 3 (2015), https://doi.org/10.1111/obr.12358.

32. Kelly Jakubek, "AMA Adopts New Policy Clarifying Role of BMI as a Measure in Medicine," American Medical Association, June 14, 2023, www.ama-assn.org/press-center/press-releases/ama-adopts-new-policy-clarifying-role-bmi-measure-medicine.

33. Sadiya S. Khan, Hongyan Ning, John T. Wilkins, et al., "Association of Body Mass Index with Lifetime Risk of Cardiovascular Disease and Compression of Morbidity," *JAMA Cardiology* 3, no. 4 (2018), https://doi.org/10.1001/jamacardio.2018.0022.

34. F. Norheim, Y. Hasin-Brumshtein, L. Vergnes, et al., "Gene-by-Sex Interactions in Mitochondrial Functions and Cardio-Metabolic Traits," *Cell Metabolism* 29, no. 4 (2019), https://doi.org/10.1016/j.cmet.2018.12.013.

35. E. Di Angelantonio, S. N. Bhupathiraju, D. Wormser, et al., "Body-Mass Index and All-Cause Mortality: Individual-Participant-Data Meta-Analysis of 239 Prospective Studies in Four Continents," *The Lancet* 388, no. 10046 (2016), https://doi.org/10.1016/s0140-6736(16)30175-1.

36. Christina Strack, Gundula Behrens, Sabine Sag, et al., "Gender Differences in Cardiometabolic Health and Disease in a Cross-Sectional Observational Obesity Study," *Biology of Sex Differences* 13, no. 1 (2022), https://doi.org/10.1186/s13293-022-00416-4; Anne-Marie Lundsgaard and Bente Kiens, "Gender Differences in Skeletal Muscle Substrate Metabolism: Molecular Mechanisms and Insulin Sensitivity," *Frontiers in Endocrinology* 5 (2014), https://doi.org/10.3389/fendo.2014.00195.

37. Cheehoon Ahn, Tao Zhang, Gayoung Yang, et al., "Years of Endurance Exercise Training Remodel Abdominal Subcutaneous Adipose Tissue in Adults with Overweight or Obesity," *Nature Metabolism* 6 (2024), https://doi.org/10.1038/s42255-024-01103-x.

38. Kaitlin Sullivan, "Belly Fat Isn't Always Bad: New Study Finds How to Improve Healthy Belly Fat," NBC News, September 10, 2024, www.nbcnews.com/health/health-news/regular-exercise-may-lead-healthier-belly-fat-study-finds-rcna170213.

39. Georgina Krebs, Bruce R. Clark, Tamsin J. Ford, and Argyris Stringaris, "Epidemiology of Body Dysmorphic Disorder and Appearance Preoccupation in Youth: Prevalence, Comorbidity and Psychosocial Impairment," *Journal of the American Academy of Child and Adolescent Psychiatry*, March 1, 2024, https://doi.org/10.1016/j.jaac.2024.01.017.

40. Matthew Schneier, "Ozempic Is Changing the Definition of Being Thin," *The Cut*, February 27, 2023, www.thecut.com/article/weight-loss-ozempic.html/.

41. R. Mukhopadhyay, "Essential Fatty Acids: The Work of George and Mildred Burr," *Journal of Biological Chemistry* 287, no. 42 (2012), https://doi.org/10.1074/jbc.o112.000005.

42. Sylvia Tara, *The Secret Life of Fat: The Science Behind the Body's Least Understood Organ and What It Means for You* (W. W. Norton, 2016), 25.

43. Chia-Yu Chang, Der-Shin Ke, and Jen-Yin Chen, "Essential Fatty Acids and Human Brain," *Acta Neurologica Taiwanica* 18, no. 4 (2009), pubmed.ncbi.nlm.nih.gov/20329590/.

44. "Anorexia Statistics: Gender, Race and Socioeconomics," The Bulimia Project, last updated September 22, 2022, bulimia.com/anorexia/statistics/.

45. Sidney Taiko Sheehan, "Groundbreaking Study Shows Substantial Differences in Brain Structure in People with Anorexia," press release, Keck School of Medicine of USC, June 14, 2022, keck.usc.edu/groundbreaking-study-shows-substantial-differences-in-brain-structure-in-people-with-anorexia/.

46. Nawab Qizilbash, John Gregson, Michelle E. Johnson, et al., "BMI and Risk of Dementia in Two Million People over Two Decades: A Retrospective Cohort Study," *The Lancet Diabetes and Endocrinology* 3, no. 6 (2015), https://doi.org/10.1016/s2213-8587(15)00033-9.

47. Tara, *Secret Life of Fat*, 25, 46–51.

48. Rose E. Frisch and Janet W. McArthur, "Menstrual Cycles: Fatness as a Determinant of Minimum Weight for Height Necessary for Their Maintenance or Onset," *Science* 185, no. 4155 (1974), www.science.org/doi/10.1126/science.185.4155.949.

49. Rose E. Frisch, Grace Wyshack, and Larry Vincent, "Delayed Menarche and Amenorrhea in Ballet Dancers," *New England Journal of Medicine* 303, no. 1 (1980), https://doi.org/10.1056/nejm198007033030105.

Chapter Six: Endurance, the Female Superpower

1. Hansjörg Ransmayr, "Ice Swimming Extreme in the Glacier Ice Palace," *Alpine Swimming*, April 7, 2018, www.alpine-swimming.com/en/en-swims/ice-swimming-extreme-in-the-ice-palace-of-the-tux-glacier/.

2. Jamie Monahan, interview with author, November 10, 2023.

3. "About Jaimie," Jaimie Monahan's website, accessed October 4, 2024, www.jaimiemonahan.com/aboutjaimie.

4. "Jaimie Monahan Antarctica Ice Mile Swim with Penguins," Jaimie Monahan, video, posted April 9, 2018, YouTube, www.youtube.com/watch?v=D6FnvG1Ll7w.

5. She noted that during her shorter ice swims, which max out at around thirty minutes, she has to pay much closer attention to her bodily sensations than during a multihour cold-water swim, because it's easy to get too cold without realizing it. Those swims are less meditative. "You have to be so incredibly focused on monitoring yourself, like every second, it's just a completely different animal," Monahan told me. To ensure she isn't growing dangerously cold, she

keeps an eye on the skin of her arm, noticing if it starts to turn blue, and also tests her hands.

6. Swimming instruction in Australia is taught in school, and there are free community pools in almost every town. Coastal areas often have beautiful oceanside saltwater rock pools, which is where I swim.

7. B. Knechtle, A. A. Dalamitros, T. M. Barbosa, C. V. Sousa, T. Rosemann, and P. T. Nikolaidis, "Sex Differences in Swimming Disciplines—Can Women Outperform Men in Swimming?" *International Journal of Environmental Research and Public Health* 17, no. 10 (2020), https://doi.org/10.3390/ijerph17103651.

8. Beat Knechtle, Thomas Rosemann, Romauld Lepers, and Christoph Alexander Rüst, "Women Outperform Men in Ultradistance Swimming: The Manhattan Island Marathon Swim from 1983 to 2013," *International Journal of Sports Physiology and Performance* 9, no. 6 (2014), https://doi.org/10.1123/ijspp.2013-0375.

9. Beat Knechtle, Thomas Rosemann, and Christoph Alexander Rüst, "Women Cross the 'Catalina Channel' Faster than Men," *SpringerPlus* 4, no. 1 (2015), https://doi.org/10.1186/s40064-015-1086-4.

10. Jennifer Pharr Davis, interview with author, April 2019.

11. Leigh Cowart, *Hurts So Good: The Science and Culture of Pain on Purpose* (PublicAffairs, 2021), 165.

12. V. Zarulli, J. A. Barthold Jones, A. Oksuzyan, and J. W. Vaupel, "Women Live Longer than Men Even During Severe Famines and Epidemics," *Proceedings of the National Academy of Sciences* 115, no. 4 (2018), https://doi.org/10.1073/pnas.1701535115.

13. Megan Roche, interview with author, September 29, 2023.

14. Michelle A. Kominiarek and Priya Rajan, "Nutrition Recommendations in Pregnancy and Lactation," *Medical Clinics of North America* 100, no. 6 (2016), https://doi.org/10.1016/j.mcna.2016.06.004.

15. C. Thurber, L. R. Dugas, C. Ocobock, B. Carlson, J. R. Speakman, and H. Pontzer, "Extreme Events Reveal an Alimentary Limit on Sustained Maximal Human Energy Expenditure," *Science Advances* 5, no. 6 (2019), https://doi.org/10.1126/sciadv.aaw0341.

16. Cara Ocobock, interview with author, October 27, 2023.

17. Gina Kolata, "Training Through Pregnancy to Be Marathon's Fastest Mom," *New York Times*, November 3, 2007, www.nytimes.com/2007/11/03/sports/othersports/03runner.html.

18. Lisa Salmon, "Paula Radcliffe: Getting Fit After Having a Baby," *Irish Independent*, April 29, 2011, www.independent.ie/life/family/mothers-babies/paula-radcliffe-getting-fit-after-having-a-baby/26727894.html.

19. Sophie Haydock, "Experience: I Went into Labour After Running a Marathon," *The Guardian*, June 16, 2017, www.theguardian.com/lifeandstyle/2017/jun/16/experience-went-into-labour-after-running-marathon.

20. Gerald S. Zavorsky and Lawrence D. Longo, "Viewpoint: Are There Valid Concerns for Completing a Marathon at 39 Weeks of Pregnancy?,"

Journal of Applied Physiology 113, no. 7 (2012), https://doi.org/10.1152/japplphysiol.01426.2011.

21. Francine Darroch, Audrey R. Giles, and Roisin McGettigan-Dumas, "Elite Female Distance Runners and Advice During Pregnancy: Sources, Content, and Trust," *Women in Sport and Physical Activity Journal* 24, no. 2 (2016), https://doi.org/10.1123/wspaj.2015-0040.

22. A. S. Tenforde, K. E. S. Toth, E. Langen, M. Fredericson, and K. L. Sainani, "Running Habits of Competitive Runners During Pregnancy and Breastfeeding," *Sports Health: A Multidisciplinary Approach* 7, no. 2 (2014), https://doi.org/10.1177/1941738114549542.

23. Ainslie K. Kehler and Katie M. Heinrich, "A Selective Review of Prenatal Exercise Guidelines Since the 1950s Until Present: Written for Women, Health Care Professionals, and Female Athletes," *Women and Birth* 28, no. 4 (2015), https://doi.org/10.1016/j.wombi.2015.07.004.

24. Brendan Wolfe, "Indentured Servants in Colonial Virginia," *Encyclopedia Virginia*, December 7, 2020, encyclopediavirginia.org/entries/indentured-servants-in-colonial-virginia/.

25. "Maternity Leave by Country 2020," World Population Review, 2022, worldpopulationreview.com/country-rankings/maternity-leave-by-country.

26. Lene A. H. Haakstad and Kari Bø, "The Marathon of Labour: Does Regular Exercise Training Influence Course of Labour and Mode of Delivery?," *European Journal of Obstetrics and Gynecology and Reproductive Biology* 251 (2020), https://doi.org/10.1016/j.ejogrb.2020.05.014.

27. "Exercise in Pregnancy," National Health Service—UK, March 15, 2023, www.nhs.uk/pregnancy/keeping-well/exercise/.

28. Benedikt Hallgrimsson, Rebecca M. Green, and David C. Katz, "The Developmental-Genetics of Canalization," *Seminars in Cell and Developmental Biology* 88 (2019), https://doi.org/10.1016/j.semcdb.2018.05.019.

29. Amanda Hale, interview with author, May 31, 2023.

30. Erin A. McKenney, Amanda R. Hale, Janaiya Anderson, Roxanne Larsen, Colleen Grant, and Robert R. Dunn, "Hidden Diversity: Comparative Functional Morphology of Humans and Other Species." *PeerJ* 11 (2023), https://doi.org/10.7717/peerj.15148. All the cadavers for their research were donated through the anatomical gifts program at the Duke University School of Medicine.

31. Nayoung Kim, "Sex Difference of Gut Microbiota," *Sex/Gender-Specific Medicine in the Gastrointestinal Diseases*, 2022, https://doi.org/10.1007/978-981-19-0120-1_22.

32. Aibo Gao, Junlei Su, Ruixin Liu, et al., "Sexual Dimorphism in Glucose Metabolism Is Shaped by Androgen-Driven Gut Microbiome," *Nature Communications* 12, no. 1 (2021), https://doi.org/10.1038/s41467-021-27187-7.

33. In this case, mice's similarity to humans in terms of insulin resistance, sex-based differences, hormone impacts, and other aspects of physiology means they are a reasonable analog, though as pointed out elsewhere, this is not always the case with animal models.

34. Amy Wallis, Henry Butt, Michelle Ball, Donald P. Lewis, and Dorothy Bruck, "Support for the Microgenderome: Associations in a Human Clinical Population," *Scientific Reports* 6, no. 1 (2016), https://doi.org/10.1038/srep19171. Researchers have pointed out that since gender is a cultural construct and "sex" is the biological term, it should be called the microsexome instead.

35. Matthieu Clauss, "Interplay Between Exercise and Gut Microbiome in the Context of Human Health and Performance," *Frontiers in Nutrition* 8 (2021), https://doi.org/10.3389/fnut.2021.637010.

36. Tindaro Bongiovanni, Marilyn Ong Li Yin, and Liam Heaney, "The Athlete and Gut Microbiome: Short-Chain Fatty Acids as Potential Ergogenic Aids for Exercise and Training," *International Journal of Sports Medicine* 42, no. 13 (2021), https://doi.org/10.1055/a-1524-2095.

37. Taylor Dutch, "2,189 Miles, 40 Days, and 3 Showers: How Tara Dower Destroyed the Appalachian Trail Speed Record," *Runner's World*, September 24, 2024, www.runnersworld.com/news/a62330229/tara-dower-appalachian-trail-record/.

Chapter Seven: Running on Empty

1. By the time I was in junior high, this was no longer allowed; yes, the 1980s really were a time when kids busted their spleens sledding—as my friend Chrissy did—and nobody thought that meant what we did was too dangerous. We all felt bad for her and then proceeded to create even bigger jumps for our sleds. I'm now one of those people who say, "It was a different time" and mean it.

2. Chris Heath, "Fiona: The Caged Bird Sings," *Rolling Stone*, January 22, 1998, www.rollingstone.com/feature/fiona-the-caged-bird-sings-244221/.

3. Casey Johnston, "Nothing Tastes as Good as Skinny Feels," *Ask a Swole Woman*, September 12, 2017, askaswolewoman.com/ask-a-swole-woman-archive/2017/9/12/does-nothing-taste-as-good-as-skinny-feels.

4. Sebastian A. Baldauf, Leif Engqvist, and Franz J. Weissing, "Diversifying Evolution of Competitiveness," *Nature Communications* 5, no. 1 (2014), https://doi.org/10.1038/ncomms6233.

5. J. H. Conviser, A. S. Tierney, and R. Nickols, "Essentials for Best Practice: Treatment Approaches for Athletes with Eating Disorders," *Journal of Clinical Sport Psychology* 12, no. 4 (2018), https://doi.org/10.1123/jcsp.2018-0013.

6. Michael Fredericson, Megan Roche, Michelle T. Barrack, et al., "Healthy Runner Project: A 7-Year, Multisite Nutrition Education Intervention to Reduce Bone Stress Injury Incidence in Collegiate Distance Runners," *BMJ Open Sport and Exercise Medicine* 9, no. 2 (2023), https://doi.org/10.1136/bmjsem-2023-001545.

7. Bailey Kowalczyk, interview with author, October 24, 2023.

8. Gavi Klein and Audrey Gibbs, "Tools of the Patriarchy: Diet Culture and How We All Perpetuate the Stigma," *Ms.*, July 16, 2020, https://msmagazine.com/2020/07/16/tools-of-the-patriarchy-diet-culture-and-how-we-all-perpetuate-the-stigma/.

9. Wikipedia, "100 Metres," October 25, 2020, en.wikipedia.org/wiki/100_metres.

10. Christine Yu, *Up to Speed* (Penguin, 2023), 90–91.

11. "Year by Year," Western States Endurance Run, accessed December 6, 2024, www.wser.org/history-year-by-year/.

12. Tunnel Hill 100, 2017 results, UltraSignup, accessed December 6, 2024, ultrasignup.com/results_event.aspx?did=43104.

13. Sophie Power, "Are Women Better Endurance Athletes than Men?," *Runner's World*, March 10, 2022, www.runnersworld.com/uk/news/a39393890/jasmin-paris-barkley-marathons/.

14. Owen Clarke, "Climber Babsi Zangerl Just Made History on El Capitan," Outside Online, November 26, 2024, www.outsideonline.com/outdoor-adventure/climbing/babsi-zangerl-flash-el-cap/.

15. Stephanie Case, "Ultrarunning Has a Gender Problem. Talking Won't Fix It," Outside Online, May 25, 2018, www.outsideonline.com/culture/opinion/ultrarunning-has-gender-problem/.

16. Scott Douglas Jacobsen, "Redefining Gender Equality: The Hidden Value of Unpaid Care Work," *International Policy Digest*, November 7, 2024, intpolicydigest.org/redefining-gender-equality-the-hidden-value-of-unpaid-care-work/.

17. "Factors Influencing Girls' Participation in Sports," Women's Sports Foundation, 2009, www.womenssportsfoundation.org/support-us/do-you-know-the-factors-influencing-girls-participation-in-sports/.

18. Case, "Ultrarunning Has a Gender Problem."

19. John Branch and Alex Goodlett, "Female Mountain Bikers at Rampage Backflip into Equality," *New York Times*, October 16, 2024, www.nytimes.com/2024/10/16/us/red-bull-rampage-women-mountain-biking.html.

20. S. K. Hunter, S. S. Angadi, A. Bhargava, et al., "The Biological Basis of Sex Differences in Athletic Performance: Consensus Statement for the American College of Sports Medicine," *Medicine and Science in Sports and Exercise* 55, no. 12 (2023), https://doi.org/10.1249/MSS.0000000000003300.

21. This is a well-known phrase used by men when they are beaten in a sport by a woman. As Caitlyn Pilkington wrote in a piece where she argues for taking pride in the expression, "When that guy yells at his buddy up ahead that you're 'about to get chicked, bro'—because that's happened to me!—I prefer to chuckle to myself and think, *Damn straight. Been chicking since the fourth grade*." Caitlyn Pilkington, "Why I'm Not Upset About the Term 'Getting Chicked,'" *Women's Running*, March 17, 2017, https://www.womensrunning.com/culture/upset-term-getting-chicked/.

Chapter Eight: Female Pain and Disempowerment

1. Turns out I had cracked a rear molar into several pieces, which earned me my first cavities a year later, and my right eardrum was temporarily damaged, but I was otherwise OK. The bone-deep slit in my chin needed to be well cleaned and patched back together. I opted for adhesive strips instead of stitches since the latter

were more costly and I was uninsured. The doctor, who had worked as a medic in the US Army in Afghanistan, told me cheerily that my cut looked like shrapnel damage from an improvised explosive device, and she knew just how to fix it up sans stitches. Based on her data re: flying things hitting faces, she figured I was very, very lucky that the mule's hoof hadn't hit me in my mouth, nose, eyes, or forehead, because otherwise I could have lost an eye or died from a brain hemorrhage out in the mountains. I told her I do tend to be lucky. My only car accident involved driving off a cliff and landing in a river outside Telluride, Colorado, where I exited the vehicle without a scratch, and when I was bitten by a dog in the face as a kid, it missed my eye by a millimeter. Smiling, she looked into my eyes and said, "You either have someone watching over you or really good reflexes."

2. P. Yilmaz, M. Diers, S. Diener, M. Rance, M. Wessa, and H. Flor, "Brain Correlates of Stress-Induced Analgesia," *Pain* 151, no. 2 (2010), https://doi.org/10.1016/j.pain.2010.08.016.

3. Tor Wager, interview with author, March 16, 2023.

4. Katerina Zorina-Lichtenwalter, Carmen I. Bango, Lukas Van Oudenhove, et al., "Identification and Characterization of Genetic Risk Shared Across 24 Chronic Pain Conditions in the UK Biobank," MedRxiv (Cold Spring Harbor Laboratory), June 30, 2022, https://doi.org/10.1101/2022.06.28.22277025.

5. Vincent T. Martin, "Ovarian Hormones and Pain Response: A Review of Clinical and Basic Science Studies," *Gender Medicine* 6 (2009), https://doi.org/10.1016/j.genm.2009.03.006.

6. Jeffrey J. Sherman and Linda LeResche, "Does Experimental Pain Response Vary Across the Menstrual Cycle? A Methodological Review," *American Journal of Physiology-Regulatory, Integrative and Comparative Physiology* 291, no. 2 (2006), https://doi.org/10.1152/ajpregu.00920.2005.

7. "Sexism in Mouse Research Can Lead to Medical Harm to Women, Scientists Warn," CBC, April 21, 2016, www.cbc.ca/news/science/mouse-sex-studies-1.3545486.

8. David Reby, Florence Levréro, Erik Gustafsson, and Nicolas Mathevon, "Sex Stereotypes Influence Adults' Perception of Babies' Cries," *BMC Psychology* 4, no. 1 (2016), https://doi.org/10.1186/s40359-016-0123-6.

9. Diane E. Hoffmann and Anita J. Tarzian, "The Girl Who Cried Pain: A Bias Against Women in the Treatment of Pain," *Journal of Law, Medicine and Ethics* 28, no. 4 (2001), https://doi.org/10.1111/j.1748-720x.2001.tb00037.x.

10. "Doctors Finally Confirm Period Pain Can Be as Painful as a Heart Attack," UCL News, March 1, 2018, www.ucl.ac.uk/news/headlines/2018/mar/doctors-finally-confirm-period-pain-can-be-painful-heart-attack.

11. Olivia Goldhill, "Period Pain Can Be 'Almost as Bad as a Heart Attack.' Why Aren't We Researching How to Treat It?," *Quartz*, February 15, 2016, https://qz.com/611774/period-pain-can-be-as-bad-as-a-heart-attack-so-why-arent-we-researching-how-to-treat-it.

12. Pallavi Latthe, Rita Champaneria, and Khalid Khan, "Dysmenorrhea," *American Family Physician* 85, no. 4 (2012), www.aafp.org/pubs/afp/issues/2012/0215/p386.html.

13. Camran Nezhat, Farr Nezhat, and Ceana Nezhat, "Endometriosis: Ancient Disease, Ancient Treatments," *Fertility and Sterility* 98, no. 6 (2012), https://doi.org/10.1016/j.fertnstert.2012.08.001.

14. Hugh S. Taylor, Alexander M. Kotlyar, and Valerie A. Flores, "Endometriosis Is a Chronic Systemic Disease: Clinical Challenges and Novel Innovations," *The Lancet* 397, no. 10276 (2021), https://doi.org/10.1016/s0140-6736(21)00389-5.

15. Maya Dusenbery, *Doing Harm: The Truth About How Bad Medicine and Lazy Science Leave Women Dismissed, Misdiagnosed, and Sick* (HarperOne, 2018).

16. "Pelvic Inflammatory Disease (PID)—Symptoms and Causes," Mayo Clinic, April 30, 2022, www.mayoclinic.org/diseases-conditions/pelvic-inflammatory-disease/symptoms-causes/syc-20352594.

17. Brigitte Leeners, Fabia Damaso, Nicole Ochsenbein-Kölble, and Cindy Farquhar, "The Effect of Pregnancy on Endometriosis—Facts or Fiction?," *Human Reproduction Update* 24, no. 3 (2018), https://doi.org/10.1093/humupd/dmy004.

18. Mahesh J. Fuldeore and Ahmed M. Soliman, "Prevalence and Symptomatic Burden of Diagnosed Endometriosis in the United States: National Estimates from a Cross-Sectional Survey of 59,411 Women," *Gynecologic and Obstetric Investigation* 82, no. 5 (2017), https://doi.org/10.1159/000452660.

19. Dusenbery, *Doing Harm*.

20. Rachel E. Gross, "They Call It a 'Women's Disease.' She Wants to Redefine It," *New York Times*, April 27, 2021, www.nytimes.com/2021/04/27/health/endometriosis-griffith-uterus.html.

21. Bob Cole, "'The Very Future of Our Nations': How Aboriginal Midwifery Represents a Practical Model for Utilization of Traditional Knowledge," *Journal of Integrated Studies* 10, no. 1 (2018), https://jis.athabascau.ca/index.php/jis/article/view/224.

22. Maurine R. Musie, Rafiat A. Anokwuru, Roinah N. Ngunyulu, and Sanele Lukhele, "African Indigenous Beliefs and Practices During Pregnancy, Birth and After Birth," in *Working with Indigenous Knowledge: Strategies for Health Professionals*, ed. Fhumulani M. Mulaudzi and Rachel T. Lebese (Aiosis Scholarly Books, 2022), 85–106.

23. Roger Staples, "Victoria's Secret: Chloroform and the Acceptability of Analgesia for Birth," *O&G Magazine*, August 2, 2017, www.ogmagazine.org.au/18/1-18/victorias-secret-chloroform-acceptability-analgesia-birth/.

24. Wikipedia, "Charles Delucena Meigs," October 21, 2020, https://en.wikipedia.org/wiki/Charles_Delucena_Meigs.

25. Ellen Barry, "Chloroform in Childbirth? Yes, Please, the Queen Said," *New York Times*, May 6, 2019, www.nytimes.com/2019/05/06/world/europe/uk-royal-births-labor.html.

26. "Alabama artist works to correct historical narrative around beginnings of gynecology," *PBS NewsHour*, video, posted February 27, 2023, YouTube, www.youtube.com/watch?v=_9l3Ptx4DpY.

27. Kelly M. Hoffman, Sophie Trawalter, Jordan R. Axt, and M. Norman Oliver, "Racial Bias in Pain Assessment and Treatment Recommendations,

and False Beliefs About Biological Differences Between Blacks and Whites," *Proceedings of the National Academy of Sciences* 113, no. 16 (2016), https://doi.org/10.1073/pnas.1516047113.

28. Gillian Clouser, "Blackness, Maternal Mortality, and Prenatal Birth: The Legacy of Slavery," Yale School of Medicine, May 16, 2022. https://medicine.yale.edu/news-article/blackness-maternal-mortality-and-prenatal-birth-the-legacy-of-slavery/.

29. "Country Comparisons: Maternal Mortality Ratio," The World Factbook (CIA), 2020, www.cia.gov/the-world-factbook/field/maternal-mortality-ratio/country-comparison/; Women and Equalities Committee, *Black Maternal Health* (UK Parliament, 2022), publications.parliament.uk/pa/cm5803/cmselect/cmwomeq/94/report.html.

30. "Infant Mortality and African Americans," Office of Minority Health, US Department of Heath and Human Services, last edited September 21, 2023, minorityhealth.hhs.gov/infant-mortality-and-african-americans.

31. Haben Debessai, interview with author, September 10, 2024.

32. Lee Boomer, "Life Story: Anarcha, Betsy, and Lucy," Women and the American Story, accessed October 27, 2024, https://wams.nyhistory.org/a-nation-divided/antebellum/anarcha-betsy-lucy/.

33. "Anarcha, Lucy, and Betsey: The Mothers of Gynecology," accessed October 27, 2024, www.anarchalucybetsey.org/.

34. Freeda En, "Sasheer Zamata on Periods," Facebook video, January 28, 2023, www.facebook.com/watch/?v=856861642269751.

35. Dusenbery, *Doing Harm*, 183.

36. "Why Do We Still Not Know What Causes PMS?," ResearchGate, August 12, 2016, www.researchgate.net/blog/why-do-we-still-not-know-what-causes-pms.

37. Rae Nudson, "Gynecology Has a Pain Problem," *The Cut*, June 1, 2022, www.thecut.com/2022/06/pain-in-gynecology-practice-exams.html.

38. sumsumc13, "I am seeing stars," TikTok, March 27, 2023, www.tiktok.com/@sumsumc13/video/7215361019038879022?lang=en; Summer Cartwright, "I Faced Severe Pain to Get an IUD for the Sake of My Relationship—So, Yes, I Deserve a Gift from My Partner," Well+Good, June 12, 2023, www.wellandgood.com/gift-from-partner-for-getting-iud/.

39. shortstakofpancakes, comment on post by 2ndaccountquestions, "Colposcopy pain," r/TwoXChromosomes, Reddit, 2020, www.reddit.com/r/TwoXChromosomes/comments/inwzlv/colposcopy_pain/.

40. PowerfulEast6585, "first colposcopy today. cried the entire time," r/TwoXChromosomes, Reddit, 2021, www.reddit.com/r/TwoXChromosomes/comments/ue8dgc/first_colposcopy_today_cried_the_entire_time/.

41. Melinda Wenner Moyer, "Women Are Calling Out 'Medical Gaslighting,'" *New York Times*, March 28, 2022, www.nytimes.com/2022/03/28/well/live/gaslighting-doctors-patients-health.html.

42. "Everything You Need to Know About Colposcopy, If It Hurts and Why You Need It," Nebraska Medicine, February 16, 2022, www.nebraskamed.com

/womens-health/gynecology/everything-you-need-to-know-about-colposcopy-if-it-hurts-and-why-you-need-it.

43. "Colposcopy: Biopsy, Purpose, Procedure, Risk & Results," Cleveland Clinic, May 19, 2022, my.clevelandclinic.org/health/diagnostics/4044-colposcopy; "What Is a Colposcopy," Planned Parenthood, accessed October 27, 2024, www.plannedparenthood.org/learn/cancer/cervical-cancer/what-colposcopy.

44. Nighat Arif, email correspondence with author, September 13, 2024.

45. J. H. Pope, T. P. Aufderheide, R. Ruthazer, et al., "Missed Diagnoses of Acute Cardiac Ischemia in the Emergency Department," *New England Journal of Medicine* 342, no. 16 (2000), https://doi.org/10.1056/NEJM200004203421603.

46. Sophia Antipolis, "Heart Attack Diagnosis Missed in Women More Often than in Men," press release, European Society of Cardiology, March 12, 2021, www.escardio.org/The-ESC/Press-Office/Press-releases/Heart-attack-diagnosis-missed-in-women-more-often-than-in-men.

47. Hoffmann and Tarzian, "The Girl Who Cried Pain."

48. Esther H. Chen, Frances S. Shofer, Anthony J. Dean, et al., "Gender Disparity in Analgesic Treatment of Emergency Department Patients with Acute Abdominal Pain," *Academic Emergency Medicine* 15, no. 5 (2008), https://doi.org/10.1111/j.1553-2712.2008.00100.x.

49. Raoul Daoust, Jean Paquet, Gilles Lavigne, Karine Sanogo, and Jean-Marc Chauny, "Senior Patients with Moderate to Severe Pain Wait Longer for Analgesic Medication in EDs," *American Journal of Emergency Medicine* 32, no. 4 (2014), https://doi.org/10.1016/j.ajem.2013.12.012.

50. Hoffmann, and Tarzia, "The Girl Who Cried Pain."

51. David E. Newman-Toker, Ernest Moy, Ernest Valente, Rosanna Coffey, and Anika L. Hines, "Missed Diagnosis of Stroke in the Emergency Department: A Cross-Sectional Analysis of a Large Population-Based Sample," *Diagnosis* 1, no. 2 (2014), https://doi.org/10.1515/dx-2013-0038; Benjamin Bleicken, Manfred Ventz, Marcus Quinkler, and Stefanie Hahner, "Delayed Diagnosis of Adrenal Insufficiency Is Common: A Cross-Sectional Study in 216 Patients," *American Journal of the Medical Sciences* 339, no. 6 (2010), https://doi.org/10.1097/maj.0b013e3181db6b7a; Nafees U. Din, Obioha C. Ukoumunne, Greg Rubin, et al., "Age and Gender Variations in Cancer Diagnostic Intervals in 15 Cancers: Analysis of Data from the UK Clinical Practice Research Datalink," ed. Masaru Katoh, *PLOS One* 10, no. 5 (2015), https://doi.org/10.1371/journal.pone.0127717; A. H. E. M. Maas and Y. E. A. Appelman, "Gender Differences in Coronary Heart Disease," *Netherlands Heart Journal* 18, no. 12 (2010), https://doi.org/10.1007/s12471-010-0841-y.

52. Ana Mikolić, David van Klaveren, Joost Oude Groeniger, et al., "Differences Between Men and Women in Treatment and Outcome After Traumatic Brain Injury," *Journal of Neurotrauma* 38, no. 2 (2020), https://doi.org/10.1089/neu.2020.7228.

53. Michael Sun, Tomasz Oliwa, Monica E. Peek, and Elizabeth L. Tung, "Negative Patient Descriptors: Documenting Racial Bias in the Electronic Health Record," *Health Affairs* 41, no. 2 (2022), https://doi.org/10.1377/hlthaff.2021.01423.

54. Alberto Giovanni Leone, Dario Trapani, Matthew B. Schabath, et al., "Cancer in Transgender and Gender-Diverse Persons," *JAMA Oncology* 9, no. 4 (2023), https://doi.org/10.1001/jamaoncol.2022.7173.

55. "FSRH Statement: Pain Associated with Insertion of Intrauterine Contraception," FSRH Clinical Effectiveness Unit, June 30, 2021, www.fsrh.org/Common/Uploaded%20files/documents/fsrh-clinical-statement-pain-associated-with-insertion-of-iut-jul-2021.pdf.

56. K. M. Curtis, A. T. Nguyen, N. K. Tepper, et al., "U.S. Selected Practice Recommendations for Contraceptive Use, 2024," *MMWR Recommendations and Reports* 73, no. 3 (2024), https://doi.org/10.15585/mmwr.rr7303a1.

57. Sheila K. Mody, John Paul Farala, Berenice Jimenez, Moena Nishikawa, and Lynn L. Ngo, "Paracervical Block for Intrauterine Device Placement Among Nulliparous Women," *Obstetrics and Gynecology* 132, no. 3 (2018), https://doi.org/10.1097/aog.0000000000002790.

Chapter Nine: Pain and Perseverance in Athletics

1. Bethany Brookshire, interview with author, October 30, 2023.

2. The Plastic Flamingo (plasticflamingo.ink), "Back again asking our tattoo artists what their thoughts are," TikTok.com, August 27, 2023, www.tiktok.com/@plasticflamingo.ink/video/7272111318176451883.

3. Damnit Wooddy (damnitwooddy), "Who do you think sits better?," TikTok, August, 19, 2023, www.tiktok.com/@damnitwooddy/video/7269163609811373354.

4. Joanna Witkoś and Magdalena Hartman-Petrycka, "Gender Differences in Subjective Pain Perception During and After Tattooing," *International Journal of Environmental Research and Public Health* 17, no. 24 (2020), https://doi.org/10.3390/ijerph17249466.

5. E. J. Bartley and R. B. Fillingim, "Sex Differences in Pain: A Brief Review of Clinical and Experimental Findings," *British Journal of Anaesthesia* 111, no. 1 (2013), https://doi.org/10.1093/bja/aet127.

6. Tor Wager, interview with author, March 16, 2023.

7. Bonnie Schwartz, interview with author, July 21, 2023.

8. "Guys Try Period Pain! MTV Style," MTV UK, video, posted March 7, 2019, YouTube, www.youtube.com/watch?v=Nl9LaisncLk.

9. Amanda MacMillan, "Why Men Are Much Worse at Being Sick than Women," *Time*, February 27, 2017, https://time.com/4683864/men-sick-cold-flu/.

10. Tor D. Wager, Lauren Y. Atlas, Martin A. Lindquist, Mathieu Roy, Choong-Wan Woo, and Ethan Kross, "An FMRI-Based Neurologic Signature of Physical Pain," *New England Journal of Medicine* 368, no. 15 (2013), https://doi.org/10.1056/nejmoa1204471.

11. The same study showed that, according to the brain scans, the pain of social rejection can feel just as bad as physically inflicted pain.

12. Loren J. Martin, Erinn L. Acland, Chulmin Cho, et al., "Male-Specific Conditioned Pain Hypersensitivity in Mice and Humans," *Current Biology* 29, no. 2 (2019), https://doi.org/10.1016/j.cub.2018.11.030.

13. "Men and Women Remember Pain Differently," news release, McGill Newsroom, January 10, 2019, www.mcgill.ca/newsroom/channels/news/men-and-women-remember-pain-differently-293050.

14. Yoon Frederiksen, Mimi Mehlsen, Signe Maria, Robert Zachariae, and Hans Jakob Ingerslev, "Predictors of Pain During Oocyte Retrieval," *Journal of Psychosomatic Obstetrics and Gynecology* 38, no. 1 (2017), https://doi.org/10.1080/0167482x.2016.1235558.

15. Nirit Geva, Sari Golan, Lior Pinchas, and Ruth Defrin, "Sex Effects in the Interaction of Acute Stress and Pain Perception," *Pain* 164, no. 3 (2023), https://doi.org/10.1097/j.pain.0000000000002743.

16. Carol Torgan, "Bacteria Trigger Nerve Cells to Cause Pain," National Institutes of Health, May 14, 2015, www.nih.gov/news-events/nih-research-matters/bacteria-trigger-nerve-cells-cause-pain; Aakanksha Jain, Sara Hakim, and Clifford J. Woolf, "Immune Drivers of Physiological and Pathological Pain," *Journal of Experimental Medicine* 221, no. 5 (2024), https://doi.org/10.1084/jem.20221687.

17. Robert E. Sorge, Josiane C. S. Mapplebeck, Sarah Rosen, et al., "Different Immune Cells Mediate Mechanical Pain Hypersensitivity in Male and Female Mice," *Nature Neuroscience* 18, no. 8 (2015), https://doi.org/10.1038/nn.4053.

18. "Pain for Women, Pain for Men," *Proto Magazine*, August 10, 2017, protomag.com/neurology/pain-women-pain-men/.

19. Hannah E. Braithwaite, Thomas Payne, Nicholas Duce, et al., "Impact of Female Sex on Anaesthetic Awareness, Depth, and Emergence: A Systematic Review and Meta-Analysis," *British Journal of Anaesthesia* 131, no. 3 (2023), https://doi.org/10.1016/j.bja.2023.06.042.

20. M. Rademaker, "Do Women Have More Adverse Drug Reactions?," *American Journal of Clinical Dermatology* 2, no. 6 (2001), https://doi.org/10.2165/00128071-200102060-00001.

21. A. Farkouh, C. Baumgärtel, R. Gottardi, M. Hemetsberger, M. Czejka, and A. Kautzky-Willer, "Sex-Related Differences in Drugs with Anti-Inflammatory Properties," *Journal of Clinical Medicine* 10, no. 7 (2021), https://doi.org/10.3390/jcm10071441.

22. Brooke Deal, Laura M. Reynolds, Charles Patterson, Jelena M. Janjic, and John A. Pollock, "Behavioral and Inflammatory Sex Differences Revealed by Celecoxib Nanotherapeutic Treatment of Peripheral Neuroinflammation," *Scientific Reports* 12, no. 8472 (2022), https://doi.org/10.1038/s41598-022-12248-8.

23. University of Bath, "Women Feel More Pain than Men, Research Shows," ScienceDaily, July 5, 2005, www.sciencedaily.com/releases/2005/07/050705004113.htm.

24. Bartley and Fillingim, "Sex Differences in Pain."

25. Jacob M. Vigil, "No Pain, No Social Gains: A Social-Signaling Perspective of Human Pain Behaviors," *World Journal of Anesthesiology* 3, no. 1 (2014), https://doi.org/10.5313/wja.v3.i1.18.

26. Geva et al., "Sex Effects in the Interaction of Acute Stress and Pain Perception."

27. Erika L. Manning and Roger B. Fillingim, "The Influence of Athletic Status and Gender on Experimental Pain Responses," *Journal of Pain* 3, no. 6 (2002), https://doi.org/10.1054/jpai.2002.128068; Jonas Tesarz, Alexander K. Schuster, Mechthild Hartmann, Andreas Gerhardt, and Wolfgang Eich, "Pain Perception in Athletes Compared to Normally Active Controls: A Systematic Review with Meta-Analysis," *Pain* 153, no. 6 (2012), https://doi.org/10.1016/j.pain.2012.03.005; Claire Thornton, Andrew Baird, and David Sheffield, "Athletes and Experimental Pain: A Systematic Review and Meta-Analysis," *Journal of Pain* 25, no. 6 (2024), https://doi.org/10.1016/j.jpain.2023.12.007.

28. Claire Thornton, David Sheffield, and Andrew Baird, "A Longitudinal Exploration of Pain Tolerance and Participation in Contact Sports," *Scandinavian Journal of Pain* 16, no. 1 (2017), https://doi.org/10.1016/j.sjpain.2017.02.007.

29. Susann Dahl Pettersen, Per M. Aslaksen, and Svein Arne Pettersen, "Pain Processing in Elite and High-Level Athletes Compared to Non-Athletes," *Frontiers in Psychology* 11 (2020), https://doi.org/10.3389/fpsyg.2020.01908.

30. Mollie K. Karabatsos, "The Difference in Pain Tolerance Between NCAA Division 1-A Male and Female Athletes," Marshall Digital Scholar, 2022, https://mds.marshall.edu/etd/1685/; Matt E. Jaremko, Lee Silbert, and Thomas Mann, "The Differential Ability of Athletes and Nonathletes to Cope with Two Types of Pain: A Radical Behavioral Model," *Psychological Record* 31, no. 2 (1981), https://doi.org/10.1007/bf03394739.

31. Jaremko, Silbert, and Mann, "Differential Ability of Athletes and Nonathletes."

32. Lindsay Crouse, "Opinion: Why Men Quit and Women Don't," *New York Times*, April 20, 2018, www.nytimes.com/2018/04/20/opinion/boston-marathon-women-nurse.html.

33. Adam Taylor, "Why Men and Women Can't Agree on the Perfect Temperature," The Conversation, October 26, 2016, https://theconversation.com/why-men-and-women-cant-agree-on-the-perfect-temperature-66585.

34. Megan Roche, interview with author, September 29, 2023.

35. "2024 Entrants List," Western States Endurance Run, last updated June 28, 2024, www.wser.org/2024-entrants-list/.

36. Nicolas Bouscaren, Robin Faricier, Guillaume Y. Millet, and Sébastien Racinais, "Heat Acclimatization, Cooling Strategies, and Hydration During an Ultra-Trail in Warm and Humid Conditions," *Nutrients* 13, no. 4 (2021), https://doi.org/10.3390/nu13041085.

37. R. O. Deaner, R. E. Carter, M. J. Joyner, and S. K. Hunter, "Men Are More Likely than Women to Slow in the Marathon," *Medicine and Science in Sports and Exercise* 47, no. 3 (2015), https://doi.org/10.1249/mss.0000000000000432.

38. "Are Women Tougher Runners than Men?," *Men's Running*, October 14, 2016, https://mensrunninguk.co.uk/news/women-tougher-runners-men/.

39. Crouse, "Opinion."

40. Mohamed Romdhani, Omar Hammouda, Khawla Smari, et al., "Total Sleep Deprivation and Recovery Sleep Affect the Diurnal Variation of Agility Performance: The Gender Differences," *Journal of Strength and Conditioning Research* 35, no. 1 (2021), https://doi.org/10.1519/JSC.0000000000002614; "Are Women Tougher Runners than Men?"

41. M. Corsi-Cabrera, "Effect of 38 H of Total Sleep Deprivation on the Waking EEG in Women: Sex Differences," *International Journal of Psychophysiology* 50, no. 3 (2003), https://doi.org/10.1016/s0167-8760(03)00168-5.

Chapter Ten: Held Back, Hamstrung, and Hobbled, Women Still Nail It

1. Norah Steed, interview with author, November 5, 2023.

2. The history of the 100th meridian as a dividing line is fascinating. Geologist John Wesley Powell, famous for exploring the Grand Canyon, drew a demarcation at that longitude because he found it worked as a separator between the wetter eastern half of the United States and the drier western plains. He suggested that special districts for water and land management be set up by Congress for states west of the 100th meridian due to limits in water and other resources, but then as now, the politicians ignored the suggestion of a scientist. According to the Columbia University Climate School's website State of the Planet, "Now, 140 years later, in two just-published papers, scientists examine how the 100th meridian has played out in history, and what the future may hold. They confirm that the divide has turned out to be real, as reflected by population and agriculture on opposite sides. They say also that the line appears to be slowly moving eastward, due to climate change, and that it will probably continue shifting in coming decades, expanding the arid climate of the western plains into what we think of as the Midwest. The implications for farming and other pursuits could be huge." Kevin Krajick, "The 100th Meridian, Where the Great Plains Begin, May Be Shifting," Columbia University, State of the Planet, April 11, 2018, https://news.climate.columbia.edu/2018/04/11/the-100th-meridian-where-the-great-plains-used-to-begin-now-moving-east/.

3. "Single Buck (with Assistant) Records," Stihl Timbersports Database, accessed November 2, 2024, https://data.stihl-timbersports.com/Discipline/SingleBuck?level=Intermediate&gender=Female&wood=Any&diameter=40&nation=&year=0.

4. Jay Lokegaonkar, "Gunning for Glory: Know Everything About Shooting at the Olympics," website of the Olympics, February 10, 2023,

https://olympics.com/en/news/olympic-shooting-air-rifle-3-positions-rapid-fire-air-pistol-shotgun-trap-skeet.

5. Peru, "Reminder: After a Female Shooter Won Olympic Gold in 92, Unisex Competitions Were Dropped," ResetEra, July 23, 2021, www.resetera.com/threads/reminder-after-a-female-shooter-won-olympic-gold-in-92-unisex-competitions-were-dropped.461520/.

6. "Shan Zhang," website of the Olympic Games, accessed December 15, 2024, olympics.com/en/athletes/shan-zhang.

7. Jonathan Selvaraj, "10m Air Rifle: The Olympic Sport Where Women Outgun Men," ESPN, July 18, 2021, www.espn.com/shooting/story/_/id/31828521/10m-air-rifle-sport-tokyo-olympics-where-women-outgun-men.

8. Dr. Sheree Bekker (@shereebekker), "I have been hearing more frequently the narrative . . . ," Twitter (now X), March 18, 2022 (her profile is no longer publicly available; this thread is archived at Archive.today, https://archive.is/oWe91).

9. Selvaraj, "10m Air Rifle."

10. "The IOC Takes Historic Step Forward to Advance Gender Equality Following Executive Board Approval of Bold Recommendations," International Olympic Committee, February 6, 2018, https://olympics.com/ioc/news/the-ioc-takes-historic-step-forward-to-advance-gender-equality-following-executive-board-approval-of-bold-recommendations.

11. "ISSF Rules Changes for 2018–2020 Approved," ISSF, December 18, 2017, www.issf-sports.org/news/2955.

12. Daniel Mon-López, Carlos M. Tejero-González, and Santiago Calero, "Recent Changes in Women's Olympic Shooting and Effects in Performance," *PLOS One* 14, no. 5 (2019), https://doi.org/10.1371/journal.pone.0216390.

13. "Standing Position Video271," heinz reinkemeier, video, posted December 8, 2020, YouTube, www.youtube.com/watch?v=BLZYKLm27RA.

14. Wendy Lafever, "Are Women Naturally Better Shooters than Men?," NRA Women, September 24, 2020, www.nrawomen.com/content/are-women-naturally-better-shooters-than-men/.

15. Goran Vučković, Milivoj Dopsaj, Radovan Radovanović, and Aleksandar Jovanović, "Characteristics of Shooting Efficiency During a Basic Shooting Training Program Involving Police Officers of Both Sexes," *Physical Education and Sport* 6, no. 1 (2008), https://jakov.kpu.edu.rs/bitstream/handle/123456789/190/188.pdf?sequence=1&isAllowed=y.

16. Warbow, comment on post by Humdinger, "What Separates Male/Female in the Sport of Archery?," Archery Talk, March 21, 2013, www.archerytalk.com/threads/what-separates-male-female-in-the-sport-of-archery.1975267/.

17. Adam Storer and Jimmy Grant, "Who Is World Darts Championship Star Fallon Sherrock and What Is Her Net Worth?," *US Sun*, December 10, 2023, www.the-sun.com/sport/168779/fallon-sherrock-darts-next-match-pdc-healh-scare/.

18. Wikipedia, "2021 Nordic Darts Masters," June 15, 2024, http://en.wikipedia.org/wiki/2021_Nordic_Darts_Masters.

19. Dr. Sheree Bekker (@shereebekker), "I have been hearing more frequently the narrative. . . ."

Chapter Eleven: A Well-Balanced Life and Sporting Culture

1. The pool is huge and free, and I swim there every day when I'm in Australia, even during the rain. There are a few dozen of these community pools along the coast, all unique, oriented so they can be refilled by fresh ocean water at each high tide. Most were built in the 1920s and 1930s, expanded from existing natural pools set in the native rock shelves that stretch like stages into the sea. The rock pools are often populated with small fish (and the seabirds that like to hunt them), and winds sometimes make them as choppy as the sea outside their walls.

2. Luis Madureira Pinto, "How Many Surfers Are There in the World?," SurferToday, September 19, 2020, www.surfertoday.com/surfing/how-many-surfers-are-there-in-the-world; Ryan Reft, "Riding Waves, Forging Communities: Surfing, Gender, and Feminism in 20th Century California," PBS SoCal, May 30, 2014, www.pbssocal.org/history-society/riding-waves-forging-communities-surfing-gender-and-feminism-in-20th-century-california.

3. Jake Howard, "The Sport of Surfing in America Is More Diverse than It's Ever Been," Surfing on SI, January 23, 2024, www.si.com/fannation/surfing/news/surfing-more-diverse-america-than-ever.

4. Craig Sims, "Industry Report: Changing Demography of Surfing in Australia," LinkedIn, June 20, 2023, www.linkedin.com/pulse/industry-report-changing-demography-surfing-australia-craig-sims/.

5. Mindy Pennybacker, *Surfing Sisterhood Hawaiʻi: Wahine Reclaiming the Waves* (Mutual Publishing, 2023), 9–19.

6. Gerry Vartan, interview with the author (his daughter), February 2023.

7. Pennybacker, *Surfing Sisterhood*.

8. Maggie Mertens, "This Woman Surfed the Biggest Wave of the Year," *The Atlantic*, September 12, 2020, www.theatlantic.com/culture/archive/2020/09/maya-gabeira-surfed-biggest-wave-year/616216/.

9. Like the time in Bali in 2023 when American pro surfer Sara Taylor was dropped in on (when a surfer tries to take a wave someone else is already surfing on). She pushed the guy off the wave to avoid getting hit. His friend came after Taylor, punching her in the head and pushing her underwater. Later, he again attacked. A friend of Taylor's made a video of both parts of the altercation. It went viral, and the surfing community was horrified. Still, threats against women in what some men see as "their" spaces aren't uncommon. Dashel Pierson, "Pro Surfers Respond to Viral Surfing Brawl," Surfer, April 6, 2023, www.surfer.com/news/video-viral-surfing-brawl-bali-sara-taylor.

10. Cassie Comley, "'We Have to Establish Our Territory': How Women Surfers 'Carve Out' Gendered Spaces Within Surfing," *Sport in Society* 19, no. 8–9 (2016), https://doi.org/10.1080/17430437.2015.1133603.

11. Mindy Pennybacker, email correspondence with author, May 15, 2024.

12. Comley, "'We Have to Establish Our Territory.'"

13. Joanna R. Parsonage, Josh L. Secomb, Tai T. Tran, et al., "Gender Differences in Physical Performance Characteristics of Elite Surfers," *Journal of Strength and Conditioning Research* 31, no. 9 (2017), https://doi.org/10.1519/jsc.0000000000001428.

14. Joanna Parsonage, interview with author, March 20, 2024.

15. Joanna Parsonage, Josh L. Secomb, Jeremy M. Sheppard, Brendon K. Ferrier, Rebecca A. Dowse, and Sophia Nimphius, "Upper-Body Strength Measures and Pop-Up Performance of Stronger and Weaker Surfers," *Journal of Strength and Conditioning Research* 34, no. 10 (2020), https://doi.org/10.1519/jsc.0000000000002377.

16. Lena Smirnova and Ash Tulloch, "Under the Teahupo'o Spell: Paris 2024 Surfers Explain What Makes the Olympic Venue So Special," website of the Olympics, August 15, 2023, https://olympics.com/en/news/teahupoo-tahiti-pro-what-makes-paris-2024-venue-so-special.

17. "Here's Why Men Fail at the TikTok Balance Challenge," National Academy of Sports Medicine (NASM), video, posted March 12, 2021, YouTube, www.youtube.com/watch?v=gbZN5abKBms.

18. "Men's Gymnasts Try the RAREST Balance Beam Mounts," Ash Watson, video, posted December 21, 2022, YouTube, www.youtube.com/watch?v=7TQTkGYIF48.

19. Dvora Meyers, "Female Gymnasts Used to Compete on the Rings, but the Game Changed," Jezebel, August 7, 2012, www.jezebel.com/female-gymnasts-used-to-compete-on-the-rings-but-the-g-5932478.

20. Sarah Varney, "How Older Athletes Powered Up and Transformed Women's Gymnastics," NPR, August 6, 2024, www.npr.org/sections/shots-health-news/2024/08/06/nx-s1-5064419/womens-gymnastics-ages-simone-biles-olympics#.

21. The Learning Network, "What Students Are Saying About Coed Sports," *New York Times*, March 23, 2023, www.nytimes.com/2023/03/23/learning/what-students-are-saying-about-coed-sports.html.

22. Katelyn Burns, "Why the Anti-Trans Right Is Attacking Two Female Olympic Boxers," MSNBC, August 2, 2024, www.msnbc.com/opinion/msnbc-opinion/olympics-boxer-imane-khelif-anti-trans-rcna164721.

23. Chris Mosier, "Transgender Athletes in the Olympics," Transathlete, accessed November 3, 2024, www.transathlete.com/olympics.

24. Anna Dimond, "Pro Surfing Allows Transgender Athletes to Compete. Cue the Backlash," Outside Online, May 4, 2023, www.outsideonline.com/outdoor-adventure/water-activities/surfing-transgender-rules-sasha-jane-lowerson/.

25. M. P. Schuijt, C. G. J. Sweep, R. van der Steen, et al., "'Validity of Free Testosterone Calculation in Pregnant Women," *Endocrine Connections* 8, no. 6 (2019), https://doi.org/10.1530/EC-19-0110.

26. "Polycystic Ovary Syndrome," World Health Organization, June 28, 2023, https://www.who.int/news-room/fact-sheets/detail/polycystic-ovary-syndrome; Michael T. Sheehan, "Polycystic Ovarian Syndrome: Diagnosis and

Management," *Clinical Medicine and Research* 2, no. 1 (2004), www.ncbi.nlm.nih.gov/pmc/articles/PMC1069067/.

27. Breay Paty, "Appropriate Testosterone Testing for Male Hypogonadism," This Changed My Practice, University of British Colombia, January 21, 2020, https://thischangedmypractice.com/testosterone-testing-male-hypogonadism/.

28. Cara Ocobock, "Sex in Sport: Men Don't Always Have the Advantage," *Sapiens*, June 10, 2021, www.sapiens.org/biology/female-male-athletes-differences/.

29. Melissa Cunningham, "FINA Ruling Based on 'Opinion Not Science,' Australian Researchers Say," *Sydney Morning Herald*, June 20, 2022, www.smh.com.au/national/fina-ruling-based-on-opinion-not-science-australian-researchers-say-20220620-p5av3o.html.

30. Katrina Lambert, "Tokyo Marathon Adds Nonbinary Category for 2025 Race," NBC News, June 27, 2024, www.nbcnews.com/nbc-out/out-news/tokyo-marathon-adds-nonbinary-category-2025-race-rcna159284.

31. Jonathan Selvaraj, "10m Air Rifle: The Olympic Sport Where Women Outgun Men," ESPN, July 18, 2021, www.espn.com/shooting/story/_/id/31828521/10m-air-rifle-sport-tokyo-olympics-where-women-outgun-men.

Chapter Twelve: Female Bodies: Great at Defense

1. Sharon Moalem, *The Better Half: On the Genetic Superiority of Women* (Farrar, Straus and Giroux, 2020), 49.

2. Kathryn Dovel, Sara Yeatman, Susan Watkins, and Michelle Poulin, "Men's Heightened Risk of AIDS-Related Death," *AIDS* 29, no. 10 (2015), https://doi.org/10.1097/qad.0000000000000655.

3. M. M. Addo and M. Altfeld, "Sex-Based Differences in HIV Type 1 Pathogenesis," *Journal of Infectious Diseases* 209, suppl. 3 (2014), https://doi.org/10.1093/infdis/jiu175.

4. M. Mahathir, "Women at Greater Risk of HIV Infection," *Arrows for Change* 3, no. 1 (1997), https://pubmed.ncbi.nlm.nih.gov/12292992.

5. Ross Hewitt, Nader Parsa, and Lawrence James Gugino, "The Role of Gender in HIV Progression," *Bulletin of Experimental Treatments for AIDS* 14, no. 1 (2001), www.researchgate.net/publication/11659657_The_role_of_gender_in_HIV_progression.

6. Azfar-e-Alam Siddiqi, H. Irene Hall, Xiaohong Hu, and Ruiguang Song, "Population-Based Estimates of Life Expectancy After HIV Diagnosis," *Journal of Acquired Immune Deficiency Syndromes* 72, no. 2 (2016), https://doi.org/10.1097/qai.0000000000000960.

7. Sabra L. Klein and Katie L. Flanagan, "Sex Differences in Immune Responses," *Nature Reviews Immunology* 16, no. 10 (2016), https://doi.org/10.1038/nri.2016.90.

8. Daniel J. Kruger and Randolph M. Nesse, "An Evolutionary Life-History Framework for Understanding Sex Differences in Human Mortality Rates," *Human Nature* 17, no. 1 (2006), https://doi.org/10.1007/s12110-006-1021-z.

9. Hae-In Kim, Hyesol Lim, and Aree Moon, "Sex Differences in Cancer: Epidemiology, Genetics and Therapy," *Biomolecules and Therapeutics* 26, no. 4 (2018), https://doi.org/10.4062/biomolther.2018.103.

10. S. L. Klein and A. Pekosz, "Sex-Based Biology and the Rational Design of Influenza Vaccination Strategies," *Journal of Infectious Diseases* 209, suppl. 3 (2014), https://doi.org/10.1093/infdis/jiu066.

11. Ashley L. Fink and Sabra L. Klein, "The Evolution of Greater Humoral Immunity in Females than Males: Implications for Vaccine Efficacy," *Current Opinion in Physiology* 6 (2018), https://doi.org/10.1016/j.cophys.2018.03.010.

12. Sarah E. M. Howie, "Go Girls! Efficient Female Innate Immunity," *Blood* 118, no. 22 (2011), https://doi.org/10.1182/blood-2011-09-381137.

13. Howie, "Go Girls!"

14. Roland Pongou, "Why Is Infant Mortality Higher in Boys than in Girls? A New Hypothesis Based on Preconception Environment and Evidence from a Large Sample of Twins," *Demography* 50, no. 2 (2012), https://doi.org/10.1007/s13524-012-0161-5.

15. Virginia Zarulli and Giambattista Salinari, "Gender Differences in Survival Across the Ages of Life: An Introduction." *Genus* 80, no. 1, June 24, 2024, https://doi.org/10.1186/s41118-024-00216-1.

16. Moalem, *Better Half*, 12.

17. Felix Broecker and Karin Moelling, "What Viruses Tell Us About Evolution and Immunity: Beyond Darwin?," *Annals of the New York Academy of Sciences* 1447, no. 1 (2019), https://doi.org/10.1111/nyas.14097.

18. A. Maan, J. Eales, A. Akbarov, et al., "The Y Chromosome: A Blueprint for Men's Health?," *European Journal of Human Genetics* 25 (2017), https://doi.org/10.1038/ejhg.2017.128.

19. Carl Zimmer, "Seeing X Chromosomes in a New Light," *New York Times*, January 20, 2014, www.nytimes.com/2014/01/21/science/seeing-x-chromosomes-in-a-new-light.html.

20. Hao Wu, Junjie Luo, Huimin Yu, et al., "Cellular Resolution Maps of X Chromosome Inactivation: Implications for Neural Development, Function, and Disease," *Neuron* 81, no. 1 (2014), https://doi.org/10.1016/j.neuron.2013.10.051.

21. B. J. Bain and J. M. England, "Normal Haematological Values: Sex Difference in Neutrophil Count," *British Medical Journal* (1975), https://doi.org/10.1136/bmj.1.5953.306.

22. Sarthak Gupta, Shuichiro Nakabo, Luz P. Blanco, et al., "Sex Differences in Neutrophil Biology Modulate Response to Type I Interferons and Immunometabolism," *Proceedings of the National Academy of Sciences* 117, no. 28 (2020), https://doi.org/10.1073/pnas.2003603117.

23. Henning Jacobsen and Sabra L. Klein, "Sex Differences in Immunity to Viral Infections," *Frontiers in Immunology* 12 (2021), https://doi.org/10.3389/fimmu.2021.720952.

24. I traveled to Australia, which was virtually COVID-free, to care for my sick dad at the height of the pandemic, when vaccines were rolling out in the US but only for high-risk categories of people. I was not high risk, so I left

for Australia and hoped I could get vaccinated there. Australia took the virus incredibly seriously, and I was one of thousands of people who were quarantined in a hotel for two full weeks to ensure I was not bringing COVID into the country from the US. By the time I headed back to the US four months later (my dad was much improved, thankfully), I was still unvaccinated, as Australia was behind the US and was only then vaccinating high-risk people. I flew to San Francisco and got COVID either the day I landed or the next day.

25. Moalem, *Better Half*, 50–51.

26. Fink and Klein, "The Evolution of Greater Humoral Immunity in Females than Males."

27. Jacobsen and Klein, "Sex Differences in Immunity to Viral Infections."

28. Mark Wanner, "Immune System Changes with Age Differ Between Men and Women," *JAX News*, February 10, 2020, www.jax.org/news-and-insights/2020/february/immune-system-changes-with-age-differ-between-men-and-women.

29. Eladio J. Márquez, Cheng-han Chung, Radu Marches, et al., "Sexual-Dimorphism in Human Immune System Aging," *Nature Communications* 11, no. 1 (2020), https://doi.org/10.1038/s41467-020-14396-9.

30. Moalem, *Better Half*, 78.

Chapter Thirteen: Estrogen the Pathogen Slayer

1. Caroline Duncombe, interview with author, September 11, 2023.

2. Sabra L. Klein and Katie L. Flanagan, "Sex Differences in Immune Responses," *Nature Reviews Immunology* 16, no. 10 (2016), https://doi.org/10.1038/nri.2016.90.

3. D. A. González, B. B. Díaz, M. del Cristo Rodríguez Pérez, A. González Hernández, B. N. Díaz Chico, and A. Cabrera de León, "Sex Hormones and Autoimmunity," *Immunology Letters* 133, no. 1 (2010), https://doi.org/10.1016/j.imlet.2010.07.001.

4. Klein and Flanagan, "Sex Differences in Immune Responses."

5. M. Vetrano, A. Wegman, B. Koes, S. Mehta, and C. A. King, "Serum IL-1RA Levels Increase from Follicular to Luteal Phase of the Ovarian Cycle: A Pilot Study on Human Female Immune Responses," *PLOS One* 15, no. 9 (2020), https://doi.org/10.1371/journal.pone.0238520.

6. Sean M. Hughes, Claire N. Levy, Ronit Katz, et al., "Changes in Concentrations of Cervicovaginal Immune Mediators Across the Menstrual Cycle: A Systematic Review and Meta-Analysis of Individual Patient Data," *BMC Medicine* 20, no. 1 (2022), https://doi.org/10.1186/s12916-022-02532-9; "Vaginal Immune System May Hint at Prime Vaccine Timing," news release, UW Medicine Newsroom, October 17, 2023, newsroom.uw.edu/news-releases/vaginal-immune-system-may-hint-prime-vaccine-timing.

7. Klein and Flanagan, "Sex Differences in Immune Responses."

8. "Estrogen Protective Against Flu Virus in Women but Not Men, Study Suggests," media release, Johns Hopkins Bloomberg School of Public Health, January 19, 2016, https://publichealth.jhu.edu/2016/estrogen-protective-against-flu-virus-in-women-but-not-men-study-suggests.

9. Ricardo Costeira, Karla A. Lee, Benjamin Murray, et al., "Estrogen and COVID-19 Symptoms: Associations in Women from the COVID Symptom Study," *PLOS One* 16, no. 9 (2021), https://doi.org/10.1371/journal.pone.0257051.

10. Farideh Zafari Zangeneh and Maryam Sarmast Shoushtari, "Estradiol and COVID-19: Does 17-Estradiol Have an Immune-Protective Function in Women Against Coronavirus?," *Journal of Family and Reproductive Health* 15, no. 3 (2021), https://doi.org/10.18502/jfrh.v15i3.7132.

11. Sung Wook Choi, Juhee Kim, Jae Hoon Lee, et al., "Hormone Therapy in the Era of the COVID-19 Pandemic: A Review," *Journal of Menopausal Medicine* 28, no. 1 (2022), https://doi.org/10.6118/jmm.21036.

12. Janna R. Shapiro, Sabra L. Klein, and Rosemary Morgan, "Stop 'Controlling' for Sex and Gender in Global Health Research," *BMJ Global Health* 6, no. 4 (2021), https://doi.org/10.1136/bmjgh-2021-005714.

13. "Malaria Parasite Transforms Itself to Hide from Human Immune System," Harvard T. H. Chan School of Public Health, December 13, 2012, hsph.harvard.edu/news/duraisingh-malaria-parasite-transforms/.

14. "Malaria's Impact Worldwide," US Centers for Disease Control and Prevention, April 1, 2024, www.cdc.gov/malaria/php/impact/.

15. Landon G. vom Steeg, Yevel Flores-Garcia, Fidel Zavala, and Sabra L. Klein, "Irradiated Sporozoite Vaccination Induces Sex-Specific Immune Responses and Protection Against Malaria in Mice," *Vaccine* 37, no. 32 (2019), https://doi.org/10.1016/j.vaccine.2019.06.075.

16. Caroline J. Duncombe, Felicia N. Watson, Anya C. Kalata, Melanie J. Shears, and Sean C. Murphy, "Differential Sex-Specific Immune Responses Following Prime-and-Trap Vaccination Alters Protection Against Malaria in Mice," *Journal of Immunology* 208, no. 1 suppl. (2022), https://doi.org/10.4049/jimmunol.208.supp.181.14.

17. Klein and Flanagan, "Sex Differences in Immune Responses."

18. Bruce Goldman, "In Men, High Testosterone Can Mean Weakened Immune Response, Study Finds," Stanford Medicine News Center, December 23, 2013, https://med.stanford.edu/news/all-news/2013/12/in-men-high-testosterone-can-mean-weakened-immune-response-study-finds.html.

19. Marianne Peters, Leigh W. Simmons, and Gillian Rhodes, "Testosterone Is Associated with Mating Success but Not Attractiveness or Masculinity in Human Males," *Animal Behaviour* 76, no. 2 (2008), https://doi.org/10.1016/j.anbehav.2008.02.008.

20. Kelsey A. O'Brien, Jane M. Waterman, W. Gary Anderson, and Nigel C. Bennett, "Trade-Offs Between Immunity and Testosterone in Male African Ground Squirrels," *Journal of Experimental Biology* 221, no. 16 (2018), https://doi.org/10.1242/jeb.177683.

21. O'Brien et al., "Trade-Offs Between Immunity and Testosterone in Male African Ground Squirrels."

22. Judyta Nowak, Bogusław Pawłowski, Barbara Borkowska, Daria Augustyniak, and Zuzanna Drulis-Kawa, "No Evidence for the Immunocompetence

Handicap Hypothesis in Male Humans," *Scientific Reports* 8, no. 1 (2018), https://doi.org/10.1038/s41598-018-25694-0.

23. Benjamin C. Trumble, Aaron D. Blackwell, Jonathan Stieglitz, et al., "Associations Between Male Testosterone and Immune Function in a Pathogenically Stressed Forager-Horticultural Population," *American Journal of Physical Anthropology* 161, no. 3 (2016), https://doi.org/10.1002/ajpa.23054.

24. Unaib Rabbani, Ambreen Sahito, Asaad Ahmed Nafees, Ambreen Kazi, and Zafar Fatmi, "Pulmonary Tuberculosis Is Associated with Biomass Fuel Use Among Rural Women in Pakistan: An Age- and Residence-Matched Case-Control Study," *Asia Pacific Journal of Public Health* 29, no. 3 (2017), https://doi.org/10.1177/1010539517696554.

25. Logan Stuck, Eveline Klinkenberg, Nahid Abdelgadir Ali, et al., "Prevalence of Subclinical Pulmonary Tuberculosis in Adults in Community Settings: An Individual Participant Data Meta-Analysis," *Lancet Infectious Diseases* 24, no. 7 (2024), https://doi.org/10.1016/s1473-3099(24)00011-2; Shobha Shukla, "Tuberculosis (TB) Response for Women Is Grossly Inadequate," *Pakistan Christian Post*, August 25, 2010, https://pakistanchristianpost.com/opinion-details/1121.

26. Shifa Salman Habib, Wafa Zehra Jamal, Syed Mohammad Asad Zaidi, et al., "Barriers to Access of Healthcare Services for Rural Women: Applying Gender Lens on TB in a Rural District of Sindh, Pakistan," *International Journal of Environmental Research and Public Health* 18, no. 19 (2021), https://doi.org/10.3390/ijerph181910102.

27. "Civilian Occupations with High Fatal Work Injury Rates, 2022," US Bureau of Labor Statistics, 2022, www.bls.gov/charts/census-of-fatal-occupational-injuries/civilian-occupations-with-high-fatal-work-injury-rates.htm.

28. Ian Banks, "No Man's Land: Men, Illness, and the NHS," *British Medical Journal* 323, no. 7320 (2001), www.ncbi.nlm.nih.gov/pmc/articles/PMC1121551/.

29. Heather Shattuck-Heidorn, Meredith W. Reiches, and Sarah S. Richardson, "What's Really Behind the Gender Gap in Covid-19 Deaths?," *New York Times*, June 24, 2020, www.nytimes.com/2020/06/24/opinion/sex-differences-covid.html.

30. Ann Caroline Danielsen, Katharine M. N. Lee, Marion Boulicault, et al., "Sex Disparities in COVID-19 Outcomes in the United States: Quantifying and Contextualizing Variation," *Social Science and Medicine* 294 (2022), https://doi.org/10.1016/j.socscimed.2022.114716.

31. Sarah Richardson, interview with author, July 11, 2023.

32. Oliver Ritter and Georgios Kararigas, "Sex-Biased Vulnerability of the Heart to COVID-19," *Perspective and Controversy* 95, no. 11 (2020), https://doi.org/10.1016/j.mayocp.2020.09.017.

33. F. L. Gersh, J. H. O'Keefe, and B. M. Henry, "COVID-19, the Female Immune Advantage, and Cardiovascular Impact." *Mayo Clinic Proceedings* 96, no. 3 (2021), doi: 10.1016/j.mayocp.2020.12.021. Epub 2020 Dec 29. PMID: 33673933; PMCID: PMC7832658.

34. "Coronavirus and Air Pollution," Harvard T. H. Chan School of Public Health, May 19, 2020, www.hsph.harvard.edu/c-change/subtopics/coronavirus-and-pollution.

Chapter Fourteen: Living Longer, Living Better

1. Jennifer Szalai, "The Complicated Origins of 'Having It All,'" *New York Times*, January 2, 2015, www.nytimes.com/2015/01/04/magazine/the-complicated-origins-of-having-it-all.html.

2. Adam Givens, "Okinawa: The Costs of Victory in the Last Battle," National WWII Museum, July 7, 2022, www.nationalww2museum.org/war/articles/okinawa-costs-victory-last-battle.

3. Julian Ryall, "Japan: What's Behind Okinawans' Falling Life Expectancy?," DW, December 6, 2022, www.dw.com/en/japan-whats-behind-okinawans-falling-life-expectancy/a-62088176. For comparison, in the US, the average life expectancy for women in 2020 was 79.9 and for men 74.2. In the UK, it was 82.6 for women and 78.6 for men. In Australia, it was 85.2 for women and 81.2 for men. In China, it was 80.5 for women and 74.7 for men.

4. Ryall, "Japan"; National Geographic Staff, "The 5 'Blue Zones' Where the World's Healthiest People Live," *National Geographic*, September 9, 2023, www.nationalgeographic.com/premium/article/5-blue-zones-where-the-worlds-healthiest-people-live.

5. Ruairi Robertson, "Why People in 'Blue Zones' Live Longer than the Rest of the World," Healthline, August 29, 2017, https://www.healthline.com/nutrition/blue-zones.

6. "Is Longevity Determined by Genetics?," MedlinePlus, September 18, 2020, medlineplus.gov/genetics/understanding/traits/longevity/; J. Graham Ruby, Kevin M. Wright, Kristin A. Rand, et al., "Estimates of the Heritability of Human Longevity Are Substantially Inflated Due to Assortative Mating," *Genetics* 210, no. 3 (2018), https://doi.org/10.1534/genetics.118.301613.

7. Kimiko Kamida and Mitsu Kakazu, interview with author, Naha, Okinawa, February 16, 2024. Interpreter: Christal Burnette.

8. There are about 80,000 Americans on Okinawa, including about 30,000 active-duty military personnel plus their dependents. The island's total population is about 140,000.

9. Phoebe Woei-Ni Hwang and Kathryn L. Braun, "The Effectiveness of Dance Interventions to Improve Older Adults' Health: A Systematic Literature Review," *Alternative Therapies in Health and Medicine* 21, no. 5 (2015), pmc.ncbi.nlm.nih.gov/articles/PMC5491389/; Ragavendra R. Baliga, "Sing for a Long and Healthy Life?," *European Heart Journal* 45, no. 20 (2024), https://doi.org/10.1093/eurheartj/ehad819.

10. Makoto Suzuki, interview with author, Okinawa, Japan, February 14, 2024.

11. Aislinn Kotifani, "Moai: This Tradition Is Why Okinawan People Live Longer, Better," Blue Zones, August 16, 2018, www.bluezones.com/2018/08/moai-this-tradition-is-why-okinawan-people-live-longer-better/.

12. Maya Shetty, "How Social Connection Supports Longevity," Lifestyle Medicine, December 18, 2023, longevity.stanford.edu/lifestyle/2023/12/18/how-social-connection-supports-longevity/.

13. Quinton Sanicola, Yohanna Sanicola, and Emiko Kinjo, interview with author, Okinawa, Japan, February 18, 2024. Interpreter: Christal Burnette.

14. Fernando Colchero, Roland Rau, Owen R. Jones, et al., "The Emergence of Longevous Populations," *Proceedings of the National Academy of Sciences* 113, no. 48 (2016), https://doi.org/10.1073/pnas.1612191113.

15. Stephanie Bucklin, "Women Have Always Lived Longer, Study Finds," Live Science, November 21, 2016, www.livescience.com/56956-why-women-live-longer.html.

16. Robert Young, "Supercentenarian Data—Table E," Gerontology Research Group, last updated November 13, 2024, grg.org/WSRL/TableE.aspx.

17. Steven N. Austad, "Why Women Live Longer than Men: Sex Differences in Longevity," *Gender Medicine* 3, no. 2 (2006), https://doi.org/10.1016/s1550-8579(06)80198-1.

18. Fran Baum, Connie Musolino, Hailay Abrha Gesesew, and Jennie Popay, "New Perspective on Why Women Live Longer than Men: An Exploration of Power, Gender, Social Determinants, and Capitals," *International Journal of Environmental Research and Public Health* 18, no. 2 (2021), https://doi.org/10.3390/ijerph18020661.

19. Michelle O'Donoghue, "Heart Disease in Men and Women," Mass General Brigham, February 15, 2024, www.massgeneralbrigham.org/en/about/newsroom/articles/heart-disease-in-men-and-women.

20. Kathy Katella, "Maternal Mortality Is on the Rise: 8 Things to Know," Yale Medicine, May 22, 2023, www.yalemedicine.org/news/maternal-mortality-on-the-rise.

21. Munira Gunja et al. "Insights into the U.S. Maternal Mortality Crisis: An International Comparison." Commonwealth Fund, June 4, 2024, www.commonwealthfund.org/publications/issue-briefs/2024/jun/insights-us-maternal-mortalitwww.commonwealthfund.org/publications/issue-briefs/2024/jun/insights-us-maternal-mortality-crisis-international-comparison.

22. "Physical Activity," Australian Institute of Health and Welfare, May 19, 2023, www.aihw.gov.au/reports/physical-activity/physical-activity.

23. Claudia Tamas, interview with author, January 2, 2024.

24. Belinda Beck, interview with author, January 2, 2024.

25. "What Women Need to Know," Bone Health and Osteoporosis Foundation, accessed December 16, 2024, www.bonehealthandosteoporosis.org/preventing-fractures/general-facts/what-women-need-to-know/.

26. Michele Bellantoni, "Osteoporosis Information," Johns Hopkins Arthritis Center, 2024, www.hopkinsarthritis.org/arthritis-info/osteoporosis-info/.

27. Interview with Susan Stuart-Jones, December 26, 2023.

28. Belinda Beck and Robert Marcus, "Impact of Physical Activity on Age-Related Bone Loss," in *The Aging Skeleton*, ed. Clifford J. Rosen, Julie Glowacki, and John P. Bilezikian (Academic Press, 1999), 467–478.

29. Steven L. Watson, Benjamin K. Weeks, Lisa J. Weis, Amy T. Harding, Sean A. Horan, and Belinda R. Beck, "High-Intensity Resistance and Impact Training Improves Bone Mineral Density and Physical Function in

Postmenopausal Women with Osteopenia and Osteoporosis: The LIFTMOR Randomized Controlled Trial," *Journal of Bone and Mineral Research* 33, no. 2 (2017), https://doi.org/10.1002/jbmr.3284.

30. A. Gupta, L. R. Jayes, S. Holmes, et al., "Management of Fracture Risk in Patients with Chronic Obstructive Pulmonary Disease (COPD): Building a UK Consensus Through Healthcare Professional and Patient Engagement," *International Journal of Chronic Obstructive Pulmonary Disease* 15 (2020), https://doi.org/10.2147/copd.s233398.

31. Michael Gurven and Hillard Kaplan, "Longevity Among Hunter-Gatherers: A Cross-Cultural Examination," *Population and Development Review* 33, no. 2 (2007), https://doi.org/10.1111/j.1728-4457.2007.00171.x.

32. "Redefining Knowledge of Elderly People Throughout History," news release, Australian National University via American Association for the Advancement of Science, January 3, 2018, www.eurekalert.org/news-releases/903440.

33. Aron Pinker, "The Famous but Difficult Psalm 90:10," *Old Testament Essays* 28, no. 2 (2015), http://dx.doi.org/10.17159/2312-3621/2015/V28N2A15.

34. Joanna Thompson, "Wild Chimps Shown to Undergo Menopause for the First Time," *Scientific American*, October 26, 2023, www.scientificamerican.com/article/wild-chimps-shown-to-undergo-menopause-for-the-first-time.

35. Lynnette Sievert, interview with author, March 4, 2024.

36. Kristen Hawkes, "Hadza Women's Time Allocation, Offspring Provisioning, and the Evolution of Long Postmenopausal Life Spans," *Current Anthropology* 38, no. 4 (1997), https://collections.lib.utah.edu/ark:/87278/s6j96qh8; Kristen Hawkes, "Grandmothers and the Evolution of Human Longevity," *American Journal of Human Biology* 15, no. 3 (2003), https://doi.org/10.1002/ajhb.10156.

37. Michael Gurven, Hillard Kaplan, and Maguin Gutierrez, "How Long Does It Take to Become a Proficient Hunter? Implications for the Evolution of Extended Development and Long Life Span," *Journal of Human Evolution* 51, no. 5 (2006), https://doi.org/10.1016/j.jhevol.2006.05.003.

38. K. Hawkes, J. F. O'Connell, and N. G. Blurton Jones, "Hadza Women's Time Allocation, Offspring Provisioning, and the Evolution of Long Postmenopausal Life Spans," *Current Anthropology* 38, no. 4 (1997), https://doi.org/10.1086/204646.

39. James G. Herndon, "The Grandmother Effect: Implications for Studies on Aging and Cognition," *Gerontology* 56, no. 1 (2010), https://doi.org/10.1159/000236045. Yes, Hadza grandfathers supply extra calories to their grandkids too, but not to the same extent as grandmothers.

40. Peter S. Kim, James E. Coxworth, and Kristen Hawkes, "Increased Longevity Evolves from Grandmothering," *Proceedings of the Royal Society B: Biological Sciences* 279, no. 1749 (2012), https://doi.org/10.1098/rspb.2012.1751.

41. Mirkka Lahdenperä, Virpi Lummaa, Samuli Helle, Marc Tremblay, and Andrew F. Russell, "Fitness Benefits of Prolonged Post-Reproductive Lifespan in Women," *Nature* 428, no. 6979 (2004), https://doi.org/10.1038/nature02367.

42. In fact, researchers found that the opposite may be true for men. Long-lived men often go on to have more children with new wives, which means their first (older) set of children receive less support than they would have otherwise. M. Lahdenperä, V. Lummaa, and A. F. Russell, "Selection on Male Longevity in a Monogamous Human Population: Late-Life Survival Brings No Additional Grandchildren," *Journal of Evolutionary Biology* 24, no. 5 (2011), https://doi.org/10.1111/j.1420-9101.2011.02237.x.

43. Stuart Nattrass, Darren P. Croft, Samuel Ellis, et al., "Postreproductive Killer Whale Grandmothers Improve the Survival of Their Grandoffspring," *Proceedings of the National Academy of Sciences* 116, no. 52 (2019), https://doi.org/10.1073/pnas.1903844116.

44. Roni Jacobson, "Revisiting the Grandmother Hypothesis," *Scienceline*, February 20, 2013, scienceline.org/2013/02/revisiting-the-grandmother-hypothesis/.

45. Shripad D. Tuljapurkar, Cedric O. Puleston, and Michael D. Gurven, "Why Men Matter: Mating Patterns Drive Evolution of Human Lifespan," *PLOS One* 2, no. 8 (2007), https://doi.org/10.1371/journal.pone.0000785.

46. Eric G. Hulsey, Yuan Li, Karen Hacker, Karl Williams, Kathryn Collins, and Erin Dalton, "Potential Emerging Risks Among Children Following Parental Opioid-Related Overdose Death," *JAMA Pediatrics* 174, no. 5 (2020), https://doi.org/10.1001/jamapediatrics.2020.0613; Jude Coleman, "COVID Deaths: More than 10 Million Children Lost a Parent or Carer," *Nature*, News, September 16, 2022, https://doi.org/10.1038/d41586-022-02941-z.

47. Nuria Jaumot-Pascual, Maria Jesús Monteagudo, Douglas A. Kleiber, and Jaime Cuenca, "Gender Differences in Meaningful Leisure Among Older Adults: Joint Displays of Four Phenomena," *Frontiers in Psychology* 9 (2018), https://doi.org/10.3389/fpsyg.2018.01450; Beth Almeida, "Five Facts on Older Women in the Labor Market," Center for American Progress, May 23, 2023, www.americanprogress.org/article/five-facts-on-older-women-in-the-labor-market/.

Chapter Fifteen: A More Exciting Future for Women's Bodies

1. "Yakama Nation Treaty of 1855," Yakama Nation, accessed November 11, 2024, www.yakama.com/about/treaty/.

2. Elaine Harvey, Zoom interview with author, May 7, 2024.

3. Jeremy FiveCrows, "An Interview with Elaine Harvey, CRITFC's New Watershed Manager," Columbia River Inter-Tribal Fish Commission, July 18, 2023, https://critfc.org/2023/07/18/an-interview-with-elaine-harvey-critfcs-new-watershed-manager/.

4. Vivek V. Venkataraman, "Women Were Successful Big-Game Hunters, Challenging Beliefs About Ancient Gender Roles," The Conversation, March 10, 2021, https://theconversation.com/women-were-successful-big-game-hunters-challenging-beliefs-about-ancient-gender-roles-153772. At the "Man the Hunter" conference of 1966, seventy of the seventy-five anthropologists in attendance were men.

5. Eugène Morin and Bruce Winterhalder, "Ethnography and Ethnohistory Support the Efficiency of Hunting Through Endurance Running in

Humans," *Nature Human Behaviour* 8, no. 6 (2024), https://doi.org/10.1038/s41562-024-01876-x.

6. Cara Ocobock and Sarah Lacy, "The Theory That Men Evolved to Hunt and Women Evolved to Gather Is Wrong," *Scientific American*, November 1, 2023, www.scientificamerican.com/article/the-theory-that-men-evolved-to-hunt-and-women-evolved-to-gather-is-wrong1/.

7. Emma Kemp, "How Motherhood Turned Matilda Katrina Gorry into a World Cup Starter," *Sydney Morning Herald*, July 19, 2023, www.smh.com.au/sport/soccer/how-motherhood-turned-matilda-katrina-gorry-into-a-world-cup-starter-20230719-p5dpf3.html.

8. Emma Gritt, "Breastfeeding Athletes Are Finally Taken Seriously by Paris Olympics," Yahoo! Life, March 5, 2024, https://uk.style.yahoo.com/breastfeeding-athletes-finally-taken-seriously-230400269.html.

9. Abigail Anderson, Sophia Chilczuk, Kaylie Nelson, Roxanne Ruther, and Cara M. Wall-Scheffler, "The Myth of Man the Hunter: Women's Contribution to the Hunt Across Ethnographic Contexts," *PLOS One*, June 28, 2023, https://doi.org/10.1371/journal.pone.0287101.

10. David A. Breternitz, Alan C. Swedlund, and Duane C. Anderson, "An Early Burial from Gordon Creek, Colorado," *American Antiquity* 36, no. 2 (1971), https://doi.org/10.2307/278669.

11. Amanda Foreman, "The Amazon Women: Is There Any Truth Behind the Myth?," *Smithsonian Magazine*, March 20, 2014, www.smithsonianmag.com/history/amazon-women-there-any-truth-behind-myth-180950188/.

12. Joshua Rapp Learn, "The Forgotten History of Amazon Warrior Women of Ancient Scythia," *Discover Magazine*, July 21, 2021, www.discovermagazine.com/planet-earth/the-forgotten-history-of-amazon-warrior-women-of-ancient-scythia; Derek Hawkins, "Amazons Were Long Considered a Myth. These Discoveries Show Warrior Women Were Real," *Washington Post*, December 31, 2019, www.washingtonpost.com/science/2019/12/31/amazons-were-long-considered-myth-these-discoveries-show-warrior-women-were-real/.

13. Anahit Y. Khudaverdyan, Azat A. Yengibaryan, Suren G. Hobosyan, Arshak A. Hovhanesyan, and Ani A. Saratikyan, "An Early Armenian Female Warrior of the 8–6 Century BC from Bover I Site (Armenia)," *International Journal of Osteoarchaeology* 30, no. 1 (2019), https://doi.org/10.1002/oa.2838.

14. "Archaeologists Found the Burial of Scythian Amazon with a Head Dress on Don," news release, Akson Russian Science Communication Association via the American Association for the Advancement of Science, December 25, 2019, www.eurekalert.org/news-releases/517672.

15. J. Alix-Garcia, L. Schechter, F. V. Caicedo, and S. J. Zhu, "Country of Women? Repercussions of the Triple Alliance War in Paraguay," SSRN Electronic Journal, 2020, https://doi.org/10.2139/ssrn.3598489.

16. Adrienne Zihlman, interview with author, March 3, 2023.

17. Charlotte Hedenstierna-Jonson, Anna Kjellström, Torun Zachrisson, et al., "A Female Viking Warrior Confirmed by Genomics," *American Journal of Physical Anthropology* 164, no. 4 (2017), https://onlinelibrary.wiley.com/doi/full/10.1002/ajpa.23308.

18. Hedenstierna-Jonson et al., "Female Viking Warrior."

19. Neil Price, Charlotte Hedenstierna-Jonson, Torun Zachrisson, et al., "Viking Warrior Women? Reassessing Birka Chamber Grave Bj.581," *Antiquity* 93, no. 367 (2019), https://doi.org/10.15184/aqy.2018.258. Queer theory is an interdisciplinary field that challenges traditional assumptions about gender, sexuality, and other identities. It is based on the idea that gender and sexual identities are social constructs.

20. Randall Haas, James Watson, Tammy Buonasera, et al., "Female Hunters of the Early Americas," *Science Advances* 6, no. 45 (2020), https://doi.org/10.1126/sciadv.abd0310.

21. Anderson et al., "The Myth of Man the Hunter."

22. Pat Miller, "Dog Trainers Note How Men and Women May Train Differently," *Whole Dog Journal*, July 6, 2006, www.whole-dog-journal.com/training/dog-trainers-note-how-men-and-women-may-train-differently/.

23. K. S. Singh, "Gender Roles in History: Women as Hunters," *Gender, Technology and Development* 5, no. 1 (2001), https://doi.org/10.1080/09718524.2001.11909990; Faye Flam, "The 'Man the Hunter' Myth Won't Go Away," *Noema Magazine*, October 17, 2023, www.noemamag.com/the-man-the-hunter-myth-wont-go-away/.

24. Vivek V. Venkataraman, "Women Were Successful Big-Game Hunters, Challenging Beliefs About Ancient Gender Roles," The Conversation, March 10, 2021, https://theconversation.com/women-were-successful-big-game-hunters-challenging-beliefs-about-ancient-gender-roles-153772.

25. Jan Jaap van Leusden, interview with author, May 24, 2024.

26. Jeré Longman, "'8 Players 1 Heartbeat': A Game Men and Women Play as Equals," *New York Times*, February 20, 2023, www.nytimes.com/2023/02/20/sports/korfball-gender-equality.html.

27. Quotes by Michelle Van Geffen, Merijn Mensink, and Damian Folkerts come from interviews conducted by the author, Delft, Netherlands, May 24, 2024.

28. Longman, "'8 Players 1 Heartbeat.'"

29. Interview with Sophia Nimphius, Sydney, Australia, March 25, 2024.

30. R. W. Viola, J. R. Steadman, S. D. Mair, K. K. Briggs, and W. I. Sterett, "Anterior Cruciate Ligament Injury Incidence Among Male and Female Professional Alpine Skiers," *American Journal of Sports Medicine* 27, no. 6 (1999), https://doi.org/10.1177/03635465990270061701.

31. Sophia Nimphius, Jeffrey M. McBride, Paige E. Rice, Courtney L. Goodman-Capps, and Christopher R. Capps, "Comparison of Quadriceps and Hamstring Muscle Activity During an Isometric Squat Between

Strength-Matched Men and Women," *Journal of Sports Science and Medicine* 18, no. 1 (2019), www.ncbi.nlm.nih.gov/pmc/articles/PMC6370970/.

32. Sophia Nimphius, "Exercise and Sport Science Failing by Design in Understanding Female Athletes," *International Journal of Sports Physiology and Performance* 14, no. 9 (2019), https://doi.org/10.1123/ijspp.2019-0703.

SELECTED BIBLIOGRAPHY

Angier, Natalie. *Woman: An Intimate Geography*. Virago, 2014.
Bohannon, Cat. *Eve*. Knopf, 2023.
Buckley, Thomas, and Alma Gottlieb. *Blood Magic*. University of California Press, 1988.
Butler, Judith. *Bodies That Matter: On the Discursive Limits of Sex*. Routledge, 1993.
Butler, Judith. *Who's Afraid of Gender?* Knopf Canada, 2024.
Cahn, Susan K. *Coming on Strong: Gender and Sexuality in Twentieth-Century Women's Sport*. Harvard University Press, 1994.
Cirotteau, Thomas Kerner, Jennifer Kerner, and Eric Pincas. *Lady Sapiens: Breaking Stereotypes About Prehistoric Women*. Hero Press, 2023.
Clancy, Kate. *Period*. Princeton University Press, 2024.
Cleghorn, Elinor. *Unwell Women: Misdiagnosis and Myth in a Man-Made World*. Dutton, 2021.
Clover, Carol J. *Men, Women and Chainsaws: Gender in the Modern Horror Film*. Princeton University Press, 1992.
Cooke, Lucy. *Bitch: On the Female of the Species*. Basic Books, 2022.
Criado Perez, Caroline. *Invisible Women: Exposing Data Bias in a World Designed for Men*. Vintage, 2019.
Dusenbery, Maya. *Doing Harm: The Truth About How Bad Medicine and Lazy Science Leave Women Dismissed, Misdiagnosed, and Sick*. HarperOne, 2018.
Ehrenreich, Barbara, and Deirdre English. *Complaints and Disorders*. Feminist Press at CUNY, 1973.
Gimbutas, Marija Alseikaite. *The Language of the Goddess*. Thames and Hudson, 2006.
Gimbutas, Marija Alseikaite, and Miriam Robbins Dexter. *The Living Goddesses*. University of California Press, 2001.
Goode, Starr. *Sheela Na Gig: The Dark Goddess of Sacred Power*. Inner Traditions, 2016.
Gordon, Aubrey. *What We Don't Talk About When We Talk About Fat*. Beacon Press, 2020.
Graeber, David, and David Wengrow. *The Dawn of Everything: A New History of Humanity*. Penguin Books, 2021.
Green, Monica H. *The Trotula: An English Translation of the Medieval Compendium of Women's Medicine*. University of Pennsylvania Press, 2002.

Gregory, Philippa. *Normal Women: Nine Hundred Years of Making History*. HarperCollins, 2024.
Gross, Rachel E. *Vagina Obscura: An Anatomical Voyage*. W. W. Norton, 2022.
Harari, Yuval Noah. *Sapiens: A Brief History of Humankind*. Harper Perennial, 2015.
Harrison, Da'shaun. *Belly of the Beast: The Politics of Anti-Fatness as Anti-Blackness*. North Atlantic Books, 2021.
Kaplan, Janice. *Genius of Women: From Overlooked to Changing the World*. Dutton, 2021.
Legato, Marianne J. *Principles of Gender-Specific Medicine*. Elsevier Academic Press, 2023.
Manne, Kate. *Down Girl: The Logic of Misogyny*. Oxford University Press, 2018.
Mertens, Maggie. *Better Faster Farther: How Running Changed Everything We Know About Women*. Algonquin Books, 2024.
Minchin, Louise. *Fearless*. Bloomsbury Publishing, 2023.
Moalem, Sharon. *The Better Half*. Farrar, Straus and Giroux, 2020.
Montagu, Ashley. *The Natural Superiority of Women*, 1st ed. Macmillan Company, 1953.
Natterson-Horowitz, Barbara, and Kathryn Bowers. *Zoobiquity: The Astonishing Connection Between Human and Animal Health*. Vintage Books, 2013.
O'Keane, Veronica. *A Sense of Self: Memory, the Brain, and Who We Are*. W. W. Norton, 2021.
Pennybacker, Mindy. *Surfing Sisterhood Hawaii*. Mutual Publishing, 2023.
Roughgarden, Joan. *Evolution's Rainbow: Diversity, Gender, and Sexuality in Nature and People*. University of California Press, 2013.
Sattin, Anthony. *Nomads: The Wanderers Who Shaped Our World*. W. W. Norton, 2022.
Schiff, Stacy. *The Witches: Salem, 1692*. Weidenfeld and Nicolson, 2016.
Strings, Sabrina. *Fearing the Black Body: The Racial Origins of Fat Phobia*. New York University Press, 2019.
Tara, Sylvia. *The Secret Life of Fat: The Science Behind the Body's Least Understood Organ and What It Means for You*. W. W. Norton, 2016.
Thornton, Sarah. *Tits Up: What Sex Workers, Milk Bankers, Plastic Surgeons, Bra Designers, and Witches Tell Us About Breasts*. W. W. Norton, 2024.
van Beneden, Ben, and Amy Orrock. *Rubens and Women*. Dulwich Picture Gallery, 2023.
Weitekamp, Margaret A. *Right Stuff, Wrong Sex: America's First Women in Space Program*. Johns Hopkins University Press, 2006.
Williams, Florence. *Breasts: A Natural and Unnatural History*. W. W. Norton, 2013.
Williams, Florence. *Heartbreak: A Personal and Scientific Journey*. W. W. Norton, 2022.
Wollstonecraft, Mary. *A Vindication of the Rights of Woman*. Arcturus Publishing, 1792.
Yu, Christine. *Up to Speed*. Penguin, 2023.

INDEX

abortion, 94
Abu-Salma, Ruba, 95
accuracy, in shooting, 213
ACL. *See* anterior cruciate ligament
ACOG. *See* American College of Obstetrics and Gynecology
activity levels, longevity and, 270
aging, 106, 267, 275–276. *See also* longevity
Agta people, 291, 298
AIDS, 237
air-rifle shooting, 231–232
Alan Edwards Centre for Research on Pain, McGill University, 186
Alberts, Susan, 273
all-or-nothing thinking, 194–195
Amazons, 293–294
American College of Obstetrics and Gynecology (ACOG), 61–62
American football, 18
American Medical Association, 113
analgesia, stress-induced, 155
Anand, Sonia, 111, 113
Anderson, Abigail, 292
anorexia nervosa, 116
anterior cruciate ligament (ACL), 21–22
anti-trans movement, 7, 228–229
Appalachian State University, 45
Appalachian Trail, 26, 123, 137
Apple, Fiona, 141
apps, for menstrual cycle tracking, 78–85, 94

archery, 213
Arif, Nighat, 171–172, 174
ASD. *See* autism spectrum disorder
Asher-Smith, Dina, 80
Association of Western Forestry Clubs, 202
athletes, 20–21, 143–146. *See also* female athletes
athleticism, biological sex and, 20–21
athletics, 4, 19–20, 130–131
Austad, Steven, 273
Austin, S. Bryn, 112
autism spectrum disorder (ASD), 23
autoimmune diseases, 260–261

B cells, 242–243
Baerwald, Angela, 65–66
balance, 216, 220, 223–224
balance beam, 224
Barkley Marathons, 148
Batek people, 298
The Battle of the Amazons (painting), 52
Beck, Belinda, 276, 278–279
Bekker, Sheree, 208–209, 214
Beneden, Ben van, 51
Best Practice for Youth Sport (textbook), 24–25
Better, Faster, Farther (Mertens), 24
The Better Half (Moalem), 236
Bhagwat, Anjali, 209
biased systems, elite competition preceded by, 21–22

Big Dog's Backyard Ultra (trail run), 123–124
big-game hunting, by teenagers, 297
Bindra, Abhinav, 209
biohacking, 85
biological sex, 8–9, 20–21
biology, life experience versus, 21–25, 256–261, 304
Black and brown women, health of, 101–102, 113, 161, 165, 173
Black bodies, muscles, 49
Black Girls Run, 101
"blue zones," aging in, 267
body mass index (BMI), 107–110, 112, 113
body roundness index (BRI), 111–112
body size, 100
Bohannon, Cat, 68
Bolt, Usain, 227–228
bone health
  exercise influencing, 278
  frailty and, 276
  muscular strength linked with, 280
bone strength, 276–277
bone stress injuries (BSIs), 142–143
Boston Marathon (2018), 192
Bourbon, Isabella of, 52
Brach, Tara, 121
brain, 184–188
breast milk, 129
Breternitz, David, 293
Breukheysen, Nico, 300
BRI. *See* body roundness index
British Journal of Anaesthesia, 188
Brookshire, Bethany, 177–179
Brown, Helen Gurley, 265
brown fat, 115
brown women, health of Black and, 165
Bruinvels, Georgie, 78
BSIs. *See* bone stress injuries
Buchar, Claire, 150
bulimia, 141
Burgess, Jessica, 94–95
Burnette, Christal, 265, 268
Burns, Katelyn, 229
Burr, George, 115

Burr, Mildred, 115
Butler, Judith, 7

Caber Toss event, 206
caloric needs, of male bodies contrasted with female bodies, 104–106
canalization, 132–133
Candidate Physical Ability Test (CPAT), 26–29
carbohydrates, fat contrasted with, 40, 106, 125–126
Cascades (mountains), 153
Case, Stephanie, 148–150
Cave, Christine, 281
Centers for Disease Control and Prevention (CDC), 174–175
central nervous system, 40
Cheung, Ada, 230
childbirth, pain of, 162–164, 166
chronic pain, in female athletes, 186–187
Clancy, Kate, 25
Clark, Caitlin, 24
Clarmo, Viktor, 204
Cleghorn, Elinor, 57
cold-water swimming, women dominating, 122
Cole, Bob, 163
college sports, 33–34
colposcopy, 171–172
Columbia River Inter-Tribal Fish Commission, 289
Columbia University, 17, 33
competition, 21–22, 191–196, 226–232
competitive woodchopping, 200–203
Cooper-Owens, Deirdre, 165
COVID-19, 242, 247, 250–251, 257–260
Cowart, Leigh, 124–125
CPAT. *See* Candidate Physical Ability Test
cramps, 61, 80, 82, 84, 140, 144, 159, 168
Crouse, Lindsay, 192–193
culture, metabolism and fat and, 106–110

dance, 22, 117, 225–226
  and pain, 179
darts, 213

Dauwalter, Courtney, 125, 148
Deaner, Robert, 42–43
death, cardiovascular, 113
Debessai, Haben, 165–166, 173–174, 175
Deshpande, Deepali, 231–232
DeWitte, Sharon, 9
diabetes, type-2, 41, 136
Diana and Her Nymphs Setting Out for the Hunt (painting), 52
dimorphic, humans as, 104
disabilities, 100, 169, 260
Divaraniya, Amy, 91–93
Doing Harm (Dusenbery), 160
Double Buck event, 202
Dower, Tara, 137, 148
drugs, 187–188
Duncombe, Caroline, 245–248, 251–256, 259–261
Dusenbery, Maya, 160, 169
dysmenorrhea, 159

eating disorders (ED), 101–102, 109, 140–142, 145–146, 277
Eddie Aikau Big-Wave Invitational, 220
The Egg and the Sperm (Martin), 59
eggs, 68
    battle of, 63–67
    ongoing culling events destroying, 64
    sperm chosen by, 69–70
elder-sibling advantage, 21
elite competition, biased systems preceding, 21–22
endometriosis, 72, 109, 159–161
endurance, 123–124, 192. *See also* running
    estrogen and, 129
    female advantage in, 42, 46, 147–148
    female bodies and, 40, 125–128, 290–291
    mental perseverance and, 122
    muscular, 42
    speed balanced with, 34
endurance athletes, 143–146
estrogen, 248
    endurance and, 129
    fat producing, 118
    immune systems benefiting from, 249–251, 254
    immunity increased by, 249–251
European Championships (2022), 80
European Cup, 301
Eve (Bohannon), 68
exercise. *See also* sports
    bone health influenced by, 278
    immune systems strengthened by, 275
    women were able to perform more repetitions during, 39
"Exercise and Sport Science Failing by Design in Understanding Female Athletes" (paper), 305

fat, 112
    art and, 49–50
    brown, 115
    carbohydrates contrasted with, 40, 106, 125–126
    cardiovascular death and, 113–114
    culture and metabolism and, 106–110
    estrogen produced by, 118
    female bodies strengthened by, 106–107
    menstruation and, 116–118
    subcutaneous, 110–111
    visceral, 110
fat deposition, sex-based differences in, 111
fat people, perceptions of, 49–50, 101
fat-brain connection, 115–116
fatigability, 37, 40
fatigue resistance, 38–39, 43, 126, 137
Fearing the Black Body (Strings), 49
Female Athlete Triad, 142
female athletes, 305
    ACL injured by, 21–22
    chronic pain in, 186–187
    culture discouraging, 25–26
    underfueling disempowering, 142–147
female bodies, 1, 5, 151, 280–281, 311. *See also* transgender women
    caloric needs of male bodies contrasted with, 104–106
    endurance and, 40, 125–128, 290–291

female bodies (*continued*)
   exercise science and, 24
   fat strengthening, 106–107
   generations of myths about, 56–61
   immune systems in, 237–238, 243
   male bodies compared with, 24, 35, 39–40, 104–107, 114, 126, 168, 237–238, 285
   muscles in, 17, 36, 40–41, 43, 52
   pain experienced by, 156–158, 188–189, 191
   physical stress in, 126–127
   research imbalance endangering, 187–188
   science understudying, 3–4
   sports research on, 44, 48, 80–81, 191, 221–223, 305–306
   sports-medicine research excluding, 88, 150
   testosterone levels in, 229–230
   variation between, 7–8
female buffering hypothesis, 133
"Female Genital Variation Far Exceeds That of Male Genitalia" (review study), 7–8
female immune advantage, 237–240
female longevity advantage, 273–275
female metabolisms, 103–106
female reproductive system, history of demeaning language describing, 60
feminism, 140–141, 303
"Fertilization Discovery Reveals New Role for the Egg" (article), 69
fibroids, uterine, 101
FIFA Medical Conference, 20, 22
firefighters, 26–29
fisheries science, 289
FitrWoman (app), 78, 81–84
Fitzpatrick, John, 70
fMRI. *See* functional magnetic resonance imaging
Folkerts, Damien, 303, 304
follicular waves, 65–67
food, 139–140
Foreman, Amanda, 293–294

Fortuna sports venue, 299, 303
frailty, 276, 280
Freemas, Jess, 81–82, 84
Frisch, Rose, 116–117
functional magnetic resonance imaging (fMRI), 184

Gabeira, Maya, 218–219
Gargett, Caroline, 72–74
gathering, hunting and, 36, 47, 283, 287–288
Geffen, Michelle van, 302
gender, 6–10, 19
gender equality, in surfing, 220
gender-segregation, of wrestling, 33
"The Girl Who Cried Pain" (compendium), 158–159
"A Girl's First Period" (research), 61
girls' wrestling, 31–32
"Go Girls! Efficient Female Immunity" (paper), 238
Goldhill, Olivia, 159
Goodall, Jane, 6
Gorry, Katrina, 291
Grand Raid de la Réunion (ultramarathon), 193–194
grandmother hypothesis, 283–284
grandmothers, 281–286
Griffith, Linda, 161
Gross, Rachel, 64, 71
Guillebaud, John, 159
Guinness World Records, 120
Gutin, Iliya, 108
gymnastics, 224–226
"Gynecology Has a Pain Problem" (Nudson), 170

Hadza (hunter-gatherer group), 283
Hale, Amanda, 133–134
Harvey, Elaine, 287–290
Hawkes, Kristen, 283–284
Haynes, Kenneth, 12
health, of Black and brown women, 165
health care, 191
   BMI and, 109
   pain and, 168–173

for transgender women, 173
women receiving poorer, 274
health practitioners, women not listened to by, 172
heart attacks, 274
Hedenstierna-Jonson, Charlotte, 295–296
Herodotus, 293–294
Herrick, Amethysta, 6, 8
high-intensity interval training (HIIT), 85–86
Hippocrates of Kos, 56–57
HIV, 236–237
hormone levels, during menstrual cycle, 82, 88–89
Hotel, Johannes Vermeer, 304
Hunter, Sandra, 36–38, 40, 42, 80, 210
hunters, 296–299
hunting
  big-game, 297
  gathering and, 36, 47, 283, 287–288
  by women, 290–292
Hurts So Good (Cowart), 124

IC. *See* interstitial cystitis
ICHH. *See* immunocompetence handicap hypothesis
ikigai, 267–268
immune advantage, female, 237–240
immune systems, 274
  estrogen benefiting, 249–251, 254
  exercise strengthening, 275
  in female bodies, 237–238, 243
immunocompetence handicap hypothesis (ICHH), 255
The Imperial (bar), 265–266
influencers, running, 100–101
International Surfing Association and the Australian Olympic Committee, 222
intersex people, 8–9, 35, 45, 227, 230, 307
interstitial cystitis (IC), 169
Isabella Clara Eugenia, 52
isometric muscular contractions, 37–38
ISSF Champions Trophy, 209
IUD insertion, pain of, 173–175
IUD removal, pain of, 170–172

Jackson Laboratory, 243
Jimenez, Audrey, 31–34
Johns Hopkins University, 59
Johnston, Casey, 141
Journal of Strength and Conditioning Research (Nuzzo), 39
judo, 177–179

Ka'iulani, Victoria, 217
kajimaya, 271
Kakazu, Mitsu, 265–268
Kaleokalani, Nāmāhāna'i, 217
Kamakau, Samuel Mānaiakalani, 216
Kamehameha Day surfing contest, 217
Kamida, Kimiko, 265–268
Kano, Jigoro, 178
Kansas State University, 131
Kelea (chief of Maui), 216
Keller, Evelyn Fox, 69–70
Keys, Ancel, 108
Khelif, Imane, 228
Kim, Nayoung, 135
King, Martha, 204
Kinjo, Emiko, 271–272
Klaassen, Lizet, 51
Klein, Sabra L., 250
KMSKA. *See* Royal Museum of Fine Arts Antwerp
The Knowledge (Arif), 172
korfball, 299–304
Kowalczyk, Bailey, 143–145
Kronig, Bernhardt, 167

Lacy, Sarah, 291
Lakshmi, Padma, 161–162
Lammonby, Ranae, 109
Larsen, Brianna, 87–89
Larsen, Roxanne, 133–134
Lavelle, Rose, 79
Ledecky, Katie, 228
Leeuwenhoek, Antonie van, 58
Letona, Maya, 33
life experience, biology versus, 256–261
lifting, 13, 30, 35, 279
LIFTMOR trial, 278–279

Lin Yu-ting, 228
Logging Sports Team, OSU Forestry
    Club, 200
logging-sports competitions, 200–207
long-distance swimming, 120–122, 182
longevity, 268–269, 281, 284–286
    activity levels and, 270
    female advantage in, 273–275
    on Okinawa island, 267, 272
lower-body strength, 220
Lowerson, Sasha Jane, 229
Luepp, Constance, 166
luteal phase, 85–86, 157, 249, 327n11

Magness, Steve, 194–195
Mahalingaiah, Shruthi, 81
Maher, Ilona, 112
malaria, 250–252, 253
male bodies, female bodies compared
    with, 24, 35, 39–40, 104–107, 114,
    126, 168, 185, 237–238
marathon speeds, 148–149
marathons, 99–100
Martin, Emily, 59, 69
Martin, Loren, 186
Martin, Vincent, 157
maternal mortality, 274
Mayo Clinic, 238
McDonald, Janelle, 226
McGill University, 158, 186, 187
McKenney, Erin, 133–134
Medical Physiology (Mountcastle), 58–59
Meigs, Charles, 164
Meili, Launi, 211
memory, pain impacted by, 185–186
men, women competing against, 208–214
menopause, 118, 282
Mensink, Merijn, 302–303
menstrual blood, 71–75
menstrual cycle, 58–59, 70
    adaptability and reactiveness of, 61
    history of demeaning language
        describing, 60
    hormone levels during, 82, 88–89
    mood fluctuations during, 83

menstrual cycle tracking
    apps for, 78–85, 94
    detractors of, 86–90
    hacking of, 85–86, 90
"Menstrual Cycles" (Frisch), 117
menstruation, 56, 61, 89–90, 116–118
mental perseverance, endurance and, 122
mental strength, physiology contrasted
    with, 194
Mertens, Maggie, 24
mesenchymal stem cells (MSCs), 72–75
metabolism, fat and culture and,
    106–110
microbiome, sex hormones influencing,
    135–136
Miller, Amber, 130, 131
moai, 267, 269–270
Moalem, Sharon, 236, 244
Mogil, Jeffrey, 158, 186, 187
Monahan, Jaimie, 119–123, 138
Mon-López, Daniel, 210
mood fluctuations, during menstrual
    cycle, 83
Moore, Carissa, 220
Moss, Kate, 140
Mountcastle, Vernon, 58–59
Ms. (magazine), 69, 147–148
MSCs. *See* mesenchymal stem cells
Mujika, Iñigo, 80
Murray, Alison, 48
muscle fibers, fast-twitch and
    slow-twitch, 41
muscles, 13–14, 32–33, 51
    in female bodies, 17, 36, 40–41,
        43, 52
    osteoporosis and, 277
    oxygen uptake and, 39–40
    beyond physiology, 43–46
    racism and, 49–50
Muscles (journal), 43
muscular endurance, 42
muscular power, in prepuberty, 17–18
muscular strength, 34, 280
musculoskeletal system, puberty
    changing, 22

National Institute for Allergy and Infectious Diseases, 246
National Institutes of Health (NIH), 3–4, 60, 239
National Twilight Sleep Association (NTSA), 167
National Women's Rugby League (NRLW), 25
Nature Human Behavior (journal), 290
NCAA basketball, 25–26
neonatal intensive care units (NICU), 239
New England Journal of Medicine, 184
New York City, 77, 235–236
Newtson, Andreea, 171
NICU. *See* neonatal intensive care units
NIH. *See* National Institutes of Health
Nimphius, Sophia, 19–22, 42, 45, 304–305
"No Pain, No Social Gains" (paper), 190
nociplastic pain, 156
Northwestern University, 3
nozzle reaction kickback force, 28
NRLW. *See* National Women's Rugby League
NTSA. *See* National Twilight Sleep Association
Nudson, Rae, 170
Nuzzo, James, 39

obesity epidemic, 107–108
ob-gyn care, racial aspects of, 165–166
Obstacle Pole event, 205
Ocobock, Cara, 127–129, 230, 291
Ogimi (Okinawa), 271
Okinawa island, 265–269, 272
Okinawa Research Center for Longevity Science, 268–269
Olympics, 208–210, 214, 225–226
Oova, 91–93
Oregon State University (OSU), 200, 207
Organization for the Study of Sex Differences, 187
Orrock, Amy, 51
osteoporosis, 276–279
OSU. *See* Oregon State University
ovarian follicles, 65

ovaries, 63
ovulation, 65–66, 85
Oxford English Dictionary, 191
oxygen uptake, muscles and, 39–40

paid leave, for pregnancy, 131–132
pain, 154–155, 179–180
  of childbirth, 162–164, 166
  chronic, 186–187
  communication about, 189–190
  of competition, 191–196
  dismissal of female, 158–162, 183
  female bodies experiencing, 156–158, 188–189, 191
  health care and, 168–173
  of IUD insertion, 173–175
  of IUD removal, 170–172
  memory impacting, 185–186
  nociplastic, 156
  from periods, 159
  subjectivity of, 181–184
pain relief, 164–168, 173–175
pain sensitivity, as strength, 188–191
pain tolerance, 180–181, 192
pain-brain connection, 184–188
"Painless Childbirth" (Tracy and Luepp), 166
Paré, Guillaume, 111
Paris, Jasmin, 148
Parsonage, Joanna, 25, 221–222
Parsons, Joanne, 19
PCOS. *See* polycystic ovary syndrome
Peavy Arboretum (teaching lab), 199–200
pelvic inflammatory disease (PID), 161
Pennybacker, Mindy, 216–217, 218, 221, 222
perimenopause, 92
period tech, personalized, 91–93
periods, 55–56, 61, 78, 159. *See also* menstrual cycle
Pharr Davis, Jennifer, 26, 123, 148
Phelps, Michael, 127, 228
physical labor, 47–48, 49
physical stress, in female bodies, 126–127
physiology, 43–46, 194

PID. *See* pelvic inflammatory disease
pistols, rifles contrasted with, 210–211
PMDD. *See* premenstrual dysphoric disorder
PMS. *See* premenstrual syndrome
polycystic ovary syndrome (PCOS), 91, 229
Pontzer, Herman, 105–106
poor health, fatness conflated with, 114
Pop Warner (football league), 18
post-COVID syndrome, 260
pregnancy, 104, 128, 132, 164–165, 229, 249, 274
pregnancy and athletics, 129–132
premenstrual dysphoric disorder (PMDD), 83–84
premenstrual syndrome (PMS), 170
prepuberty, muscular power in, 17–18
Prince, Sedona, 25
The Prodigal Son (painting), 51
progesterone, 82–87, 249, 254
prolapse, 73–74
puberty, musculoskeletal system changed by, 22

Quanta Magazine, 69
Quetelet, Adolphe, 107–108

Race Across America, 127
racism, muscles and, 49–50
Radcliffe, Paula, 129, 131
Rapinoe, Megan, 77, 309–311
Ratner, Vicki, 169
Red Bull Rampage (mountain-bike competition), 150
Reiches, Meredith, 89–90
Reinkemeier, Heinz, 209, 211–212
reproductive strategies, 67–68
research imbalance, female bodies endangered by, 187–188
Richardson, Sarah, 258–259
Richmond, Tracy K., 112
rifles, pistols contrasted with, 210–211
Roar (Sims), 79

Roche, Megan, 126, 136–137, 142, 145, 193, 196
Roe v. Wade, 95
Rowling, J. K., 228
Royal Museum of Fine Arts Antwerp (KMSKA), 50–51
Rubens, Peter Paul, 50, 51–52
Rubens & Women (Beneden and Orrock), 51
running
    endurance, 290–291
    trail, 123–124
    ultra, 125, 149–150
running influencers, 100–101

Sanicola, Quinton, 271
Sanicola, Yohanna, 271
Sauromatians, 295
Schwartz, Bonnie, 181–184, 195
science, 10, 268–269
    biases in, 44
    exercise, 24
    female bodies understudied by, 3–4
    fisheries, 289
    sport, 306
Scott, Dawn, 78
Seattle Pacific University, 292
Seattle Reign (soccer team), 309–310
The Secret Life of Fat (Tara), 117
separate-gender sports, 18
sex. *See specific topics*
sex binary, 9
sex hormones, 135–136, 248–250
sex-based differences, in fat deposition, 111
sexism, women's events and, 204–208
Shan, Zhang, 208–209
Shaw, Natalie, 61
Sherrock, Fallon, 213
Shirur, Suma, 208
shooting. *See also* hunting
    accuracy in, 213
    air-rifle, 231–232
    at Olympics, 208–210
Shukla, Shobha, 257

Sievert, Lynnette, 67–68, 282, 284–285
Sims, James Marion, 165–166
Sims, Stacy, 78–79, 80–81, 105–106
Single Buck event, 202–203, 204
skills, gender and, 19
sledgehammer skill test, 27
sleep deprivation, 195
Smith College, 3
Snell, Latoya Shauntay, 99–103
spatial ability, 23–24
speed, endurance balanced with, 34
sperm, 63–64, 69–70
sports, 19, 149, 206–207, 307. *See also* female athletes; hunting; running; shooting; surfing
    college, 33–34
    separate-gender, 18
    third sex category for, 230–231
    transgender women and, 228–230
sports coverage, 77–78
sports-medicine research, female bodies excluded from, 88
Stanton, Elizabeth Cady, 164
Steed, Norah, 199–207
stem cells, 72–75
strength. *See specific topics*
stress injuries, 144
stress-induced analgesia, 155
Strings, Sabrina, 49
Stuart-Jones, Susan, 277–278
subcutaneous fat, 110–111
supercentenarians, 273
surfing, 215–216, 222–223, 229
    gender equality in, 220
    strength of women and, 217–221
    women excluded from, 218–219
Surfing Australia, 222
Suzuki, Makoto, 268–270
swimming, 122, 182
Syers, Madge, 214
Syracuse University, 7, 16, 64, 108

Tamas, Claudia, 275–277, 279
Tara, Sylvia, 117
tattoos, 179–181
TB. *See* tuberculosis
TBI. *See* traumatic brain injury
teenagers, big-game hunting by, 297
testosterone, 251–256
testosterone levels, in female bodies, 229–230
"The Theory That Men Evolved to Hunt and Women to Gather Is Wrong" (article), 291
throwing ability, 23
Title IX, 18
Tolentino, Jia, 94
tonnage, 20
Tracy, Marguerite, 166
trail running, 123–124
training volume, 20
Transgender Health (journal), 110
transgender women
    health care for, 110, 173
    inclusion and, 6–7
    sports and, 228–230
transgender men, 14, 110. 170, 180, 296
traumatic brain injury (TBI), 172–173
tuberculosis (TB), 257
Tuljapurkar, Shripad D., 285
twilight sleep, 166–167
type-2 diabetes, 41, 136

ultra running, 125, 149–150
underfueling, female athletes disempowered by, 142–147
Underhand Chop event, 201
University of California, 3
University of Cambridge, 48
University of Colorado, 9
University of New South Wales, 3
University of Waterloo, 39
Unwell Women (Cleghorn), 57
Up to Speed (Yu), 147–148
urology, men dominating, 169
US women's national soccer team, World Cup won by, 77–79
uterine fibroids, 101
uterine lining, 71–72
uterus, 71–75

vaccine, for malaria, 251, 253
Vagina Obscura (Gross), 64, 71
Vartan, Gerry, 218–219
vasectomy, 173–174
Victoria (queen), 164
Vincent, Lawrence, 117
visceral fat, 110
Vora, Kalini, 70

Wager, Tor, 155–156, 181, 184–185, 186–187, 191
Wall-Scheffler, Cara, 292, 298
Wambach, Abby, 21
War of the Triple Alliance, 294–295
warriors, women as, 292–296
Waters, Alice, 272
weightlifting, 13, 30, 279
Western States Endurance Run, 148, 193
"Why Are Great Athletes More Likely to Be Younger Siblings?" (article), 20–21
"Why Men Are Much Worse at Being Sick than Women" (article), 182
Willard, Huntington F., 241
women. *See specific topics*
"Women Are Calling Out 'Medical Gaslighting'" (article), 171
Women in Sports Conference, 19
"Women Live Longer than Men Even During Severe Famines and Epidemics" (paper), 126

women's events, sexism and, 204–208
Women's National Basketball Association, 24
Women's Sports Foundation, 149
woodchopping, competitive, 200–203
Wooddy, Adam, 180
Working with Indigenous Knowledge (book), 163
World Cup (2019), US women's national soccer team winning, 77–79
World Journal of Stem Cells, 74
World Surf League, 220
World Surfing League, 229
wrestling, 31–32, 33

X chromosome, 128–129
XX chromosomes, 240–244
XY chromosomes, 240

Yakama Nation, 287, 289
Yale School of Medicine, 74
Yu, Christine, 147–148

Zamata, Sasheer, 168
Zangerl, Barbara "Babsi," 148
Zihlman, Adrienne, 295

Starre Vartan is a science writer who was raised in a family of creatives and medical professionals. She grew up in New York and now splits her time between the Pacific Northwest and Sydney, Australia. She contributes regularly to *Scientific American* and *National Geographic* and has written for CNN, *The Washington Post*, *Slate*, and *New York* magazine, among many other outlets.